A Different Sort of Time

Jerrold R. Zacharias (1905–1986)

A Different Sort of Time

The Life of Jerrold R. Zacharias
Scientist, Engineer, Educator

Jack S. Goldstein

The MIT Press
Cambridge, Massachusetts
London, England

© 1992 Massachusetts Institute of Technology

This book was set in Bembo by the MIT Press.

Library of Congress Cataloging-in-Publication Data

Goldstein, Jack S., 1925–
    A different sort of time : the life of Jerrold R. Zacharias, scientist, engineer, educator / Jack S. Goldstein.
        p. cm.
    Includes bibliographical references and index.
    ISBN 978-0-262-07138-3 (hc.: alk. paper) — 978-0-262-51909-0 (pb. : alk. paper)
    1. Zacharias, Jerrold Reinach, b. 1905   2. Physicists—United States—Biography.   I. Title.
    QC16.Z23G65   1992
    530'.092—dc20
    [B]   91-37934
                                                                                        CIP

MIT Press is pleased to keep this title available in print by manufacturing single copies, on demand, via digital printing technology.

To Absent Friends

Contents

Acknowledgments

No doubt some people live their lives in the expectation that some day biographies will be written about them. Mindful of the demands of the historian, they preserve the letters they receive and copies of the letters they write; thoughtful of their own place in history, they leave few loose ends and few unanswered questions. Jerrold Zacharias was surely not one of these people.

Much about Jerrold's life can nevertheless be discovered or inferred from the many documents that do survive and from the lively recollections of many who were part of the events of his life. Some of Jerrold's own reminiscences are fortunately preserved in several interviews made late in his life, and the memories of those among whom he lived and worked have often turned out to be rich indeed. I interviewed sixty-five individuals in the course of my research, all at length and some several times. I am deeply grateful to them, not only for patiently sharing their recollections of Jerrold with me but for helping me to understand better that wonderful but often difficult period spanned by Jerrold's life: a time in which optimism competed with anxiety, in which ordeals and trials and wartime hardships coexisted with invention and discovery. Jerrold's friends helped me to sense what it may have been like, for example, to have worked in physics in the 1930s, or to be among those who had important scientific skills to contribute to the nation in the time of the Second World War, the Korean War, and Vietnam; what it was like to be a responsible and engaged citizen in the times of Eisenhower, Kennedy, Johnson, and beyond.

Those years seem to me to have constituted a different sort of time, as the title of this book suggests—a time when the inequities and imbalances of our society seemed somehow correctable, provided only that we wished to correct them; when a hopeful individual might seek to change the world through effort and persuasion and example; when one person might dream of making a difference. Jerrold could become discouraged now and then,

like anyone else, but overall, he was a terrific optimist, a man of great faith
in the ability of people to help themselves if given a chance. His colleague
Martin Deutsch told me that, "for Jerrold, every day was a new day." That
remark was right on the money.

In alphabetical order, those who were interviewed are James Aldrich,
Albert V. Baez, Rebecca Brown, Edwin C. Campbell, J. M. B. Churchill,
Peter Demos, Martin Deutsch, Eleanor Duckworth, Bernard Feld, Herman
Feshbach, David Frisch, Lawrence Fuchs, Myles Gordon, Paul Gray, Joseph
Griffith, Uri Haber-Schaim, David Harmon, George Hein, Albert Hill,
Malcolm Hubbard, Carl Kaysen, Arthur Kerman, James R. Killian, John
King, Edwin H. Land, Douglas Lapp, Mitchell Lazarus, Margaret MacVicar,
William T. Martin, Sidney Millman, Jean Morin, Philip Morrison, Phylis
Morrison, Adeline Naiman, Barbara S. Nelson, Charles Oberdorfer, E. R.
Pariser, Alan Pifer, Emmanuel Piore, Edward Purcell, Helen Rabi, I. I. Rabi,
Norman Ramsey, George Rathjens, Berol Robinson, Hartley Rogers,
Emily Romney, Walter Rosenblith, Bruno Rossi, Patricia Sacco, M. B. R.
Savage, Donald Schon, Judah Schwartz, Arthur Singer, Benson Snyder,
Edwin F. Taylor, Robert Vessot, Rainer Weiss, Victor F. Weisskopf,
Jacqueline G. Wexler, Stephen White, Janet Whitla, Jerome B. Wiesner,
Johanna Zacharias, Leona Zacharias, Susan Zacharias. Others, referred to in
the text, have written letters, sometimes at considerable length. I thank each
of them; the hours spent with them have constituted one of the principal
fringe benefits of the work.

It is appropriate that I record here both my affection for and my
particular debt to the late Leona Zacharias. She allowed me free rein to
rummage around in what most people would regard as her private business.
She was a valuable storehouse of recollection and made both the memory
and the record of her own life, as well as Jerrold's, freely accessible to me,
offering her opinions and judgments without attempting to influence what
I might write. I am glad that she had the opportunity to read the entire book
in manuscript, and I wish that she could have seen it in print.

It is appropriate as well that I single out Albert Hill for special thanks;
he has devoted many more hours to filling in background and setting me
straight than he could have anticipated would be necessary when we began,
and he has done so with cheerful patience and forbearance. He should not
be held responsible for the errors that remain; they probably result from my
occasional failure to take his advice.

I thank both Arthur Singer and the Alfred P. Sloan Foundation for a
grant in support of this work and for continued encouragement in the course
of it. Their attempts to capture some of the history of the last fifty years

through the development of a videotape archive deserve more attention than they have received.

Warm thanks are due to S. S. Schweber, of Brandeis University, for encouraging me to think that I might be able to carry out this project and for valuable advice along the way. My other colleagues in the physics department at Brandeis have shown a friendly tolerance for and interest in this work, which I greatly appreciated. Paul Forman, of the Smithsonian Institution, and Myles Gordon, of Education Development Center, have been especially generous in sharing materials relating to Jerrold Zacharias, and I am grateful to them. Emily Meyer, of the University of Massachusetts, has done an extraordinary editorial job converting many of the convoluted, highly reflexive sentences that, like many other scientists, I seem to enjoy into simple declarative statements. The reader will have little trouble spotting where she has failed to reform me. I thank her warmly.

Friends and family put up with a great deal, I see, while a book of this sort is being written. That has not gone unnoticed or unappreciated.

It is usual in acknowledgments of this sort for the author to accept full responsibility for errors and misunderstandings, which I earnestly do; but I think I have not known until now how easy it can be to slip. Every remark not followed up, every event not placed accurately in perspective, every comment taken out of context constitutes a possible source of error. A biographer is obliged to choose as well as he or she can what is important and what may be left aside, but at the very real risk of distorting what really happened. I hope I have not misquoted anybody, and I think I have not; but that doesn't let me off the hook, for both accuracy and meaning depend on context as well as getting the words right. That is what is meant, I now understand, when an author accepts the responsibility for errors and misunderstanding. It is not simply a matter of routine politeness; rather, it is the last expression of the author's fading hope that he has not garbled what he has been told.

Finally, my very special thanks, as always, go to Nita, who makes life possible. How do single folks ever manage to write books?

Prologue

Kano, Northern Nigeria, February 1965

Along with about a dozen other visitors, I somehow managed to cram myself into the narrow space at the back of an African fifth-grade classroom, between the last row of seats and the rear wall. Some of us were scientists, others were educators and senior administrators; we had all traveled from distant places—America, England, other parts of Africa—and we felt somewhat awkward and out of place, as no doubt we were. We had come to Kano to attend a remarkable conference on African science education organized by Jerrold Zacharias, which would begin that afternoon. Zacharias had arranged to start us off a few hours early with a classroom science demonstration.

Kano is an exotic place and certainly an unlikely spot for such a conference. For that matter, Zacharias seemed an unlikely character to organize it. He was a professor of physics at MIT who I knew had early specialized in sophisticated experiments with molecular beams and had later involved himself with a well-known and well-regarded high school physics program in the United States. Neither of these accomplishments seemed to have much relevance to rural Africa. He had a reputation as a shaker and mover, someone who got things done but he seemed ordinary enough in the flesh. I knew little else about him.

At ten o'clock, the morning was already hot. Kano is on the southern edge of the Sahara and is warm even in winter. I was still uncomfortable from the abrupt changes in time and climate I had experienced, as well as feeling a bit stunned at finding myself in Africa, a continent I had never expected to visit. But the strangeness of the locale seemed to add to my sense that perhaps I was about to participate in something important. There was a sense of excitement among us, and I was susceptible to it.

Zacharias had arranged for us to come to this school to witness something that was actually more of an experiment than a demonstration.

A local African teacher, after only two weeks of special training, was going to teach that morning by a method new to him: he would introduce his pupils to science not by telling them about it but by putting some simple materials of science—in this case, flashlight bulbs and batteries—into their hands, letting the children manipulate them and make their own experiments. Boxes of scraps containing bits of wire, paper, string, wood, paper clips, rubber bands, anything that came to hand, were available in the room. The teacher would simply ask, "Can you light the bulb?"

The children were extremely self-conscious at first, aware of us at the back of the room and also finding it difficult to accept an unfamiliar premise: instead of being required to sit in one place, hands folded in respectful silence, they were free to talk, move around the room, confer with each other, visit each other's work spaces if they wished. Instead of being told what to do, they would be free to invent what to do. For the teacher, too, the premise was unfamiliar, and he seemed not quite certain of himself as he set about the activities.

Their self-consciousness and uncertainty evaporated quickly. One by one, flashlight bulbs lit up, first in one part of the room, then another, illuminating briefly each time the intense absorption and the delight in the children's faces. Each of us watching, scientist or not, had felt that same sense of wondering delight as young children, in rare and lucky moments of discovery. We were thrilled to recognize it again now. The teacher moved quietly among the groups of students, offering encouragement here, asking provocative questions there: "Can you make the bulb give a brighter light? Can you light two bulbs? Will string work in place of the wire?"

The class probably did not last longer than an hour. By the end of it, the children were well on the way to developing for themselves the simple logic of circuits and had managed in the process to experience something about the electrical properties of different materials. For those of us watching in fascination, the class had seemed to last only a few minutes. By its end we had learned, as Zacharias had expected we would, that the new, hands-on method of teaching science to children could work in Africa as well as it had in the United States or in England. Kano, rural and poor, isolated, still closer to medieval times than to the present, had provided a severe test. Kano's children could learn in this way, we saw, and so could children everywhere.

In all, sixty of us had gathered in Kano to help decide how to begin a new African science education program. Each of us had come at the invitation of Jerrold Zacharias. Many of us who were professional physicists, chemists, biologists, and physicians had little prior experience either with children's science education or with Africa; yet here we were, filled with a contagious enthusiasm, ready to learn, ready to go forward. I did not understand then, and I am not sure I understand now, by what magic Zacharias persuaded hard-headed scientists (as we supposed ourselves to be) that we might be useful in such an adventure; but he did, and we were willing.

He had found a brass cowherd's bell somewhere in the town, and with it he convened the sessions of our meeting. Those events are unforgettable for me, for they altered the course of my own life. At the end of the Kano conference, I accepted Zacharias' invitation to organize the development of what became the African Primary Science Program. I remained deeply involved with it in one way or another for more than a decade. It is thus easy for me to recall Jerrold Zacharias at Kano; I see him standing at the door to the courtyard where we all met, bell in hand, calling us to work, the meeting place a bright space of sunshine and shadow behind him. He was then an attractive gray-haired man of middle height and middle age (he had turned sixty only a month earlier, in fact) with a youthful manner; he seemed light on his feet, mild spoken in spite of his enthusiasm for what was under way, and pleasant. He kept a questioning tone in his voice and had a ready laugh; he wore his customary pale blue shirt, with rolled-up sleeves and no tie, as he tolled his bell, bringing us back from our coffee, our arguments and discussions. "Let's go," said the bell. "Let's go," said Zacharias.

Twenty-five years later, the bell now sits on a bookshelf in my study. It is a curious-looking object and often arouses visitors' comment. It was fashioned, no doubt, entirely from utilitarian sources: melted-down water taps and the like. It is roughly made; the practice of metal casting was developed to a fine art in southern Nigeria, but this bell comes from the north. Sand-cast by an amateur hand, the bell lacks the fine symmetry of its commercial or artistic counterparts. A small hole is patched with solder; the bell bears simple decorations incised in its surface, which is otherwise unpolished. Unexpectedly, given its manufacture, the bell has a clear, rich and resonant tone, pleasant to the ear and easily heard. The bell is certainly one of a kind.

So was Jerrold Zacharias. His message also was resonant and easily heard. The bell that called us together at Kano and the man who held it spoke with the same clarity of voice.

When I came to write this book, I became deeply interested again in understanding how he had managed to attract so many people with so many varied and diverse abilities to his side. How had he done that? How had he become such a leader? Philip Morrison, an MIT physicist who had also been in that classroom at Kano and had been deeply involved in many other

Zacharias projects as well, once remarked on Zacharias's skill in catching so
many people in his net. I asked Morrison how he thought Zacharias had done
it. "Ask yourself," said Morrison. "You were one of those he caught."

In part, at least, that is what this book is about: an attempt to understand
the remarkable attractiveness—seductiveness might be a better word—of
Jerrold Zacharias, who was able to make everything he was involved in seem
like a wonderful adventure. He had not always been a leader, it turned out;
he discovered his penchant for leadership relatively late in a life that at first
had been mostly quiet and private. His early career had been happily spent
in what was then obscure and esoteric physics research and in teaching. It
is clear, on the evidence of what came later, that his entire philosophy,
indeed, his character itself, was determined in those years when he had been
most exposed to the rigorous demands of science. It had been his great good
fortune, he said, to have come of age just when the great experiments in
quantum mechanics were being done. He was one of a remarkable group
of young physicists who gathered around I. I. Rabi at Columbia University
in the 1930s, and he played an essential part in a remarkable series of
experiments that contributed importantly to the dramatic development of
quantum mechanics.

He was already thirty-five when the United States entered World War
II, forty when it was over. Somehow, his wartime service at the MIT
Radiation Lab and at Los Alamos brought about a major change in him, and
he emerged from the war ready to play a role of authority and influence.
His base was the Massachusetts Institute of Technology, which became his
professional home; most often he found there a remarkable degree of support
and encouragement, but sometimes it was also the arena where he had to
fight for what he believed. As MIT went about reestablishing itself in the
postwar period, he was one of those who gave it the direction and the
leadership it needed to turn it into a great university.

He interacted strongly with the events of his time; his life serves as a
window on that time, a particular vantage point from which much of this
nation's recent history may be viewed. In part, therefore, this book is about
the view from Jerrold's window, for he was an active participant in the
nation's history. During much of his postwar life, especially during the years
of the cold war, his advice, like that of a small number of other physicists,
was in demand and highly valued. For the decade or so after the war, scientists
frequently served as advisers to government and industry and as consultants
to the military. A new kind of relationship emerged between scientists and
government; new national laboratories of unprecedented scope, such as
Brookhaven and the Lincoln Laboratory, were born. Jerrold Zacharias was
one of the architects of those new relationships, shuttling without a hitch

between Washington and Boston, between classroom and conference room, between laboratory and military base.

The cold war influenced him strongly, reinforcing his unembarrassed love for democracy, decency, and fair play. He felt a scientist's abhorrence of dogmatism, especially the forms of it that he encountered: McCarthyism, communism, know-nothingism, radicalism of any sort.

He came to believe, ultimately, that education was the most effective way to address the ills of the world. "In order to save our democracy," he said, "we've got to educate the people who vote. There's no question in my mind about that." And so, convinced of the inadequacy of existing science education, he initiated programs of educational reform; these developed into a movement that many people termed a revolution. It was as a small part of that revolutionary movement that we had come to Africa. We had responded enthusiastically to him; he was the acknowledged leader of the movement and we were a good example of the diverse collection of people he persuaded to follow him, whether to high school classrooms in the United States or to elementary school classrooms in Africa or to any of the other places to which he might lead us. His appeal resonated equally among scientists and writers and filmmakers, Nobel laureates and ordinary schoolteachers. By his own account, he had hardly opened his mouth until he was thirty-five, but afterward he became transformed into a modern day pied piper, leading enchanted followers across three hectic decades, often as surprised at what they found themselves doing as we were at Kano.

His remarkable drive and energy, his sense of humor and his enjoyment of everything, came from within. He was really an old-fashioned sort of man; like the Kano bell, he had a functional integrity. He stood for honesty and courage, and, although he could be sophisticated, he was seldom subtle. He had started out in an earlier epoch; he had been born and raised in an age that seems, at least in retrospect, to have been simpler and more forthright than our own, and he thought that "getting your head straight" was the most important thing you could do.

He lived until he was almost eighty-two, never ceasing to work on the things he thought important for the country and the world. On his eightieth birthday, among the good wishes and congratulations he received from his many well wishers was a letter from I. I. Rabi, welcoming him to the ranks of octogenarians and reflecting on their friendship of more than sixty years' standing. "I cannot think of any time," Rabi wrote, "in which the world has not been a better place because of Jerrold Reinach Zacharias."

It was an appropriately admiring sentiment from one fine and distinguished old man to another, and it was certainly reciprocal, but it was not what either would have written at any earlier time in their lives. They had been young together, full of drive and intelligence and energy, with a fair measure of mischief and fun. They had been much more interested in getting important things done than in hearing compliments from each other.

But in fact the world did become a better place because of Jerrold Reinach Zacharias. Here is his story.

A Different Sort of Time

1

Early Years

Origins

Where did you come from?
Where will you go?
Where did you come from,
My cotton-eye'd Jo?[1]

Jerrold Reinach Zacharias was born on January 23, 1905, in Jacksonville, Florida, in a comfortable two-story house that had recently been built by his father, Isadore.[2] Theodore Roosevelt had just been elected to his first full term in the White House, a populist president presiding over a growing nation whose forty-five states and territories claimed a population of nearly 80 million. The country had largely recovered from both a serious depression in the 1890s and a bitter and divisive political dispute made famous by an impassioned debate on the gold standard and the purchase of silver. A brief military engagement with Spain over Cuba in 1898 had led to a speedy victory for the United States, and the country was beginning to feel its muscle. Jacksonville had been a principal depot for troops and supplies headed for Cuba and had prospered in the war. A spirit of optimism and self-confidence permeated virtually every group and every class, save, of course, the southern blacks, whose prospects remained bleak.

Business was good in 1905. In Jacksonville, the largest city in a predominantly agricultural southern state, you could have taken a trolley ride for two and a half cents, and for a nickel you might have refreshed yourself either with a bottle of Coca-Cola or one of a local rival, Koca-Nola—take your pick.

By 1905, the Zacharias family was already long established in Jacksonville. Jerrold's paternal grandfather, Aaron Zacharias, a tobacco dealer and general merchant, was listed in the earliest city directory, dated 1870, as doing business in the downtown area. He had been born in either New York City or in Prussia in 1845[3], of German-born parents who had come to the United States in the first great tide of Jewish immigrants that flowed from various parts of Germany, beginning in about 1835. These were Jews who were escaping the restrictive laws that came into effect in post-Napoleonic Europe, severely limiting the occupations in which they might engage and even the number of marriages that might take place in the Jewish community.

Aaron Zacharias and his younger brother, Abraham, moved to the American South immediately after the Civil War, as did many other German Jews. Those who made that migration typically found opportunity, becoming tradesmen, merchants, and settlers—pillars of the community. Like them, Aaron and Abraham fulfilled for themselves the immigrant's dream. There were perhaps 200 Jewish families in Jacksonville at that time, and they formed an important part of the city. By 1882, Aaron was the secretary of the city's oldest Jewish congregation, Ahavath Chesed, and the Reporter (the recording secretary) of the fraternal organization Knights of Honor.

Aaron Zacharias married Theresa Budwig, who had been born in Germany in 1855 but about whose origin little else is known, except that a branch of the Budwig family had settled in Ohio before 1866. Aaron and Theresa had seven children, all born in Florida; their second child, Isadore August Zacharias, born in 1876, was Jerrold's father.[4]

Jerrold's maternal grandparents were Julius Kaufman, born in New Orleans in 1854, and Leah Reinach Kaufman, born in Memphis, Tennessee, in 1857, the daughter of Abraham and Elise Reinach. The names of two German towns are written on the flyleaf of Abraham Reinach's bible: Essingen and Liest. Presumably their roots were in those places. Julius and Leah Kaufman had three children, the oldest of whom was Irma, Jerrold's mother. She was born in New Orleans in 1879, the eldest of three children. Sometime before 1895, the Kaufman family moved to Jacksonville, where Julius went into the scrap metal business and where, in 1901, Irma Kaufman married Isadore Zacharias.

By that time Isadore was a promising young attorney, in business for himself and beginning to invest in real estate. He and Irma had two children: Dorothea, born in 1901, and four years later, Jerrold.

Jacksonville in 1905 was small, but attractive and lively. The St. John River, a wide and particularly beautiful waterway, lent it much charm, and most of the city lay within an easy walk of its banks. Jacksonville was an essentially southern city, which had changed hands three times in the Civil War, with great suffering and damage. Memories of that conflict, which had ended only four decades earlier, were still sharp in the minds of its citizens. A contemporary description of Jacksonville records "many handsome buildings . . . its residential streets are shaded with live-oaks, water oaks and bitter orange trees."[5] Another assessment, somewhat less flattering, described it as "a one-horse town, divided by Hogan's creek."[6] It was then Florida's principal city, with a population of about 50,000; the population had doubled twice in the preceding thirty years and was still growing rapidly. The railroad connecting Florida with the North ended at Jacksonville; the city, fresh from the prosperity of the Spanish-American War, was known as the Gateway to Florida. It was a major port as well as the rail center of the region.

In May 1901, a disastrous fire in the city had destroyed nearly all of the commercial district. Nearly 150 blocks of buildings burned down. By the time Jerrold was born in 1905, however, more than 5,400 new buildings had been erected, and the rebuilding was far from over; by 1910, more than 9,000 new buildings had been constructed. In a time of such explosive growth, it would not have been hard for an intelligent young lawyer, interested in real estate, to do well, and Isadore Zacharias did. By the time Jerrold came along, the family was prosperous.

They had not always been comfortably off. While Isadore read for the law during the depression years as clerk in the Jacksonville firm of Walker and Stripling, he had been obliged to support himself in part by playing the violin in the small orchestra of a local theater. But he had been glad to do that; both he and Irma shared a deep interest in music. Years later Jerrold would describe his parents as "fiddle-struck." Irma was a highly gifted musician and eventually became a serious and well-trained violinist. Such musical accomplishment and interest cannot have been common in boom-time Jacksonville. The city directory for 1905 lists some seventy five attorneys-at-law, but only one piano tuner.

Irma Zacharias was small and slender, barely five feet tall. She was a striking woman with dark, intense eyes and an imperious manner. She had been born during Ulysses Grant's presidency into a society that considered the only role suitable for women, especially southern women, to be in the home. Constrained by custom as firmly as by whalebone, Irma Zacharias could not vote, could not, once married, own property in her own right, and, at least as far as the outside world was concerned, was not expected to have independent opinions worth considering. There was never a time in all her years as a young girl or young matron when she did not have several servants to attend to her. Nevertheless, she developed into a remarkably independent individual, remembered as "a little Napoleon," gender and relative size notwithstanding.

Isadore, by contrast, seems to have been a more genial sort. He was above average height, with impressive good looks. He was fond of cigars and good company, and he was an admired raconteur. He was a skilled businessman and a knowledgeable investor. People tended to rely on him; he managed investments for a number of friends and relatives. He evidently adored his wife and let her make many of the critical decisions affecting their personal lives, including those that involved the family in the world of music performance and would eventually move them all to New York.

As a young married woman, Irma Zacharias proved to be enterprising and decisive. As her musical ability developed, her ambitions grew accordingly. The musical isolation of Jacksonville was clearly not to her liking, and, characteristically, she dealt with it directly and effectively. She became an amateur impresario, bringing a series of professional musicians, such as the young Mischa Elman, to perform in Jacksonville. Her initiatives brought her many important and abiding friendships with professional musicians.

Sometime around 1911, her schedule of performers included Beryl Rubinstein, a brilliant twelve-year-old pianist who was already a five-year veteran of the concert circuit. Earlier, in 1910, managed by his father, an orthodox rabbi from Athens, Georgia, Beryl had made his New York debut with the Metropolitan Opera Orchestra and had toured hard thereafter. By the time they reached Jacksonville, the young performer was ready to rebel. It seemed apparent to the Zachariases that the boy was being tyrannized and exploited by his father. As Jerrold later recalled, the boy was so miserable that he confessed planning to run away and never again play the piano. The Zachariases agreed to adopt him, and Jerrold, at age six, suddenly had an older brother.[7]

Irma and Isadore took seriously their responsibilities as patrons of the young prodigy. Irma decided to take him to Berlin for appropriate training, and, with Dorothea and Jerrold in tow, they traveled there later in 1911. They made additional trips to Berlin in 1912 and 1913, and Jerrold could remember years later how he had noticed that the streets seemed full of soldiers. Irma found a distinguished violin teacher in Berlin for herself as well. She was naturally much preoccupied with music on these trips, and Jerrold spent a good deal of time waiting, not always patiently; he later recalled that his patience and good behavior would be solicited by the promise of a visit to the zoo. He also remembered that the promise was not always fulfilled unless he insisted on it.

Irma Zacharias had decided that Jerrold had no particular talent for music, and in the months and years that followed she was deeply occupied with Beryl. Jerrold remembered that Beryl and his mother often played Beethoven sonatas together, while he, more or less content to be left alone, played on the floor.

His sister, Dorothea, was not a very satisfactory companion for him, then or later. She was four years older than he and was herself busily involved with the piano. Jerrold used to say that living surrounded by so much music taught him to "shut his earlids." The experience put him off music for a long time, he said. Fortunately, he rediscovered music as a young adult and became an appreciative and knowledgeable listener, especially enjoying Mozart. For most of his adult life, he took great pleasure from music.

In any event, there was someone else in the picture who did provide real companionship for Jerrold: his nurse, Anna Liza Johnson, who had come to work for the Zacharias family when Jerrold was six months old. She was an illiterate black girl who in 1905 had walked the nearly fifty miles from Woodbine, Georgia, to Jacksonville, somehow arriving at Grandmother Leah Kaufman's kitchen. Leah Kaufman sent Anna to her daughter, Irma, who hired her to take care of young Jerrold. Anna gave her age as sixteen when she was hired but many years later confessed to having been only fourteen.

Anna stayed with the Zacharias family for the rest of her life, remaining close even after her retirement. Although she never learned to read and write, she developed into a strong, competent and intelligent woman. She adored Jerrold, to whom she referred as "my boy," and he spoke of her as "my real mother." Anna's influence on him was strong and lasting. Many years later he reflected bitterly that when Anna was growing up in Georgia, there was no school in the entire state that she could have attended. Jerrold noted a cruel irony in the fact that when he eventually achieved national attention and praise for his innovations in education, Anna could not read the magazine and newspaper articles describing his accomplishments.

When Jerrold married and had children of his own, Anna came to help take care of them, and they too loved her deeply, taking the same pleasure in her inexhaustible store of invented stories as had their father in his time.

Growing Up

Almost no one now is left to recall the days of Jerrold's boyhood and youth. Rebecca Brown, a few years older than Jerrold, lived as a young girl just around the corner from the Zachariases. She knew the family and remembered Jerrold as a bright and lively youngster who liked to run after fire engines and who with his friends occasionally got into trouble digging holes in the neighbors' gardens.

Much can be surmised about Jerrold's youth, which seems to have been normal and healthy in all respects. He was certainly intelligent and able although not always in ways that earned him much credit at home, where musical ability was the paramount virtue. He was a good but not outstanding student from the beginning, and he continued to be so in high school and college. He was far from being a solitary type; he became a Boy Scout when he was twelve, and later, in his teens, was a member of the fraternal Order of De Molay, the youths' affiliate of the Masonic Order. These were important associations in the Jacksonville of that time. Jerrold was evidently very likeable. Photographs from the period display a grin that is engaging even across seven decades.

One naturally looks for early signs of interest or aptitude in science, and they are not lacking. Years later, writing in 1972, Jerrold reminisced about his early experiences:

None of my scientific colleagues has ever disagreed with me about how and when he became a professional scientist. The details are different but the general plan is the same. For me, it started when I was four, and the images are still clear. There was an automobile, a 1909 White Gas, belonging to my grandfather; the chauffeur showed me the gears, how to change the tires, how the engine worked. Later, with tools and trees and lumber and the abundance of stuff that nature supplies in a small town, it was a full experimental life. For reading, there were Jules Verne, Tom Swift, the Motor Boys and the Rover Boys. . . .

A friend, Dr. Helene Deutsch, the psychoanalyst, has told me that most grownups recall their first deep thoughts as occurring at age nine. Dr. Helene, I know you are right; I was nine, and I know where I was. My family and I were spending the summer of 1914 in Asheville, North Carolina . . . I asked myself various hard questions. For one, if I shine this flashlight out into the night sky, how far does the beam go? How far is it out there? How could time have begun? If everything is made of atoms, why does this piece of stone or wood have such a hard surface or a sharp edge? And on and on.[8]

He became especially interested in cameras and photography in his early teens and fooled around with crystal-set radios. Long after, when he talked about the introduction of sophisticated silicon crystal rectifiers in radar during World War II, he recalled: "I started working with crystal receivers at the age of twelve or fourteen, putting a needle or a pin on a piece of galena . . . and I could receive signals from Norfolk, Virginia, so I knew about semi-conductor rectifiers."

He was twelve years old when the United States entered World War I, and he remembered that his father had great difficulties at the time finding people to work in his office. He and Dorothea were drafted to do clerical work—one in the morning and the other in the afternoon. As a result, he learned how to type; later, as a member of a rather high-powered physics research group, he found that he was the only one who knew how, so that the job of preparing texts for publication generally fell to him.

His affection for the gasoline engine, begun early with that 1909 White Gas, lasted all his life. He liked to say later that you could learn more good physics and more practical know-how from a Briggs & Stratton one-cylinder lawn mower engine than from any other single device he could think of. He learned to drive the family car almost as soon as he could see over the steering wheel; Florida did not require drivers' licenses in those days.

He started high school in 1918. Duval High School in downtown Jacksonville was then the only high school for whites in the fully segregated city. His high school career was successful, both academically and socially. His report cards have survived, and apart from a small difficulty with spelling in his freshman year, which never recurred, all of his grades were excellent.

He was evidently popular, judging from his club memberships and recorded school activities. He was a member of the group that edited the class yearbook and the first president of the French club, which he appears to have organized; subsequently, his applications for various positions always listed French as a language he could read and write easily. He seems generally to have sported a little bow tie, judging from his yearbook photographs. He was a "tap member" of the Senior Fellows Club, meaning that he had been chosen in his junior year, another sign of popularity. He had a leading part in the senior class production, and his characterization of a successful but unscrupulous businessman won him a reviewer's praise.

His contribution to the senior yearbook, curiously, was a piece complimenting the school's manual training program, singling out mechanical drawing for particular commendation. Two years later, in college, he would switch from an engineering program into physics because he disliked mechanical drawing so much that he did not want to go on with it. But as a senior in high school, the subject apparently still had some appeal.

He graduated Duval High School in 1922, when he was seventeen. He studied physics in his senior year, choosing it over chemistry, which he did not like very much; he explained to his mother that physics was just like chemistry, but it did not smell as bad. He did well in physics but recalled that he did not think it was interesting. Physics in those days meant learning the types of pulleys and the classes of levers, and if the beginning texts were very modern, they would have included some material on direct current electricity. Such matters as radio, including Jerrold's crystal set, would have only appeared in extracurricular activities, if at all. But there is absolutely no indication in the senior yearbook of anything like that—no radio club, no photography club, no science club.

It is unlikely that Duval High School provided any but the simplest sort of laboratory work of the most routine variety. In fact, it would have been hard for any one to generate a passion for physics in any ordinary American high school around 1922, and most of those who did generate such feelings did so on the basis of extracurricular activities. But that did not happen to Jerrold. There is no evidence that his crystal sets and cameras provoked much continuing excitement in him at that time, even if later he remembered moments of wonder.

All in all, he did not find schoolwork difficult, and he performed well, but when he assessed it later, he did not think he had learned much—as he put it, "that education never laid a glove on me." Perhaps not, but neither did it kill his curiosity. He recalled something about that in a talk he gave to college undergraduates fifty years later:

Let me start as a little boy in Jacksonville, Florida. There was an electric generating station called the Waterworks, and every time I passed by the Waterworks and saw steam and smoke coming out of the smokestack . . . a steam engine driving a generator, I would say, Now wait a minute, that's kind of silly. You can run a motor with an electric current, so why don't they just use the motor to run the generator that generates the electricity that runs the motor? Then you'd have the current free. Why do you waste all this coal and stuff?

There was a young fellow, who was an electrical engineering student at MIT—this was a long time ago, 1920, or earlier—who said, Oh, there's something in the second law of thermodynamics. I said, That's mysterious. I don't know what it means, but I believe you. Maybe I should understand something about this, because Tom Swift and his Electric Rifle, and the Motor Boys and Jules Verne could always do quite unusual things . . . it was clear that there was something to this education business.[10]

There was a major scientific revolution going on during the first decades of this century, a revolution as great as the one that had changed the world in the time of Newton and Galileo. But there was hardly any way Jerrold could have heard much about it in a small town and surely no way that he could have guessed that in only a few years, he himself would be caught up in it. He was, after all, only a bright young kid in Jacksonville, living a pleasant, unchallenging life in a world very far from the world of Einstein and Bohr, of Heisenberg and Pauli, of Rutherford, of Planck, of the Curies. And yet he was headed for that very world.

In fact, until the mid-1920s, one would have been hard put to discover much trace of the physics revolution even in the best American universities. The exciting developments had been happening mostly in Europe until then, but the change was on its way to America. Jerrold Zacharias, born near the beginning of this scientific revolution, came of age just about when it arrived in the United States. When he eventually got into modern physics, the field was intensely active, at the flood of change and discovery. Thinking back over his life as a physicist, he marveled at what had happened. "I have had the great luck," he said, "to be living through and participating in some of the few great revolutions in physics, such as the quantum mechanics."[11]

In the first two decades of the century, the outstanding achievements of American physicists for the most part were those of precise measurement; the triumphs of theoretical understanding largely belonged to the Europeans. In America, pride of place went to experiments that were often considerably elegant, but nevertheless were more practical than conceptual. It was an important distinction, for when American physics finally embraced the new theory in the mid-1920s, it did not abandon its dedication to precise measurement. Rather, America was able to contribute something of its own. A new kind of physicist emerged, one who was able to combine high skill in the laboratory with deep theoretical understanding. Such scientists were precisely what was needed to build on the largely theoretical European triumphs.

But these triumphs were not widely known. For the public in those years, American or European, scientific achievement meant things like the first radio transmission across the Atlantic by Marconi in 1903 or the first phonographic recording of Caruso; the year 1915 was probably more notable in the United States for the first transcontinental telephone call than it was for the theory of general relativity. The public heroes of technology were those who built bridges and tall buildings. In 1905, the year of Jerrold's birth, the tallest building in America was the Philadelphia City Hall, whose spire rose to the then astonishing height of 505 feet. By the time Jerrold graduated Duval High, that record had been broken many times. Those were the engineering feats that a high school student would have heard about. Not surprisingly, when Jerrold Zacharias graduated Duval High at age seventeen, he decided to become an engineer. He had no reason at all to consider being a physicist or even to think about it.

New York

As a young man, Isadore Zacharias had wanted to go to New York to study law at Columbia University. However, the family's financial circumstances at the time of his graduation from Duval High School did not permit him to attend either Columbia or any other law school, and so, following a common practice of the day, he prepared for the bar by reading law as a clerk in the private offices of a Jacksonville law firm.

In trying to make an arrangement with Columbia for himself, however, Isadore somehow was able to establish that graduates of Duval High School would be permitted to matriculate without taking College Board Entrance examinations, and in 1922, Jerrold was able to enroll at Columbia under this dispensation as a member of the freshman class. He was not, however, exempt from having to take an IQ test in order to matriculate, and he recalled, years later, one of the questions: "How can you keep a wagon horse's hooves from slipping on the ice?" It was a curious question for a southerner. "The horse is not on the ice," answered the kid from Jacksonville; "the ice is in the wagon." No doubt Jerrold's tongue was in his cheek as well. But equally clearly, he had even then the beginnings of the iconoclastic attitude that decades later, in the 1960s and 1970s, would bring him squarely into the controversies on IQ testing. In 1922, however, iconoclasm notwithstanding, he managed to pass the examination and became a Columbia undergraduate.

The decision that Jerrold would attend Columbia turned out to be far more than a simple matter of academic choice. Irma Zacharias made it the occasion to move the family, including Anna, to New York, leaving Jacksonville and their southern roots permanently behind. Probably she had had this in mind for some time, but in any event, by 1922 she could no longer be content with life in Jacksonville. She had social ambitions for her daughter, Dorothea, and she required for herself much more contact with the world of music. The family took up residence in a large, handsome apartment at 114th Street and Riverside Drive, only a short walk from Columbia University , and they settled in to their new lives. Isadore maintained his law and real estate interests in Florida all during the 1920s, however, and commuted almost weekly to Jacksonville.

Irma maintained a sort of musical salon in the New York apartment. Isadore, who continued to idolize her, purchased for her a Stradivarius violin, the Berou, dated 1714. Irma renewed her contacts with Mischa Elman and with other musicians whom she had brought to Jacksonville in her impresario days. She entertained often and splendidly in these years. Jerrold remembered that the young George Gershwin was a frequent caller.

Irma Zacharias was a complicated woman. By the time the family moved to New York, she had developed into an excellent violinist, outstanding both as a teacher and as a chamber music player. Some years later she became first violinist in the Dorian String Quartet, a group that achieved some success in New York concert halls. She had many friends, who along with her pupils admired her for her energy, style, and talent. But she was also a difficult woman—highly prejudiced, snobbish, domineering, and opinionated. People were rarely indifferent to her: they found her admirable or impossible. Increasingly, Jerrold had trouble reconciling his mother's obvious abilities and talents with what he came to see were her serious faults.

He was keenly aware of how his mother saw him: a poor third after the eligible Dorothea and the talented Beryl Rubinstein. He saw that it was important, even essential for him to achieve some separation from the family, from Irma's dominating personality. If he was ever to accomplish anything, it would have to be on his own. Quietly and gradually, without inviting discussion or argument, he became less and less involved with the activities of the family and more involved with friends and activities of his own choosing. It was a pattern that became permanent.

Judging by external appearances, it would have been reasonable at that time to think of Jerrold as blessed by fortune. As an undergraduate at Columbia, he had a substantial allowance; he belonged to a fraternity, and he enjoyed tea dances. He had his own Packard touring car nicknamed "Blue-y," and he sported a raccoon coat. His own idea of himself when he later spoke of this part of his life was that he was a playboy.

He was nevertheless a fairly good student, as he had been in high school, although rather an easygoing one. In his first year at Columbia, he enrolled in the appropriate engineering courses and found them not difficult but not overly interesting. On the advice of his professors, he deferred physics courses until his sophomore year, but when he finally got around to them, something clicked, as it had not done in high school. He could do extremely well in those courses with little strain or effort, and he took all the physics courses he could manage. In the end, the required course in mechanical drawing drove him out of engineering; he had lost patience with such things in the time since he had written about them so favorably in his high school yearbook. It was not possible to major in physics at Columbia at that time—few American universities provided such an opportunity—so he majored in mathematics with a physics minor.

In 1925, as a Columbia junior, still seeking to distance himself from his family, he met two people who were to have a profound effect on him: Leona Hurwitz, whom he married in 1927, and I. I. Rabi, then a physics graduate student, who became Jerrold's mentor and lifelong friend.

Leona

Leona Hurwitz was just eighteen when Jerrold met her. She was a sophomore at Barnard College and a serious student of biology, looking forward to a career. She was very popular, beautiful, and clearly capable and bright. They had met when Jerrold was brought along as a blind date for one of Leona's visiting friends and again when Leona was taken by a mutual friend to one of Irma Zacharias's musical parties.

They began to see each other more frequently. He found her charming and good to be with; as she put it, she saw that he would always be entertaining, never dull. They were married on June 23, 1927, about two weeks after they received their degrees: Leona, the A.B. from Barnard College and Jerrold, the A.M. from Columbia. He was planning to continue as a graduate student in the fall and suggested to Leona that she do the same. Ultimately, she did enter graduate school, but not until a few years later, and by that time their circumstances had changed considerably.

Although they had moved in the same circles before they were married, their backgrounds were rather different. Leona was the daughter of Joseph Hurwitz, a New York City high school teacher of mathematics, and his wife, Edith Gottfried Hurwitz. Joseph had been born in New York City, the child of Russian immigrants; Edith had come to the United States as a four year old, accompanying her widowed mother, Leah, from Iasi, Rumania. Both had grown up on the Lower East Side of Manhattan; Joseph Hurwitz had managed to attend the·City College of New York, graduating in the class of 1900.

Economically, Leona's family circumstances in 1927 were much more modest than those of Jerrold's family, and a considerable social distance separated the Hurwitzes and the Zachariases as well. Leona's grandparents had come from Eastern Europe rather than Germany, something that Irma Zacharias was unlikely to overlook. She had a remarkable mimetic talent and enjoyed imitating foreign accents, but she found it distasteful to imitate the Yiddish accent that denoted Eastern European origins and could not be persuaded to do it.

Joseph Hurwitz had a wonderful sense of humor and a great affection for puns and word play, which he shared with his son-in-law, for Jerrold too always loved aphorisms and puns. In time, he would be guilty of many of them. Joseph thought very highly of Jerrold, and they got along extremely well.

Leona remembered being taken by her parents to plays, concerts, operas, and dance recitals. Typically they went to Saturday matinees, even when it meant climbing up to the cheapest seats just under the roof. She was sent to the Ethical Culture School in New York for her secondary schooling, and there she encountered modern trends in education and social thought. She was encouraged there and at home to think in terms of a career for herself although that was not common at the time. But she recalled that there was never any question about whether she would attend college. As far as she knew, it was simply assumed that she would.

Both before and after marriage, Jerrold seems to have been conscientious about his studies, but outside the classroom and laboratory he was still very much the playboy. Leona remembered that when they were married, he had fifty-two pairs of socks and a dozen dress shirts. He had the raccoon coat, of course, which is remembered still, and his blue Packard touring car. He maintained a membership in a golf club right up to the time of their marriage. Nevertheless, there were some signs of a growing seriousness. In the summer of 1926, for example, while he and Leona were engaged, he managed to devote some time to the Leiden Summer Conference on Physics while on an otherwise carefree vacation trip to Europe.

He was just twenty-two years old when they were married, and Leona was only twenty. Their honeymoon started brightly enough; they sailed to Europe tourist class on what Leona remembered as a "miserable ship," the *S.S. Carmania,* which Cunard Lines scrapped the following year. Their plan was to bicycle around France, but their program ended abruptly in Strasbourg when Jerrold came down with a serious case of hepatitis. Irma Zacharias, in Europe with Isadore and Dorothea at the time, marched dramatically to her son's bedside and took over. Leona found herself peremptorily displaced.

Jerrold and Leona returned to New York as soon as he could travel again, and they took up residence—rather reluctantly, Leona recalled—in the spacious Zacharias apartment on Riverside Drive. They had a living room, bedroom, and bath to themselves but were expected to take meals with the family; dinners were formal and elegant. They had not yet achieved the full separation from his family that Jerrold wanted; they did not yet have the full independence or the full responsibility that comes with setting up one's own household. They were, of course, still very young, and in Irma's view there was still much for them to learn. In fairness, the arrangement cannot have been easy for the senior Zachariases either. Leona remembered that both Irma and Isadore were often generous and forbearing.

In the fall, Jerrold was well enough to return to his studies at Columbia, and Leona found a job as a research assistant in the Herpetology Department of the American Museum of Natural History.

Apprenticeship of a Physicist

There were only two members in the Columbia College Class of 1926 who were physics majors. And if it had not been for Rabi, I would very likely not have stayed the course.

—J.R. Zacharias[12]

Undergraduate physics education at Columbia at that time was mostly mathematics, some theory, and a set of canned laboratory exercises, that is, preplanned experiments with known outcomes. There were no real labs to work in, Jerrold recalled; he said that "there was no one who insisted or made a point to link theory to practice."

In graduate school, he was drawn to Shirley Quimby, a specialist in what would later be called solid state physics, who agreed to direct his dissertation research. Quimby was a man of great precision. He taught an elegant course in electricity and magnetism, each lecture beginning exactly where the previous one had ended. Proof followed didactic proof in logical order rather than in the more human way in which any subject actually develops. Unfortunately, physics continues to be taught along such lines even today—unfortunate, because it suggests that the scientist never deviates, is never in error, never follows a false trail. In time, Jerrold came to understand that while physics might be sometimes taught that way, real research follows a more complicated path. Many years later, when trying to sum things up for an interviewer, he said: "The shortest distance between two points is a straight line. But you rarely get there as the crow flies. Even science, which to the naive observer seems to move right foot, left foot, doesn't work that way. If only it were that straightforward."[13]

Quimby was, first, a laboratory man, a measurer, and he was attractive to Jerrold for that reason. He was relatively young; he had just been promoted to assistant professor in 1926 at the age of thirty-three. But he was a man of parts. He was a Californian who in addition to his academic career became an accomplished amateur magician, the head of the New York Assembly of the Society of American Magicians, and editor of its national publication, *Sphinx*. He was the official measurer of the New York Yacht Club, and Jerrold often accompanied him to the docks, measuring tape in hand.

Quimby gave Jerrold a laboratory to work in and a problem to work on. Quimby had earlier worked out an idea for a new way to measure the stress-strain relationship in a solid—that is, how much force it takes to produce a prescribed distortion of a given type. A single number, known as Young's modulus, is enough to describe this relationship; it is the ratio of the force to the distortion it produces.

The object was not simply to measure this number, which was already known approximately, but to do it with high precision. Quimby's idea was a simple one. Every solid object vibrates at natural or characteristic frequencies, which are sharply defined if the internal friction in the material is low; thus, a bell emits well-defined characteristic tones when it is struck, and so does a metal rod. Softer materials, like clay or putty, have high internal friction and so do not ring well.

The natural frequencies of an object depend on its shape, as might be expected, and on the value of Young's modulus as well. Since frequencies can be measured with great accuracy, a means was at hand for determining Young's modulus with greater precision than ever before. Jerrold took on the problem of measuring Young's modulus for single crystals of nickel using Quimby's idea. The choice was one Jerrold apparently made for himself, and it proved not to be an easy task. It took him more than three years to do it.

Single crystals of nickel, of a size sufficient to permit careful shaping, cannot be purchased off the shelf. Purified nickel can be obtained commercially, but it is normally in a polycrystalline form. To get it into the form of a single crystal, the nickel had to be melted at incandescent temperatures in a special furnace he had built for the project and then allowed to cool slowly and recrystallize under very carefully controlled conditions. Great care was necessary to avoid even the slightest impurities, which would cause the delicate crystal structure to fragment. Tiny local fluctuations in temperature would also cause the crystalline structure to be imperfect, so that the whole process would have to be done over again. Even small, uncontrollable vibrations in the building itself sometimes caused imperfections in the crystal structure, and sometimes imperfections would result for no apparent cause whatever. It was painstaking work, and Jerrold often came home disappointed after a wasted effort.

Quimby was a demanding supervisor with stern ideas about what constituted a suitable apprenticeship in physics. No short-cuts were permitted to a young experimenter, nor was there ever a suggestion that the goal might be redefined or that any sort of failure might be acceptable. If the work was painstaking and demanding, well, that was the nature of physics, at least according to Quimby; you did it that way or not at all.

Eventually Jerrold completed the work and obtained his doctoral degree, but it left him unsatisfied. He had enjoyed the actual work; he had learned to grow the large single crystals himself, and he had built both the special furnace and the precise frequency equipment that the experiment demanded. He liked doing that and was capable of the patience and persistence required to make it all work. But something was missing, he thought. The measurements were useful, even valuable, but they also revealed puzzling properties of the metal for which no one seemed to have an explanation. That bothered him.

What he had discovered in his experiments was that as the temperature of his single-crystal nickel samples was raised, the internal friction continued to increase; the material got more and more puttylike. Then, when the temperature reached a value known as the Curie temperature, all internal friction abruptly disappeared, and the sample rang like a bell again. The Curie temperature originally had been singled out for other reasons in studies of ferromagnetism: it is the temperature at which the magnetic properties of the material change dramatically. But what could the magnetic properties of a material have to do with its response to stretching and pulling? It was perhaps an important discovery, but no one could understand it. The theory of metals was still principally the simple free-electron theory set forth by P. Drude and H. A. Lorentz around the turn of the century,[14] and the new quantum theory was only just being applied to this problem by A. Sommerfeld and W. Pauli in Europe. Understanding of these phenomena was still years in the future.

Jerrold decided that there were more interesting problems to pursue. For example, in 1925, while Jerrold was still an undergraduate, C. E. Davisson of Bell Labs had given a Friday evening colloquium talk at Columbia on the recent work he had done in collaboration with L. H. Germer, showing that electrons are particles that sometimes behave like waves. That talk was Jerrold's first exposure to the new and revolutionary quantum mechanics, and years later, he remembered that he had felt a sense of outrage at what seemed so self-contradictory. "And I went to Rabi after the colloquium," he recalled, ". . . he was seven years older, even then . . . and he said, look, look at the data, that's what the data says [sic]. It's going to take some understanding, but don't reject it. And that was 1925."[15]

Jerrold had been attracted to Quimby's lab by the sense of precision that he found there and had greatly enjoyed the work. But now he wondered what to do next. He had become a true physicist, after all. He was no longer satisfied simply to make things work; he needed to understand why.

Physics in the Thirties

Getting Started

Jerrold finished his dissertation research and began the task of writing it up in 1931. The depression was near its lowest point and his family circumstances had become greatly altered. Much of the money that his father had accumulated had disappeared in the great crash of 1929, and the family had had to accommodate to a more modest style.

Earlier, in 1929, Jerrold and Leona had moved to a small apartment of their own on New York's East Side. In the wake of the stock market crash, his parents' apartment on Riverside Drive was unnecessarily large and expensive, and the senior Zachariases also moved to a smaller place, on West 71st Street. The beloved Stradivarius (Leona remembers that it had been thought of as the most important member of the household) was sold. Jerrold's fellowships had run their course. Leona had several jobs at once, but she was also in graduate school and was expecting their first child. Jerrold needed a job.

The job market for physicists had not totally vanished, although in 1931 things were beginning to be difficult and by a year or so later were bleak. The physical chemist Linus Pauling wrote to his physicist friend Samuel Goudsmit in May 1933: "I haven't the faintest idea as to where [your former student] can get a job. Caltech is filled with our own PhD's and former National Research Fellows hoping for a small stipend. It is a shame these able men should be without positions. We have had only a 10% [salary] cut, a year ago, but may well have another. I am hoping that conditions will improve soon."[1]

There were still some industrial jobs among the scarce opportunities that remained in 1931, and the possibility of an academic job, though remote, could not be ruled out. Columbia was out of the question; Jerrold's friend

Rabi had been appointed lecturer there after returning from two years in Europe in 1929, an apostle bringing with him the gospel of the new physics.[2] The Columbia physics department, having so recently hired its first Jew, was not ready to hire a second. That possibility was not even considered.

It may be difficult today to realize the prevalence of antisemitic attitudes at that time, to understand how they were exacerbated by the worsening economic conditions, and to recall how persistent and pervasive these attitudes were, at least until World War II. Complete exclusion of Jews had been the rule in some physics departments in the 1920s and in some industries as well.[3] "Bell Labs didn't take any Jews," Jerrold recalled, "no University took Jews, no Jewish engineers got jobs anywhere. . . . Rabi was hired by Columbia primarily because Heisenberg said 'you've got one man here on fellowship who's wonderful.' So they swallowed their pride and hired him. I mean not MIT, not at Harvard, not at no place."[4]

The antisemitism offended Jerrold deeply. He remembered an interview at that time with a representative of a major instrument company, in which the company declined to make an offer once it was realized that Jerrold was Jewish. "We had a Jew once," he was told. "It didn't work out."

Even after he had obtained an academic position (he felt that the department that hired him needed an "exhibit Jew"), he found that these attitudes persisted. Once, not many years later, when the members of his department were invited to the country home of a senior colleague for a picnic, he and Leona were not included. "The neighbors wouldn't understand," he was told. Such experiences angered him and left a bitter residue, but he did not let the antisemitism, prevalent though it was, discourage him. He had a practical attitude about things that were hurtful, a way of putting them aside and getting on with his life. But the hurts were nonetheless real, and he never forgot them.

Jerrold eventually received an offer from the Sperry Company but by that time had decided that he would prefer to teach, if possible, and to find some way to continue doing research. The choice was not based on financial expectations, of that one may be certain. The median salary for physics instructors in 1931 was a little less than $3,000 per year and only twice that for full professors.[5] Academic life clearly did not promise luxury. Nevertheless, in spite of a growing national investment in industrial research and its rapidly growing popularity among new graduates, the common understanding was that pure research was done principally in the universities, and Jerrold hoped to find a way in. Others had managed to combine teaching at local colleges with research at Columbia; this would be his route as well.

When a teaching position opened up at Hunter College, the municipal college for women in New York, he applied for it and was appointed tutor in physics at a salary of $2,000 per year. He recalled in 1985 what that job had been like:

Being a lecturer in physics at Hunter was no doubt typical of physics teaching at many small colleges in the '30s, probably in the '80s as well. My teaching load was sixteen contact hours per week, to which one must add all the student, faculty and administrative contact hours. I had a desk in a room fitted as a laboratory for undergraduates. I had neither the time, the colleagues, the space, nor the stuff (cheap labor in the form of graduate students and apparatus in the form of existing equipment or a skilled machine shop) to carry out my own work. A small college in a large city may be a fine place for working on literature, history, or any kind of solitary research that requires only a brain, a library, a pencil and a pad for a lab. But it's no place for doing experimental physics.[6]

He became a very successful teacher. Rebecca Salant Skydell, one of his students in a junior year class in electricity and optics, recalled him as "sweet, very gentlemanly, very understanding. He was very clear. Everybody liked him. He never put anyone down, as people tend to do in girls' schools. He was just happy to have girl students interested in the sciences." She remembers him as having "dark, straight hair, a round face, kind of boyish, but already with a slight professional stoop. An adorable-looking man." And Rosalyn Sussman Yalow, who shared the Nobel Prize in Medicine in 1977, took Jerrold's optics course when she was an undergraduate at Hunter. She recalls the strong interaction between the small class of students and the teacher.

Jerrold spent his first year at Hunter, he said, "settling in, learning what it takes to teach physics to young women, and putting the finishing touches on my thesis." The dissertation was to be published in the *Physical Review* after the usual lengthy peer review process.[7] This was the desirable way to proceed, not only for the considerable prestige attached to publication in that journal but because Columbia, like most other universities at the time, required that one hundred printed copies of the dissertation be presented. It was far cheaper to order one hundred reprints of a published article than to have them printed privately, but the process seemed interminable. Overall, it took two additional years; Jerrold finally received his degree at Columbia's commencement in June 1933, one of only 133 individuals nationwide who received the Ph.D. in physics that year.

The year 1931 thus saw Jerrold with a new job and with fatherhood imminent. He and Leona moved yet again to a larger apartment, more suitable for a family with a child, on the West Side of the city. In October, their daughter, Susan, was born, and Anna, who had taken care of Jerrold when he was an infant, came to cook for the family and to take care of their baby. Things were settling down comfortably, and by 1932 Jerrold was ready to get back into a research laboratory.

It would have been natural at this point for Jerrold to look for research opportunities in the area of solid state physics, since he had done his dissertation work in that field. He could have anticipated help from Quimby had he done so; dissertation supervisors are more or less expected to give that sort of assistance, and Quimby, who liked Jerrold, would have been glad to oblige. Instead, Jerrold jumped ship, as he put it; he went to see his friend Rabi and asked for a place in his lab. "There was something very appealing, very attractive about him," Rabi later said. "I had known him for a while, a few years. He used to stop and watch the baseball team on the way to the lab. He had a coat, a raccoon coat, I think it was. Oh, he was very capable. He had a lot of energy." For his part, Jerrold said he struck a deal with Rabi: "I told him that I would, without pay, spend my spare time—afternoons, nights, weekends and holidays—working in his lab, on one condition: we work on atomic hydrogen."[8]

Atomic hydrogen is the simplest atom there is. Jerrold had learned his lesson, he said, about systems that could be measured but were too complicated to understand. He wanted no more of that.

Hydrogen

Rabi had just begun to set up his own laboratory at Columbia in 1931. He had spent the years 1927 to 1929 in various laboratories in Europe, and there he had immersed himself in the new quantum mechanics in a way that had not been possible for him in the United States. Rabi said:

I first went to Zurich [to be] with Schrodinger, and I arrived the day he left. Then I went to Sommerfeld, where I met Bethe, he was a student at that time. And to Copenhagen. These are all by accident. From Copenhagen I was told to go to work with Pauli, which I did. . . .
I found when I got there [Europe] I knew a lot of theoretical physics. In fact, more than most of the German students, except the very good ones. Certainly far above the German experimental physicists.[9]

Rabi made a considerable impression on the European physics community during his stay there, which, as Jerrold had noted, helped him afterward to obtain his position at Columbia. The German physicist Werner Heisenberg, another of the principal developers of the new physics, had urged his appointment on the Columbia people so strongly that they had felt obliged to put aside whatever reservations they may have felt for any other reasons.

Among the great European research centers at which Rabi had worked was Otto Stern's laboratory in Hamburg. There, under Stern's influence, Rabi had become interested in and challenged by the problem of the magnetic properties of the atom and of its nucleus. These properties were what he now proposed to study in his own laboratory in New York.

In 1931, Rabi joined forces for a time with Gregory Breit, a theoretical physicist at New York University, and together they published a short but very important article in which they derived a formula for the various energy states available to an atom in a magnetic field.[10] By the time Jerrold approached him, Rabi was already working with his first graduate student, Victor Cohen, on the magnetic properties of the sodium atom.[11] Their idea was to test the applicability of the Breit-Rabi formula.

The sodium atom is not particularly simple, however. The neutral sodium atom contains eleven electrons that move in complicated orbits about a nucleus which is itself complex: the nucleus consists of twenty-three particles—eleven protons and twelve neutrons—very tightly bound together and interacting strongly with each other. Almost nothing was known in 1931 about the forces that hold the nucleus together. In fact, not until the year following was the neutron actually detected experimentally, although the possibility of its existence had been suggested. Most of the neutron's properties were still unknown. All in all, it seemed to Jerrold that the sodium atom would turn out to be about as hard to understand as his nickel crystals had been. Satisfactory methods for understanding the behavior of aggregates of strongly interacting particles simply did not yet exist. He wanted to work on a set of problems that could be disentangled from each other and dealt with one at a time, with results that could be interpreted. Discussing it later, he said "hydrogen was as complicated a beast as I could understand."[12]

In contrast to an atom such as sodium, hydrogen has a particularly simple structure that permits one to get right at the fundamentals. It had been the simplicity of hydrogen, for example, that enabled Bohr to make the great conceptual breakthroughs that gave quantum mechanics its start. The nucleus of hydrogen consists of only a single elementary particle, a proton, around which orbits a single electron. It is as uncomplicated a system as one can find, and yet, as Bohr and others had shown, it displays all of the rich complexity of atomic physics.

Deuterium is chemically identical to hydrogen and almost as simple. Its nucleus is the second in the table of elements, containing one proton and one neutron, tightly bound together. Like hydrogen, it has only one electron; since the chemistry of an atom depends on its electronic structure, deuterium behaves chemically exactly like hydrogen. The experimental techniques that would serve for hydrogen might serve for deuterium as well, and experiments on deuterium might even turn out to reveal something about the newly discovered neutron.

Heavy hydrogen, as deuterium was also called, had been discovered only the year before by a Columbia man, Harold Urey.[13] That had been one of a series of brilliant and exciting discoveries during the short interval between late 1931 and roughly the middle of 1933, in both the United States and Europe, that later led people to refer to that period as the year of miracles. Urey was ready to provide them with samples of heavy water, that is, water some of whose molecules contained deuterium instead of hydrogen. If Rabi and his group wanted it, he could supply deuterium in gaseous form as well. The availability of deuterium was thus another good and practical reason for working on hydrogen.

There was yet another compelling reason for studying hydrogen, and it was not convenience or even simplicity. Very recent work by Estermann and Stern in Hamburg, in the very laboratory where Rabi had worked, had shown that the magnetic properties of the proton were not those that theory predicted for it if the proton were really an unstructured elementary particle.[14] The proton promised interesting surprises.

Rabi readily agreed with the proposal to study hydrogen and invited Jerrold to join forces with Jerome M. B. Kellogg, a recent Ph.D. from the University of Iowa, who had just arrived at Columbia as an instructor. Rabi was able to obtain laboratory space for them in Columbia's Pupin Laboratories building, and they went to work.

Jerrold's pattern of holding down a full-time teaching job in one of the municipal colleges while doing research in Columbia's laboratories was unusual, but not unprecedented. Rabi himself had followed that pattern for a time before he had gone to Europe on his fellowship, and others were to follow it later. Speaking about it afterward, Jerrold said that he figured he could meet all of his commitments at Hunter in about a thousand hours a year. He estimated that this left him with at least another thousand or perhaps fifteen hundred hours for research, maybe even more; and the Pupin Laboratories is where he spent them for the next seven years.

Rabi's Lab

Isidor Isaac Rabi was a pleasant-looking man who stood only about five feet four inches tall and wore round horn-rimmed glasses. There was nothing about him that would have caused any New Yorker's head to turn as he passed, except perhaps his tendency to hum operatic arias under his breath. Nevertheless, for nearly all of his long and fruitful life, he was a major figure in the world of physics. He was a leader, an innovator, and for many, including Jerrold, a dependable and resourceful mentor. During the 1930s, some of the most talented and able men in science started distinguished careers under his guidance—men such as Victor Cohen, Donald Hamilton, Jerome Kellogg, Polykarp Kusch, Sidney Millman, Norman Ramsey, Julian Schwinger, and Jerrold Zacharias.[15]

It was hard work. They often kept at it seven days a week, long hours into the night, although some of them—Zacharias, for one, and Millman, for another—had full-time teaching jobs elsewhere at the same time, involving some fifteen or sixteen classroom hours a week plus the associated time for preparing lectures, grading papers, and so on. Norman Ramsey, reflecting on those days in a conversation with Jerrold years later said, "I was always impressed, because I was working with you at the time you were working this amount at Hunter College. Nevertheless I thought of you as a full-time person around the Columbia beam lab. I mean I would have to remind myself, well, you had to be away at certain times, but evenings when I was working, you'd be in. Weekends you'd be in fully." And in the same conversation, Jerrold recalled, "I remember specifically Norman Ramsey and I were in the lab about one o'clock of a morning, and the apparatus was working beautifully. And the question was, should we shut down and go home? And I remember saying to Norman, look, the apparatus is willing to talk, let's explore."[16]

One cannot doubt either the enormous talent or the enormous commitment that these young men brought to their work, nor can one doubt that Rabi was their unchallenged leader. When Jerrold was asked many years later how it had been possible for Rabi to remain the dominant intellectual figure in such an extraordinary group, he replied, "He was just smarter than anyone else. He was the guy you went to for ideas."

There was always an extraordinary feeling of excitement in Rabi's laboratory, which arose from the group's sense of being at the forefront. It arose also from working side by side with bright and capable young physicists who were succeeding in new ways. Sidney Millman put it this way: "And yet I can't remember any period when there was more real fun working in the lab. Perhaps it was the excitement of new results—signs of moments, the indium [summer] moment,[17] and the first nuclear resonance curve. Perhaps it was the relaxed feeling of having a field of research all to yourself. Perhaps even more important was working in the lab without technical aids, without 'associates' but with just colleagues."[18]

Primarily the sense of excitement was generated by Rabi himself—from his own energy and his own sense of wonder. It was not that Rabi was a good experimentalist in the conventional sense; Zacharias and Kellogg often would not let him touch the equipment for fear he would upset some delicate arrangement or other. But he could produce ideas no one else would think of, and he led the way to interpret and understand the experimental results. He had a keen sense of what was important, and he was rarely wrong.

The difference between Quimby's lab, from which Jerrold had just come, and Rabi's lab was striking. Quimby's lab, his experiments, his style of work, all connected with the past; they were solid, traditional, and pedestrian, and they maintained a sharp distinction between experiment and theory. Rabi's lab was part of the future of physics; it was one of the important places where the future was beginning to happen. For a young man in Jerrold's position, Rabi's lab was exactly the right place to be, and he had chosen exactly the right time to be there.

Molecular and Atomic Beams

A casual passer-by at 120th Street and Broadway on a summer's evening in 1933 might have been startled to see a bright violet-blue light shining from a tenth-floor window of the Pupin Physics Laboratories at Columbia. No doubt it would have meant little to him, but to Jerrold it meant that the Wood's discharge tube he had built was working. He claimed to be able to tell all the way from the drugstore on the corner whether it was working right, from the color alone.

The Wood's tube apparatus was the source of monatomic hydrogen for the Zacharias-Kellogg experiments. Hydrogen normally exists as a diatomic molecule: two atoms linked together by the sharing of their electrons. In Wood's apparatus, which consisted of about 10 feet of folded glass tubing and resembled nothing so much as a blue "neon" sign, molecular hydrogen was torn apart in an electric discharge. R. W. Wood, a physicist at Johns Hopkins, had discovered that under usual circumstances, the hydrogen atoms would recombine at the walls of the tube to re-form ordinary molecules, but if a little water vapor was introduced into the tube, this kind of recombination would be prevented and the monatomic form would survive. How this worked was not then very clear; the interaction between the electrons of hydrogen and the electrons of the glass walls of the tubing is complex, and it was not yet understood how the presence of water vapor affected that interaction. But as Jerrold had learned earlier in his work on single crystals of nickel, an experimenter frequently has to proceed without full theoretical understanding of some aspect of the technique, sometimes relying on the physicist's analog of folk wisdom. There were large handbooks full of useful hints and tricks of the trade, and in the 1930s, at least, no other area of physics was more dependent on such folk wisdom than high-vacuum research with gaseous discharges.

The experimental arrangements were simple enough conceptually. Hydrogen atoms were to emerge from the Wood's tube apparatus, passing through a narrow slit; the only atoms that could emerge into the experimental region were those that happened to be moving in the right direction for getting through the slit. The atoms would thus form a beam which would then pass into a space containing a carefully designed magnetic field.[19] The atoms would be deflected by the magnetic field, each by an amount that depended on the magnetic state of the atom. Thus, the field sorted out the atoms according to their magnetic states. After being deflected, the atoms would be detected at the other end of the apparatus. By measuring where they wound up and by knowing the precise configuration of the magnetic field that had sorted them out, the experimenter would be able to infer the magnetic states of the atoms themselves.

The difference between conceiving such an experiment and actually carrying it out is enormous. The art of experimentation lies in understanding more or less intuitively what can be done and what cannot. Rabi had a rare ability to extend his intuition into the atomic domain—to guess how atoms would behave in this or that circumstance. Jerrold's talent was a different one; he had an almost instinctive understanding of machine tolerances and vacuums and what could and could not be measured.

For example, the inhomogeneous magnetic fields Stern and his co-workers used in their Hamburg experiments were produced by carefully shaped magnets with extremely small gaps through which the beam was made to pass. Jerrold felt strongly that the magnetic fields could not be measured in those gaps with the precision necessary for a satisfactory result. He described later what he had thought: "Stern and Estermann had made a measurement of the magnetic moment of the proton which depended on their having measured magnetic fields in very small spaces. So I tried to make measurements of magnetic field gradients in very small spaces and I decided they couldn't do it. I couldn't do it and I figured they couldn't do it—we'd better do something different."[20]

Perhaps that was arrogant, given the reputations of Stern and Estermann, but more likely Jerrold's fine experimental intuition was at work, distinguishing what could be measured from what could not. He thought it would be necessary to design a different way to produce the fields. Since measuring the fields would always be difficult, he proposed that the fields be produced not by magnets but by electric currents. If the geometry of the currents could be kept simple, the fields could be calculated rather than measured, and the difficult measurement could thus be avoided. This was the procedure he adopted in the first experiments.

The experiment had to be carried out in a high vacuum, for several clear reasons. First, in order for the hydrogen to be as pure as possible, they would have to eliminate practically all contaminants, at least to the extent possible. Second, if the atoms were to be formed into a beam and were to stay in the beam, anything that could get in the way would have to be eliminated. By pushing the current technology hard, the experimenters could reduce the pressure to something like one one-hundred-millionth of atmospheric pressure. But achieving and maintaining such a pressure for the course of the experiment would always take a significant portion of their time.

The heart of the apparatus was contained in a brass can—a piece of thick-walled brass tubing about 5 or 6 inches in diameter and about 2 feet long in the early versions. The can enclosed the space in which the magnetic field was produced and the region in which the beam interacted with it. The detector was located at the far end of the can. To the can was joined the Wood's discharge tube, the vacuum pumps, and a McLeod gauge (mercury and glass and an ingenious arrangement of tubing and stopcocks) for measuring very low pressure.

Two vacuum pumps were required. The first was the fore-pump, a commercial device capable of reducing the pressure by a factor of between 100,000 and 1 million. The second was an oil diffusion pump, which they had to make themselves, out of glass and a special low vapor-pressure oil. When everything was working well—when there were no leaks and no unidentified sources of contaminants—the diffusion pump could slowly take the pressure down about another factor of 1000.

Most of the glass parts were made of Pyrex glass, fairly new at the time—but some of it had to be made of a softer glass called lime glass. The difference between the two is important. Pyrex is a low-expansion glass, so its dimensions do not change much as it cools. Stresses in the glass remain small, and no great skill is required to keep it from cracking or shattering. Even inexperienced graduate students can often work directly with it.

Lime glass is not so forgiving. It must be slowly annealed in ovens or by continuously flaming the glass joints for many hours as the glass cools slowly. Stresses are particularly severe where electrical conductors pass through the glass walls of the system, but stresses can show up anywhere. Francis Bitter, a good friend of Jerrold in the Columbia physics department, noted later, "Even when we had taken as many precautions as we knew how, it was as likely as not that, on returning the morning after having created a masterpiece, we would find it cracked at some vital point."[21]

The working substance of the diffusion pump was a special oil, the surface of which had to be exposed to the vacuum to some extent. This was also true of the mercury in the McLeod gauge. Various stopcocks and valves were necessary, and they had to be lubricated with stopcock grease. Joints and seams often had to be overlaid with glyptol or putty to seal them. As all vacuum experimenters sadly knew, these substances continually emit vapor into the vacuum, and even solid surfaces may have gases adsorbed on or dissolved into them which are released into the vacuum. This phenomenon, known as *outgassing*, is the bane of the experimenter's existence. Some of the vapor could be driven off by local heating and then pumped away, but oil and mercury vapors were continually present. Cold traps were introduced—U-shaped sections of tubing kept at liquid nitrogen temperature. Some of the vapor could be condensed and trapped in such places and kept out of the way. Pinhole leaks might occur anywhere in the system; these had to be tracked down with soapsuds and a shaving brush. When a layer of soap film was brushed over an otherwise invisible hole, the pressure in the system would alter momentarily as the tiny flow of air through the hole was modified. The hole might be located and repaired, but the task was time-consuming. There never seemed to be opportunity to clean away the dried soap films, and after a while the apparatus became spotty and disreputable looking. The external appearance was deceptive, however. The precision lay inside, and to achieve it, the experimenters were ready to take all the care necessary, no matter how long it took.

Working at high vacuum with the materials and techniques then available was a never-ending battle, and that was only the prerequisite to the experiment, not the experiment itself. Jerrold likened the process to keeping a car running: "In many ways, every apparatus I've ever worked with reminds me of the 1927 Dupont, serial number 2850, that I bought in 1931. . . . It had a four cylinder Stutz engine with electric fuel injection, four forward speeds, balloon tires, hydraulic brakes and a hand crank to supplement an occasionally failing electric starter. Two tons of automobile and 17 miles per gallon when a gallon was 12 cents. It had been raced on the roads of Belgium. A marvel of engineering; my fingernails were never clean. There was always something that needed tending to."[22]

Everything about a new kind of experiment was a problem to be solved in some fresh way. The slit that formed the beam would have to be as narrow as a few hundredths of a millimeter—much less than the thickness of a human hair. Nothing like that was available from laboratory suppliers; how could such a slit be made? In order to create sufficiently intense magnetic fields, the conductors would have to carry very large currents—as large as 100 amperes. What would be the source of these currents? How could the conductors be kept cool? Once the beam had been sorted out by the magnetic field, how was it to be detected?[23]

Virtually no component of the apparatus could be acquired commercially even if money had been available for it. The Columbia physics department had only one machinist, so the experimenters had to do most of the actual fabrication and construction themselves. Willy-nilly, Jerrold and the others turned themselves into first-class machinists, welders, plumbers, electricians, and repairmen as the experiments progressed.

No detail of the experiments could be ignored. The alignment of all of the parts required exquisite care, so much so that existing techniques of optical alignment were insufficient. They had to make use of the hydrogen beam itself to do the alignment; they estimated that an error in alignment of the slit by only one-tenth of a millimeter would result in a loss of 90 percent of the beam.

In the end, it was the equipment that always called the tune. There would come a rare moment, after much tinkering and adjusting, when everything would be working perfectly, but it might come at ten in the evening or even at one in the morning. Whenever the moment came became the time to do the experiment, to take data. Who could tell when another opportunity would come?

Finally, they made their first measurements of the strength of the proton magnetic moment, and they followed that up immediately with a measurement of the magnetic moment of the deuteron (then called the deuton). The results were highly significant; they showed that the value of the proton magnetic moment was indeed anomalous, as Stern and Estermann had found, and they showed that the neutron magnetic moment is not zero as one would have expected from the fact that it had no electric charge. They estimated their overall accuracy at about ten percent, and a substantial quantitative disagreement with the measurements of Stern and Estermann remained. These differences would not be cleared up until new methods of measurement were introduced by Rabi's group several years later.

The proton magnetic moment was much larger than could be accounted for by the theory of an elementary particle.[24] This was the essential discovery, the fact that didn't fit, the surprise that hydrogen provided. Stern and Estermann had found the anomaly first, and it was now confirmed by Rabi's group; the numerical discrepancy that remained between the results of the two laboratories was certainly less important than the large anomaly itself. The numerical difference would be resolved by more precise experiments, but the anomalous values of the magnetic moments would remain to demand explanation. Rabi, Kellogg, and Zacharias, reporting their results in the *Physical Review* in August 1934, stated, "The substantially fair agreement of the two results must be regarded as more important than the difference. . . . It can be taken as certain that the proton is not describable by the simple type of Dirac wave equation which describes the electron."[25]

The Resonance Method

The first measurements of the proton moment provided great satisfaction to Rabi, Kellogg, and Zacharias and attracted considerable attention as well. It was clear to them, however, that they should try to improve the technique. There were straightforward improvements they could incorporate into future measurements, such as developing a new detector, that would be more efficient and easier to read than the screen of molybdenum oxide soot they had been using, and steps of that sort could readily be taken. But there were some inherent difficulties in that first experiment, and to remedy these they would have to redesign the apparatus. In a very real sense, this willingness to start over was what distinguished Rabi's group. Others might have settled for the small improvements that would have resulted from becoming more painstaking, or from developing a detailed theory of the apparatus, or from simply collecting more data to gain some statistical advantage. Their technique had not yet been pushed to its limits, after all. Instead, Rabi and his group preferred to give free rein to their ingenuity, always seeking a new design or a new experiment that would give unequivocal answers quickly. New technique rapidly followed new technique, and new experiment followed new experiment. Rabi's laboratory set both a new style and an impressive pace.

As Kellogg, Rabi, and Zacharias went about improving the measurement technique, the first problem they sought to solve was that the atoms did not emerge from the source with a single speed but rather with a distribution of speeds, so that they took different lengths of time to pass through the magnetic field. The beam of atomic hydrogen was expected to be sorted out by the magnetic field into four separate beams, according to the rules of quantum mechanics. But since the actual deflection of an atom depended on how long it took to pass through the sorting field, the four beams spread out and became blurry, losing the original sharp definition for which the slits had been so carefully designed. Instead of well-resolved traces on the molybdenum screen, the experimenters saw overlapping smears; the overlapping was so severe that they could resolve only two separate smeared-out traces on the screen. They then had to estimate by eye where the centers of the smears were. Each made his own independent estimate, peering through a traveling microscope, and then they averaged their observations. The whole procedure was unsatisfactory, and while it was in order to design a detector better than the molybdenum screen, that would not solve the problem of beam spreading, which would have remained a serious limitation.

Following the Rabi philosophy, they set about designing a new experiment that would be insensitive to the time it took for the atoms to move through the field. Their idea was to introduce a second magnetic field into the experiment; it would be stronger than the first one but would occupy a shorter region. The first field, now called the A-field, would be set to split the beam into four streams exactly as before; the second, the B-field, would operate selectively and could be set to refocus each of the beams one at a time. For each segment of the beam—each "beamlet"—the B-field precisely undid the effect of the A-field, while the other three beamlets remained unfocused. The beamlets could be studied separately, then, and the results would not depend on the transit time. The spread in the velocities of the atoms would have no effect on the results. The new refocusing method was an elegant solution and gave far more precise results, but, as it turned out, it was only a stepping-stone along the way.

The equipment, of course, now had to be completely redesigned to make room for two magnetic fields instead of one, as well as to accommodate other changes. A new type of pressure gauge known as a Pirani gauge was to be used in place of the molybdenum screen and provision had to be made for it. The Pirani gauge technique was very slow, requiring long intervals of uninterrupted time for taking data. Since the only such periods occurred at night, Zacharias and Kellogg, and often Rabi as well, became nocturnal.

With the newly built equipment came new problems. Jerrold's notebooks from that time report over and over again the long hours spent trying to correct problems with the vacuum. "Discovered system had sprung a leak. Hg [mercury] would not go in capillary tube. Ran 'buster' [the vacuum pump] from 9 to 10:10 PM and in meantime closed the leak. Whereabouts unknown. Phooey. Interruption. Supper." "Ever since Aug. 1 have been horsing around with vacuum. . . . On August 7 we installed JMBK's ion gauge & finally after pumping for several hours found leak in solder . . . did two outgassings."

Other new design problems arose as well. In addition to the many pages in Jerrold's notebooks devoted to the calculation of the atomic angular momentum rules are many filled with design calculations relating to the magnetic fields. In order to get stronger magnetic fields in the later experiments, he and Kellogg began to combine the use of magnets with the use of currents. The magnets were carefully designed to give the same field configurations that had been achieved with currents alone, so that the fields and their rates of change could still be calculated rather than requiring measurement. Details of magnet design are also scattered throughout the notebooks, together with notes about selection of appropriate materials and improvements in the electronic circuits associated with the Pirani gauges.

It was important to measure the sign of the magnetic moment of the proton, as well as its magnitude. This was tricky. Quantum mechanics established clearly that the magnetic moment of any system must point along the direction of its angular momentum axis, but it could point either way along that axis. Which way was it for the proton? If the magnitude of the moment could not be predicted, then neither could the sign; it would have to be measured.

They developed a technique for measuring the sign of the moment that was a natural outgrowth of the refocusing method, and it turned out to have a feature that would become of transcendent importance. Each of the four beamlets produced by the experiment consisted of atoms in a particular magnetic state, different from those in the other beamlets, and each magnetic state of the atom had a different energy in the magnetic field. Once an atom had gotten into a particular state, it would remain in it unless disturbed sufficiently to make a quantum jump. Very weak disturbances could not affect its state. For a disturbance to be effective, it had to be strong enough to provide for the necessary energy transfer. Only then could an atom change its state.

According to the rules of quantum mechanics, the four magnetic states of the hydrogen atom are grouped into a triplet set and a singlet. The three triplet states all have nearly the same energy in a weak magnetic field, but the energy of the singlet state differs by a significant amount called the *hyperfine structure energy*. This was the situation that Rabi now proposed to exploit.

The Breit-Rabi theory predicted that if the proton moment was positive, then the atoms of the singlet state would be deflected toward the stronger part of the inhomogeneous field. As Jerrold put it, "So the question we have to ask those atoms that fall into the stronger field is 'are you from the singlet state or from the triplet state?' . . . Atoms in the triplet state can easily pick up enough energy to flop into one of the adjacent states within the triplet. But atoms in the singlet state can't go anywhere . . . Flopping a singlet atom into one of the triplet states requires adding or subtracting the hyperfine structure energy that separates the singlet from the triplet."[27]

The trick was to use the refocusing device, introducing yet a third magnetic field—the "tickler" field, or simply the T-field—into the space between the A- and B-fields. This was done by means of four vertical wires just below the beam, which could be placed to one side or the other of the beam or astride it. Passing a direct current through the wires would produce a weak magnetic field; an atom passing through this region would feel a small rotation in the direction of the magnetic field, a sudden small disturbance. If it were an atom in a triplet state, this might be enough to cause a flop in its orientation; the atom could wind up in a different triplet state. If that happened, the refocusing magnet would not refocus it; it would be lost, and the detector would record a drop in intensity. But if the atom were in the singlet state, it could not flop; there would not be enough energy. The detector then would show no effect.

Rabi was able to get an exact mathematical solution to the problem of flopping, which pleased him very much. He said later, "I wrote a paper, a theoretical paper, an exact solution. I was always happy with an exact solution. You have a spin in a gyrating magnetic field."[28] Of course, that was just Rabi's manner of speaking, a kind of shorthand. What he meant was that for an atom moving through the T-field, that field would look like a gyrating magnetic field. The ability to sense what the atom would feel was a Rabi specialty. His paper, which appeared in 1936, became famous. Here he explained more carefully what he meant:

"An atom moving with constant velocity through a magnetic field varying in strength and direction along its path is equivalent, for these questions, to an atom at rest subject to a field varying in time in the same manner."[29] Rabi had here recognized the equivalence for the atom of spatial and time variations of the magnetic field. Before long this would lead to further refinements of the technique, new experiments, and new achievements.

Once more the apparatus had to be redesigned. By this time, there were several apparatuses in operation; the tenth floor apparatus was devoted to hydrogen, where Jerrold collaborated with Kellogg, and there was another on the fifth floor, which Jerrold had helped build and where he collaborated with Millman and Kusch. It was the fifth floor apparatus that they now redesigned. The can had to be lengthened once again to make room for the T-field, but the design had to be very much ad hoc; there was no theory to say what would be the best way to produce the T-field. The decision to produce it with four vertical wires was dictated by what was practical to build, and the precise location of the wires for maximum effectiveness was a matter of trial and error. Again, the notebooks show long hours of adjusting the design to make the apparatus work. In the end, the atom yielded. In Jerrold's words, "We had found a way to ask an atom a question and to have it answer."[30] The magnetic moment of the proton was unquestionably positive.

Finally, in 1937, the next step came, suggested by Rabi's 1936 paper. It seemed to them like a very small step at the time, but it was tremendously important. Jerrold described how it happened: "I remember it was Rabi who said, you know—this atom doesn't know whether it's moving or not, what it sees is a varying magnetic field, and that's nothing but a frequency. And that's how that thing came about. I'm sure I didn't say it. I'm sure Rabi and I were in the same room. And one of us said it."[31]

There is a well-known theorem in mathematical physics, known as Fourier's theorem, that says that any time variation can be represented as a sum of harmonic (perfectly periodic) variations, each having a different frequency and added together in different amounts. For example, if the lid of a piano is suddenly slammed down, that can be thought of as a complicated time-varying disturbance to which all of the strings of the piano will resonate (if the dampers are off so that the strings are free to vibrate). In effect, the piano becomes a device that analyzes the abrupt disturbance, and reveals which frequencies are present in it. By analogy, the atom passing through the T-field would experience what it would "see" as a time-varying disturbance. If the natural frequencies of the atom matched the frequencies in the disturbance, the atom too would resonate.

It was also well known to physicists since Niels Bohr and the very earliest developments in quantum theory that to every frequency there corresponds an energy difference, and vice versa. Rabi now pointed out that as the atom entered the T-region and saw a collection of frequencies, one of those frequencies might correspond exactly to the energy necessary for it to make a transition. But that is just what is meant by resonance in quantum mechanics: when a stimulating frequency matches the frequency corresponding to a quantum jump, the quantum jump can take place.

Rabi said later that it had not been such a major insight (although few would have agreed): "We all knew Bohr's relationship. We were all thoroughly imbued with the physics of this. . . . The whole theory of it was well-known, I think, to everyone there. . . . Actually, things of that sort had been done, beginning in 1925 at least, by Fermi, and Breit had worked on things of that sort, and . . . Gorter, so no, there was nothing really new. The real important thing was to do it, and discover the precision which you get out of it."[32]

In fact, although Rabi had sensed the important point in his 1936 paper, it had still required a comment from a visitor to push them over the threshold and to get them moving toward a frequency approach. C. J. Gorter was a young Dutch physicist who visited the laboratory at the end of the summer of 1937. Earlier, in 1932, Gorter had tried unsuccessfully to detect nuclear magnetic moments by imposing an oscillating magnetic field on a sample of material and varying the frequency. When resonance occurred, the sample would absorb energy from the field as the atoms made their quantum jumps. The temperature of the sample should show a sudden small increase that he hoped might be detectable. Unfortunately, the experiment failed because competing effects masked the tiny indication that Gorter was looking for. Gorter's technique was crude compared to using the molecular beams method, but he had in fact come up with the idea of probing the atom with a single resonant frequency.

Now, on his brief visit to the laboratory at Columbia, Gorter suggested to Rabi that the T-field might be replaced with a radiofrequency magnetic field, and everything seemed to fall into place at once. The idea was an obvious extension of Rabi's paper, and Rabi, Millman, Kusch, and Zacharias immediately began to implement it. Instead of using the T-field to produce a change in the field direction, the T-field itself would be modulated at a single chosen frequency. If that frequency matched the energy of a particular transition, something would happen, and the detector would register a change. If the frequency did not match a transition energy, nothing would happen. The scheme was beautiful, it was simple, and, because frequencies can be measured far more accurately than anything else, it promised an improvement in precision of orders of magnitude.

Results came quickly. The equipment had to be modified once again, but they were getting to be old hands at that, although they now had to master new techniques involving radio oscillators and resonant circuits. They acquired a copy of the *Radio Amateur's Handbook* and went to work. In January 1938, Rabi, Zacharias, Millman, and Kusch published a short note, with a brief acknowledgment to Gorter, describing the new method as applied to lithium chloride but without yet offering a definitive number. "The effects are very striking," they wrote, "and the resonances are very sharp."[33]

Even before they had obtained a numerical result, they could tell that they had a surgically precise tool in their hands and over the next several years they used it with remarkable effect. One after another, they measured dozens of nuclear moments with an accuracy they could only have dreamed of a few years earlier.

Unexpectedly, a new result came out of this precision: a measurement of the *quadrupole* moment of the deuteron. If the nucleus were a perfect little electrical sphere spinning around its axis, then the magnetic moment they had already measured would be the end of the story. But it had already been suggested from spectroscopic evidence that the electric charge on the nucleus is not always spherically symmetric, and the quadrupole moment was the measure of its asymmetry. It was expected to be tiny.

In 1939 and 1940, Kellogg, Rabi, Ramsey, and Zacharias published two papers reporting what is now regarded as a classic experiment.[34] They had remeasured the magnetic moment of the deuteron using the resonance method. With the precision that the new method provided, they could see that a small bump appeared in the graph of the data, off to one side. They could not have detected it in their earlier, less precise experiments, but it was real and it was reproducible. Jerrold recalled that it was one of those middle-of-the-night discoveries, made between one and four in the morning; made at that hour because of the rule he applied to himself and that he imposed on the others: never walk away from an apparatus when it's running and giving data.

At first they didn't know what the bump in the data meant. It was not expected, and the meaning was certainly not obvious. Rabi was away at the time, as a visiting professor at Stanford, and they communicated with him by telegram. Rabi recalled, "The first thing I thought about was, here's something and we should improve the resolution. Which they did in quick time, it was amazing . . . that it was a quadrupole moment wasn't immediately apparent, but that there was something there, was."[35]

But before very long they realized that the anomaly must be due to a nuclear quadrupole moment, and with some calculational help from A. Nordsieck,[36] they were able to determine an experimental value. The importance was immediately apparent. The simple structure of the deuteron had allowed an essential feature of the nuclear force to be revealed: the neutron and proton attract each other, but the force is not precisely along the line joining them.

The result was so important that no fewer than eleven theoretical papers interpreting the result appeared in the *Physical Review* in the nine months or so between the first and second publications on the subject by the Kellogg-Rabi-Ramsey-Zacharias group.

The first magnetic moment paper, which gave a relatively imprecise value for the moment of the proton, had appeared in August 1934, and the paper on the deuteron quadrupole moment was published in April 1940. In the five and a half intervening years, Rabi and his co-workers had entered a new domain of precision. By dint of incredibly hard work, they had invented new techniques, producing and mastering them as fast as they needed them. They had built new equipment, used it, modified, it and frequently abandoned it for still newer equipment, all at a furious pace. There was no precedent in physics for such a whirlwind of combined theoretical and experimental activity. What had begun with a straightforward attempt to improve on someone else's measurement had developed into a whole new field of physics, radiofrequency spectroscopy, and had led directly to nuclear magnetic resonance techniques. The application of these techniques in chemistry, medicine, and biology continues to expand every day.

The 1943 Nobel Prize in physics was not awarded until 1944 because of the war, and it was awarded to Otto Stern for the experiments that had initiated the field. Simultaneously, the 1944 Nobel Prize was awarded to Rabi for the invention of the resonance method.

For the others, the rewards were much less tangible. Some, like Millman and Ramsey, got their Ph.D.s out of bits of it; for others, like Kellogg and Kusch, it was their job. But fundamentally, they were mostly like Jerrold, who was there entirely for the pleasure of doing good physics. There was no other reward and none other that he expected.

Outside the Lab

Jerrold's industriousness in these years was closely matched by Leona's. As they had planned before they were married, Leona enrolled as a graduate student at Columbia in the fall of 1927. She was awarded a master's degree in biology in 1928 and promptly began the long pull toward a Ph.D. She and Jerrold took for granted that she would pursue her education as far as possible and that she would have a career. When she and Jerrold were married, that idea formed part of their expectations, and there is no evidence that it was ever questioned by either of them, nor does Leona recall that it ever was, not even when their daughter, Susan, was born in 1931.

In any case, they needed two incomes in those early years of marriage. In one busy interval, Leona held three jobs simultaneously: research assistant at the American Museum of Natural History, assistant in neurology at the College of Physicians and Surgeons (P&S, Columbia), and instructor at Hunter College in the evenings. In 1937, she received the Ph.D. degree in embryology from Columbia University and accepted a position in the department of optometry at P&S as instructor of anatomy, physiology, and embryology. Her achievement was at the least uncommon; to have accomplished it at that time in combination with being a wife and a mother was surely remarkable.

Jerrold and Leona were young and enthusiastic enough not to find fifty- and sixty-hour work weeks oppressive. There was thus time for other things, and fortunately, considering how hard they both worked, they had energy and appetite for them as well. But they did not find themselves unusual. The 1930s was a vigorous and creative time in New York, perhaps its finest time in many ways. The city suited Jerrold and Leona very well. In spite of the depression, or perhaps because of it, New York marked the period with a burst of creativity in music, literature, and the theater. There were new plays and concerts to attend, new kinds of magazines and books to read.

In the depression, the difference between having a job and not having one was enormous, for a little money went a long way. Jerrold and Leona were able to entertain their friends with small dinners and to be entertained in turn. They rented a small summer cottage on a farm in Yorktown Heights for several months each year, commuting the forty miles or so to the city each day. On weekends their friends and relatives generally came to visit. The Rabis came, as did the Quimbys and other friends. When the great European physicist Wolfgang Pauli came to visit Columbia, he was brought to the Yorktown Heights cottage for a summer's day excursion. Jerrold found welcome the chance to deal with the routine problems of having such a place, a refreshing change from the problems of the laboratory. Such time as he took away from work at Columbia or at Hunter could be spent at such domestic tasks as laying a new attic floor, tinkering with whatever car he owned at the time, or picking fruit in the orchard surrounding the house.

In 1937 Jerrold was promoted to the rank of assistant professor, an advancement that he felt was overdue. He had greater ambitions by then, in any case. The chairmanship of the Hunter physics department became vacant in that year, and he decided to apply for the post. He got Rabi and Harold Urey (the Columbia Nobelist who had discovered the deuteron) to write letters of recommendation. They wrote strong letters of support, but Jerrold could not be sure that the authorities at Hunter would recognize those names. He decided that his case would be stronger with a supporting letter from Albert Einstein: everyone knew of Einstein. Here is Jerrold's own account of what he did:

Through H. P. Robertson, a physicist at Princeton and a close friend of Einstein, the meeting was arranged. I drove down to Princeton, prepared at Robertson's suggestion to talk about our work on the quadrupole moments of nuclei. I stood at the blackboard drawing diagrams and formulas, describing the procedures and our results, and then the kid from Jacksonville turned from the blackboard, looked Einstein in the eye, and said, "Is that clear?" After I recovered from hearing myself say that, we chatted for a bit and Einstein then dictated a letter in German to his secretary. I had the letter translated, submitted it to the powers that be at Hunter, and someone else got the job.[37]

Clearly a good research record was not enough to set aside the departmental order of succession. Perhaps in the fullness of time Jerrold would be in line for the chairmanship but not yet. The disappointment was not serious, however, for he had not actually expected to get the job. His promotion was in hand, and he could look forward to a sabbatical leave in 1939–1940. Life seemed full and rewarding for the Zachariases. Their careers were on track, and as the country gradually emerged from the depression, whatever worries they had had over money and jobs diminished considerably.

If they had thought to wonder about it, Jerrold and Leona might have believed at that point that they could see what shape their lives would take. But the situation in Europe would soon explode into war, and the war would spread to engulf them and everyone they knew. Quite soon their careers would be permanently altered by events beyond anyone's control.

The Trials of War

War Comes to Europe

Physics research flourished in the United States during the 1930s, but elsewhere the rise of the dictatorships brought scholars and scientists into terrible difficulties. While Rabi and his co-workers conceived new experiments and sought ways to improve on Otto Stern's techniques, Stern himself was forced to abandon his laboratory and to start anew in another place. Because of his reputation, Stern might have lasted in Germany a little while longer, but when his colleague Immanuel Estermann was driven from his position at the University of Hamburg, Stern decided to resign. The *New York Times* of October 10, 1933 carried a small article on page 11 under the headline "Jewish Professors Here From Germany." The sub-heading read "One Expelled by University, Other Quit as Protest—Going to Carnegie Tech."[1]

Stern and Estermann were only two among the latest victims of nazism; the expulsion of Jews from the German universities had been widespread and abrupt. Jewish professors, together with those few non-Jews who stood up for them, were systematically deprived of their positions at universities all over Germany. The Times article, in retrospect, was not only the obituary of the Hamburg laboratory but it was also an omen of all that was to come. Europe had started its descent towards 1939, towards calamity and war.

The arrival of Stern and Estermann in the United States occurred early in what became known as the "intellectual migration," a migration as important to America as that which had brought Jerrold's forebears to this country a hundred years earlier. The number of those reaching sanctuary here was small, for only a tiny proportion of the victims of fascism managed to escape. For those who did not or could not, and for Jews especially, the circumstances prevailing in Europe were cruel and harsh, eventually to become more brutal than anyone could have guessed. The *Times* quoted

Estermann as saying that many of his colleagues "were wandering about Europe with nowhere to go." In May 1933, the *Manchester Guardian* published a list of more than two hundred names of those known to have been dismissed from their posts at German universities.[2]

Of those who made it to the United States, some were lucky and found places right away, but many did not. Success was not necessarily a question of skill or ability. Leo Szilard wrote of his own escape, "you don't have to be much cleverer than other people, you just have to be one day earlier than other people."[3] The universities of the United States could provide only a few places for the best-known refugee scientists. Enrico Fermi, whose wife was Jewish, took advantage of the award ceremonies in Sweden where he was awarded the Nobel Prize: he and his family simply did not return to Italy, and he obtained a position at Columbia. Leo Szilard was not so well known and found no such opportunity until much later. Hans Bethe, whom Rabi had known in Europe, was able to locate at Cornell; Victor Weisskopf, a fellow student of Bethe, managed to find a place at the University of Rochester. Bruno Rossi and Edward Teller were other examples of refugees who found places, Rossi arriving at Cornell in 1940 and Teller having come to George Washington University as early as 1935. Each would play singular and important roles in the coming war; for each, the opportunity simply to do physics in the calm and reflective way he enjoyed was a long way off.

Naturally, such matters were discussed by the physicists in the molecular beams laboratory, although while they were working they generally talked little of anything other than the experiment at hand. But they were all very much aware of what was going on. New York was the principal port to which refugees came, and for the physicists among them, New York meant Columbia. A steady stream of visitors arrived at the university during the late thirties; colloquium talks were given, and ideas exchanged. The hottest topic was nuclear physics. At first the neutron work being carried out in Rome by Fermi, Segrè, Amaldi, and others held center stage. Then in 1939, with war already a certainty but before secrecy clamped down on everything, the discovery of nuclear fission in Germany aroused alarm. The molecular beams work also evoked strong interest among the visiting physicists, particularly because those experiments might shed some light on the nature of the force that holds the nucleus together.

But if the physics news was exhilarating, the daily political news was not. There was an ominous feeling about it, a sense of disaster looming, and worst of all, a sense of helplessness. As Jerrold put it, "Now, we read the newspapers. We were responsible people. We said, Lord help us, what can we do?"[4] But there was nothing yet any of them could do.

As he had planned, Jerrold spent the academic year 1939–1940 on sabbatical leave. Hunter College supplied half his salary for that period, and Rabi was able to locate some fellowship money for him at Columbia for the other half. The money meant that he could concentrate fully on research for a whole year, and he was eager to do that. He had spent nearly seven years "in Rabi's shadow," as he said, and he felt keenly the need to do something entirely on his own. He decided to start a new experiment to measure the nuclear magnetic moment of a radioactive isotope of potassium, K^{40}. This presented a difficult challenge but held uncommon interest.

K^{40} has a particularly long lifetime, and nuclear theory suggested that its long life might be explained if the nucleus turned out to have a large magnetic moment. Rabi and Millman had attempted to measure the moment but had not succeeded; no one, in fact, had yet measured its value, principally because of the isotope's low abundance: only one atom in about 10,000 of naturally occurring potassium is K^{40}. Thus, the measurement promised to be satisfyingly difficult, but Jerrold felt that he could accomplish it within the allotted year. With no need to spend a thousand hours elsewhere, he looked forward to working steadily in the lab.

As it turned out, the experiment was even more difficult than he expected, and he needed all the time. He carefully redesigned the resonance apparatus to work in the so-called flop-in mode; that is, so that only K^{40} atoms that resonated with the flopping frequency would be focused on the detector, while those of the more abundant types, K^{39} and K^{41}, would be deflected elsewhere. It was a straightforward scheme, but it required that the apparatus work nearly perfectly. Unfortunately, it is never possible to make an ideal apparatus. Stray atoms would be scattered by residual gas molecules onto the detector wire, or might bounce off the inside of the can, or even off the magnets themselves, and wind up striking the detector. If even one or two out of every 10,000 of the non-radioactive atoms strayed to the detector, the K^{40} signal would be masked.

Jerrold succeeded only by exercising the most painstaking care, operating his equipment only when it was working well, and devoting much time to keeping it carefully tuned and adjusted. After the experiment was completed, he realized that he ought to have made use of the fact that different isotopes have different masses. The mass of K^{40} differs from that of its neighboring isotopes by one part in forty, or 2.5%. The difference was large enough to have allowed the K^{40} to be separated at the detector by mass spectroscopic techniques, but this did not occur to him, he admitted, until it was too late.

The result was important nevertheless, and he was invited to report on it in the spring of 1940 at the summer meetings of the American Physical

Society in Seattle. His colleague Norman Ramsey, with whom he had worked on the deuteron quadrupole moment, was also invited to give a paper at the meeting. Then, as now, it was considered an important honor to give an invited paper before the society, and they were very pleased. The plan was that Norman and his bride, Eleanor, would drive across the country with Jerrold (Ramsey described it as "three of us going on our honeymoon") in Jerrold's green 1936 Ford touring car. Leona had to remain in New York for some weeks longer because of her job and Susan's school and preparation for summer camp. She could only join the others later, when they reached Glacier National Park.

On June 5, 1940, Jerrold entered into his laboratory notebook, which generally contained little of a personal nature, the remark, "Vacation to west from June 8 to July 24. Longest in history." He and the Ramseys drove across the northern part of the United States at about the same speed, Jerrold said, as the Germans were moving into France. The thought was alarming and made them realize how serious the European situation had become. If the new blitzkrieg meant that the German army could move at a speed that they, unencumbered, could just about manage, then the outcome of the war in Europe looked very uncertain. There would be no easy resolution of it, they thought.

Still on the way to Seattle, they stopped in Coeur d'Alene, Idaho, Ramsey remembered, and there, in a cabin overlooking a beautiful lake, prepared their papers for the meeting. In due course they presented the papers, and their reports received satisfying attention and praise. Jerrold recalled that they encountered Robert Oppenheimer at the meeting and fell into conversation with him; Oppenheimer and his girlfriend then joined them for a three-day trip down the coast.

Jerrold had met Oppenheimer previously at the Leiden meeting he had attended in 1926 while still an undergraduate. By 1940 Oppenheimer had achieved a formidable reputation as a theoretician and was an important figure in California physics circles. He had become one of the principal early importers to the West Coast of the European advances in quantum mechanics, a function that in good part had fallen to Rabi on the East Coast. Oppenheimer was also considered by many to be a bon vivant, and, as Jerrold recalled, he offered to prove during the trip that California wines were every bit as good as French wines. Jerrold reported that Oppie at least made a valiant effort.

They parted company with Oppenheimer in Berkeley and set off for Glacier National Park where Leona joined them. They drove up the Columbia Ice Field Highway to Jasper, in the Canadian Rockies, and then down to Yellowstone National Park. On the journey eastward they stopped in Rock Island, Illinois, to visit Ramsey's parents. General Norman Ramsey, Sr. was then Commandant of the Rock Island Arsenal, where U.S. Army tanks were being manufactured. Jerrold recalled that they were given a cross-country ride in a tank. It was Jerrold's first, but far from his last experience with military technology.

By August 1, 1940, Jerrold was back in the laboratory, seeking to finish the measurements on K^{40}; his notebook picks up the thread on that date in an all-too-familiar way by reporting a small vacuum leak in the apparatus. He finished the experiment on September 4 although did not manage to get the experiment submitted for publication for another eighteen months.[5] The last entry in his notebook gives the observed value of the magnetic moment of the K^{40} nucleus. The remainder of the notebook is blank.

That part of his life was over, although he did not yet know it. Within weeks, he would be caught up the country's preparations for war. Nothing would ever be the same again.

Science Mobilizes for War

There were many who felt in 1939 and 1940 that the United States was moving painfully slowly, if indeed at all, toward the useful mobilization of its scientific manpower for defense[6]. The country was not yet at war, in spite of its posture of "belligerent neutrality." As late as 1941, lend-lease agreements, embargoes on scrap iron and the sale of fifty superannuated destroyers to Britain were about as far as isolationist sentiment would allow the Roosevelt administration to go in providing support for the Allies. It was not yet clear to all that the United States would inevitably be drawn into the fighting, and many opposed the idea passionately. The mood of isolationism was gradually diminishing, but it was not yet gone.

But in fact there was movement, slow at first and then more rapid. It was a small triumph that anything moved at all. The military authorities, by and large, had no great expectations from the scientific community. The National Research Council (NRC), formed during World War I, had been hampered by political problems and had managed to contribute but little until the last year of that war. Those contributions had been useful but neither definitive nor dramatic. The development under NRC auspices of the first versions of sonar for submarine detection was undoubtedly helpful but not as effective as the use of convoys had been. Flash and sound ranging had been a decided help to the Allied artillery forces, and the development of the tank promised to alter forever the use of infantry in battle. But these technological successes had taken too long to develop and had come along too late to alter the course of the war. The general view of the military was that it required about four years from the time a new weapon was developed to the time it could be deployed on the battlefield. Therefore, military planning following World War I, did not envisage an important role for the possible contributions of science. As World War II approached, military budgets for scientific research remained meager. Such war-directed research as there was took place mostly in the military laboratories, such as the Naval Research Laboratory, the Picatinny Arsenal, and the Aberdeen Proving Ground.

The example of the NRC had not been without effect, however. Many private industrial enterprises were motivated in the postwar period to establish or augment their own research laboratories. Some wartime developments, while perhaps not of major importance during the war, had turned out to have significant commercial value in peacetime. For example, sound-ranging techniques, developed for submarine detection, were of considerable value as navigational aids to commercial shipping. Achievements in infrared, ultraviolet, and supersonic signaling held promise for commercial application in the communication industries, and the new discoveries in telephonic amplification were clearly of great importance. Bell Telephone Laboratories, Westinghouse, General Electric, and RCA are examples of corporations—particularly those in the electrical, radio and telephone industries—that established a significant research base during the twenties and thirties, and there were many others. But there was no coordination among any of these research efforts, whether governmental or private; there was no regular exchange of ideas, for example, between the Signal Corps and the Naval Research Laboratories, although they were working on similar problems. There was no overarching structure, no effective organization that could decide about the nation's priorities. The NRC, which might have coordinated the activities in the government laboratories, lacked both the authority and the funds to do so. In any case, most of the nation's scientists were not in government laboratories and could not be reached easily by the NRC.

There existed other committees, however, with more specialized tasks. One of the oldest of these committees, established during World War I, was the National Advisory Committee on Aeronautics (NACA). It had the authority to initiate research on aeronautical matters; one of its functions was to provide information requested by any government agency. The committee had become effective and highly respected; under its guidance the research facilities at Langley Field and elsewhere had been developed, and it had sponsored many useful studies concerning aerodynamics. This committee had had only one chairman for twenty-four years; this was Joseph S.

Ames, who died in 1939. He was succeeded by a newly arrived member of the committee, Vannevar Bush, who had just become the president of the Carnegie Institution of Washington.

Before coming to Washington, Bush had had a highly successful career at MIT during the presidency of K. T. Compton—first as professor of engineering, then as dean of engineering for many years, and finally as vice-president of MIT. He and Compton, himself a distinguished physicist, were very close friends.

Bush and Compton were among the first of a new generation of scientist-advisers just beginning to emerge in Washington, men of considerable authority and standing, who would have a major part in determining the interaction of science and government during the coming war and for years thereafter. With others like themselves, they formed a sort of inner circle, never very large, and succeeded in establishing a new kind of discourse between the nation's scientists and its government. These men were well connected to existing universities and laboratories, such as MIT, Bell Labs, Chicago, Harvard, and Berkeley, and these connections would have much to do with the future growth and importance of those institutions. When they needed to get something done quickly and effectively, they naturally turned to the places they knew best, where they had the influence to get things done. Those to whom they would turn would be selected not so much by design or by partiality as by the urgency of war and the exigency of always being in a hurry. Bush and those he recruited were motivated principally or entirely by the imminence of war. They sought answers to the same question that everyone was asking: "What can we do?" Perhaps it was the free-wheeling approach adopted by the Roosevelt Administration during the depression that made it possible for them to operate effectively; perhaps it was simply the growing sense of emergency. In any case, they were profoundly effective.

Bush and Compton were members of the NRC Committee on Scientific Aids to Learning, on which they served with James B. Conant, president of Harvard University and a respected chemist, and Frank B. Jewett, a physicist, director of Bell Laboratories, and president of the National Academy of Sciences. Meeting privately, they discussed the need for a new kind of agency, capable of mobilizing the full scientific resources of the country for the war they believed to be certain. Following Bush's lead, they became convinced that a new federal agency was required, to function with the authority of the executive branch and with the resources of government behind it. They felt that the NACA, of which Bush was the new chairman, would be a suitable model.

With the help of Roosevelt's aide, Harry Hopkins, Bush made a proposal to the president, which he approved immediately. A few weeks later, at the end of June 1940, the National Defense Research Committee (NDRC) came into existence under the Bush chairmanship with an assurance of presidential support and as much money as it needed.

Bush moved the NDRC into action at once. Four divisions were set up—one to deal with problems of ordnance, one with problems of chemistry, one with problems of communication and transport, and one, under Compton, to deal with "miscellaneous projects." Compton, in turn, immediately set up Section D-1, which became known as the Microwave Committee, under the direction of a most unusual man: Alfred L. Loomis, a wealthy New York lawyer and investment banker who managed at the same time to be an effective researcher in microwave technology, maintaining his own private laboratory at his estate in Tuxedo Park, New York.

The timing of these events was fortuitous. The NDRC was activated on June 27, 1940; on July 8, an official communication from the British government was delivered to President Roosevelt, indicating their desire for "an immediate and general interchange of secret technical information with the United States, particularly in the ultra short wave radio field."[7] The British wanted to exchange information about radio ranging (soon to be known as radar), and Section D-1 had been created just in time to be part of this all-important exchange.

The formation of the NDRC altered permanently the structure of the scientific establishment in this country. The change did not take place all at once, of course, but the central policy set by Bush and his committee remains in place today, fifty years later. The policy was simply this: rather than creating costly new Federal research centers, contracts with existing research enterprises such as universities and industrial laboratories should be utilized as much as possible. The result was an impressive strengthening of the nation's institutions that has lasted well beyond the immediate needs that gave rise to the policy itself.

Historians agree that it was a remarkably foresighted policy. These decisions clearly arose from the fact that the members of the committee were nearly all research scientists themselves. They understood well how the scientific community could best function, and they spelled out the mechanisms that would make it possible.

Radar

The idea of radar was not new in 1939.[8] Its development had been incremental over at least two decades, with important advances made by a number of early workers such as R. Watson-Watt, R. V. Appleton, and M.

A. F. Barnett in England and A. H. Taylor, L. Young, and L. A. Gebhard in the United States. Radar is not the invention of any single individual or group. As early as 1922, Guglielmo Marconi himself had suggested that beamed radio waves might be used to detect ships at night, or in fog.[9] In 1925, G. Breit (the same Breit who a few years later collaborated with Rabi on the magnetic fine structure of atoms) and M. A. Tuve had determined the height of the ionospheric layer of the atmosphere by measuring how long it took for a radio pulse to make the round trip from ground to ionosphere and back.[10] By the time war broke out in Europe in 1939, England and Germany had active radar development programs under way, and in the United States there were separate research programs on radar at the Naval Research Laboratory and the Army Signal Corps Laboratories.

Advances in radar were not even always secret. The French luxury liner Normandie had a well-publicized system on board capable of detecting large obstacles such as icebergs (but not giving their range and direction) as early as 1935.[11] By 1939, the British had deployed an extensive radar system known as the Home Chain, consisting of some twenty-five radar stations along the south and east coasts of England, to provide early warning of the approach of enemy aircraft. Linking the individual stations into a system was very much Watson-Watt's achievement; he had originated and promoted the idea that aircraft could be tracked from station to station, from radar set to radar set.

The Home Chain stations operated at relatively long wavelengths and provided coverage out to more than 100 miles. Depending somewhat on weather, the system could be effective in detecting aircraft at such ranges if the aircraft were at altitudes of 15,000 feet or more. These installations were decisive in the Battle of Britain in 1940; without the advantage the Home Chain gave to the British, the outcome might well have gone the other way.

The long wavelengths at which the Home Chain system operated required that the transmitting antennas be on masts some 360 feet high, with receiving antennas at a height of 240 feet in order to avoid excessive reflections from the ground. Both sets of antennas were permanently mounted at the edge of the sea. They worked well as designed, but it was clear before war broke out that what was needed was a mobile radar system capable of being mounted on an airplane and usable under airborne battle conditions. Weight and size would have to be reduced drastically; wavelengths, instead of being measured in meters, would have to be much shorter, in the centimeter or microwave range. In addition, a single antenna would be required to do double duty, switching back and forth, first as powerful transmitter and then as extremely sensitive receiver. The difficulties of switching between two such different modes were enormous. The Home

Chain equipment, although it was immeasurably useful in those critical first months of the war, was a far cry from what had to be designed.

There were good reasons beside those of weight and size for going to shorter wavelengths. It is a general rule of thumb that in order to "see" an object with a wave signal, the wavelength must be significantly smaller than the object; otherwise the break in the wave pattern caused by the object simply heals itself, and no substantial reflection takes place. This would turn out later to be of major importance in the use of radar against submarines, relatively small targets even when surfaced and nearly invisible to long-wave radar. Another reason for preferring short wavelength is that such radar can be made highly directional even with small antennas; it was clearly of primary importance to know in what direction lay the enemy target and short wavelength would make that possible. Furthermore, without such directionality, there would be confusing reflections from many things other than the target. For example, the ground or the sea also reflects energy back to the receiver, providing a bright background called ground clutter. Just as it is difficult to see stars in the daytime because of the background of scattered daylight, it becomes extremely difficult to separate out the radar target signal from everything else in its background. In the engineers' language, the problem is to separate the signal from the noise, that is, from any electric disturbance that is not the signal.

Above all, there was the question of power. Any radar beam, no matter how large the antenna, must have some angular width and must therefore spread out as the signal propagates outward. Thus, only a fraction of the emitted pulse is actually incident on the target; the narrower the beam is the larger is the fraction. The target then reflects only a small fraction of what is incident, but reflects it in all directions, so that in the end, only a tiny part of the emitted energy is returned to the receiving antenna. The energy returned in a radar echo from a distant target might be less than a million-million-millionth part of that emitted.[12]

This stringent arithmetic presented a formidable problem at the beginning of the war, for there were no generators in the microwave range sufficiently powerful to give detectable echoes at a useful range. Research and development were proceeding at a frustratingly slow pace. The lack of a powerful generator placed a severe restriction on the effectiveness of any microwave radar system.

Radically new equipment had to be developed for microwave radar, suitable for wavelengths shorter than about 10 centimeters, and perhaps as short as 3 centimeters. This need was widely understood; the experiments that the Germans had been making were in the microwave region, and the radar that had been mounted on the Normandie had operated at 16 centimeters, a respectably short wavelength. The U.S. Naval Research Laboratory and the Signal Corps Laboratories as well had been working hard on airborne radar for some years, but the problem of obtaining sufficient microwave power had defeated everyone's efforts.

In the early autumn of 1940, a British mission came to the United States to exchange technical and scientific information. What they brought with them would change the radar situation dramatically.

The mission was headed by Sir Henry Tizard, an informal science adviser to Winston Churchill. What they brought with them, in a specially escorted black metal deed box, was a new invention, the resonant cavity magnetron. This tiny device, barely five inches across, eventually had a profound impact on the outcome of the war. Its inventors were Harry Boot and John Randall of the University of Birmingham. It was so new that at the time of the Tizard mission, only twelve production models were in existence. The cavity magnetron was capable of an astonishing 10 kilowatts of pulsed power at a wavelength of 10 centimeters and it was introduced to the Microwave Committee with dramatic effect.

The first meeting of the British group with members of the Microwave Committee took place on September 19 in Alfred Loomis's New York apartment. Within ten days, the committee had decided to get the magnetron into production, and Bell Telephone Laboratories was given a contract to begin producing it. To make centimeter radar practical, development of appropriate systems to use the new magnetrons would have to be accelerated. A special laboratory was needed, and by early October, Loomis and the committee had made several important policy decisions: the laboratory should be a civilian establishment, and in order not to weaken already existing war work, its staff should be recruited from the universities rather than from industry.

An important open question remaining for the committee was where to locate the new laboratory. When Karl Compton was able to offer space at MIT, including access to an airplane hangar at what was then called East Boston Airport, the committee leaped at the suggestion and accepted it without ceremony.

On October 16, Lee DuBridge, dean of the faculty and a highly respected professor of physics at the University of Rochester, accepted the committee's invitation to head the new laboratory. DuBridge then invited I. I. Rabi to leave Columbia for the duration and to become the associate director. By the end of the first week in November, Rabi was at Cambridge. The next week he asked Jerrold Zacharias and Norman Ramsey to join him.

All he could tell them was that it was secret war work, but that was enough for the moment; they went. The creation of the MIT Radiation Laboratory was quickly under way, and Jerrold's career now took a sharp turn in a new direction.

The Radiation Laboratory, I

Jerrold arrived at the Radiation Laboratory in the third week of November 1940—"I arrived three weeks late," he said—and reported for work on November 26, as soon as security clearance could be arranged. He had no idea what the work would be, or the commitment it would require. Planning was impossible, and Leona and Susan stayed in New York. He took a temporary leave from Hunter, without any idea of how long it would last; he had every intention of returning to New York when he could. The green Ford touring car went up on blocks in a New Jersey garage since gasoline rationing made it impractical to keep it on the road. Jerrold found a furnished room in Boston's Back Bay, and he and Leona resigned themselves to commuting one way or the other on weekends, by plane or train, whenever the pressure of work allowed it.

Officially the war had not yet started for America; the attack on Pearl Harbor was still more than a year off. The Selective Service Act had been passed in September 1940, however, and young men were beginning to find themselves arbitrarily moved about the country in the name of "defense mobilization." Jerrold was thirty-five, with a wife and child, but it never occurred either to him or to Leona that he would have been eligible for draft deferment, that he might have stayed put in New York, teaching elementary physics to new draftees. It was almost normal in 1940 for individuals to be dislocated and for families to be separated; the state of emergency was beginning to intrude everywhere.

When Jerrold arrived at the laboratory, not more than a couple of dozen people had been hired. According to the official history of the laboratory, "By the middle of December [1940], the total staff numbered only 35, consisting of 30 physicists, three guards, two men in charge of the stockroom and the purchase of supplies, and one secretary."[13] Nevertheless, work started immediately. In September the Microwave Committee had rounded up a number of industrial laboratories, including Bell Labs, Sperry, RCA, and Bendix, and they had pledged to deliver to MIT the components of a 10-centimeter radar system, magnetron and all, by the first of the year. On December 18, with all laboratory activities going forward at a furious rate, Jerrold was put in charge of a group building a workspace on the roof of

MIT's Building 6.[14] Their task was to install the first Radiation Lab microwave radar system and to get it operating well enough to detect signals reflected from distant objects. Their target date was the January 6, 1941.

The pace, Jerrold recalled, was terrific. "We hit the ground running," Jerrold said. "[It] was a far cry from any academic laboratory."[15] In fact, none of the scientists had any real experience with the high frequencies that the magnetron made possible. Back in Rabi's lab, Jerrold and his co-workers had been able to rely on the *Radio Amateur's Handbook* for the information they had needed, but the handbook offered them nothing they could use now.[16]

Incredibly, on January 4, 1941, two days ahead of schedule, the first experimental roof system was operating, at a wavelength of 10-centimeters. It had to be a two-antenna system because they had not yet solved the problem of switching rapidly and safely between transmission and reception using a single antenna, but they succeeded in detecting radar pulses bounced off the buildings of Boston across the Charles River[17]. Some six days later, they were able to find a partial solution to the duplexer problem using existing electronics components, and a single-antenna system was made to operate.[18] Jerrold recalled that they sent a somewhat cryptic telegram to DuBridge, who was in Washington at the time: "Mother Church [of the Christian Science church in Back Bay] on one eye."[19]

It was only the beginning. On February 7, the scientists reported the first tracking of an aircraft with the rooftop system. The laboratory began to get requests from the navy for systems capable of air-to-surface vessel detection, which offered a whole new class of problems. On March 15, Jerrold was assigned to go with a very small group to supervise the installation of the first experimental shipboard microwave radar system on the destroyer USS *Semmes*, a four-stacker out of New London, Connecticut. He said later that he did not know why he had been sent, since he had no particular skills for the job, but it is clear in retrospect that he did. Certainly, he adapted quickly and the task was completed successfully before the end of May. He recalled the general attitude of the military people at that time, as expressed by one naval officer: "I remember the Admiral of the Atlantic Fleet—I forget his name—asked me what I was doing, and I told him. He said, 'We need something for this war, not the next one.'"20 The *Semmes* installation became the prototype for a production model, and in June a contract was signed with the Raytheon Company. The first production set was installed on the USS *Augusta* in April 1942.

It is worth noting the difference between an experimental laboratory model of a complex system and a production version of the same thing. Jerrold made a point of it:

How do you keep the moisture out? Moisture was a pest. Because in a laboratory you think about something working with a laboratory humidity in a uniform temperature, simply because human beings don't work in cold weather. If the laboratory is cold, you turn up the heat . . . but that's not the way it is in the field. It's always a nuisance to look at the military specifications, going from very low to very high temperature, and from very low to very high humidity, and back and forth . . . These [things] are very prosaic, but understanding them makes all the difference between a reliable device and an unreliable device.[21]

Experimental models are to be operated by their designers, people whose skills are high and whose understanding of the system is virtually complete. Stopgap measures involving vacuum wax, shaving soap, and string can be relied on to solve unexpected problems in an experimental system, but in a production model, unexpected problems cannot be allowed to arise, for no one on hand has the knowledge or the skill to make such improvisations, and there is never enough time to allow much tinkering, adjusting, or fine tuning. The radar that was installed on the USS *Augusta* was expected to work under the extreme conditions of battle. To have managed a production model of an operating radar system within ten months was a remarkable accomplishment.

From May 26, 1941, until January 26, 1942, Jerrold served at the Bell Laboratories in Whippany, New Jersey, as the Radiation Laboratory representative working alongside a group engineering the production versions of S-band radar for use in night fighters.[22] Albert Hill, a colleague at the Radiation Lab at that time and later one of Jerrold's closest friends, thought that Jerrold had probably been selected for that job because he could "stand up to the Bell people." This might not have been a simple matter. Bell Labs had been in the business of industrial electronics for many years and had some of the best electronics engineers in the world. There was what Jerrold called a "wholesome jealousy"[23] between them and the Radiation Lab, and they were not likely to welcome advice or criticism from a bunch of bright and enthusiastic amateurs, not yet well organized, and with little experience in microwaves or in problems of production engineering. Jerrold no doubt had his work cut out for him.

He found it exhilarating. "It was a great experience," he said, "working shoulder to shoulder with real engineers."[24] His experience in the molecular beams laboratory had in any case required him to learn something about engineering a system as opposed to simply inventing or building a device. It had taken some doing at Columbia to match the disparate components of the apparatus to each other, to get everything working at once and to make it all reliable. Jerrold had had to learn to see in the mind's eye all the different parts functioning together as a single system and to anticipate what additional

problems would arise from that. It was excellent preparation for the work
he was now doing.

There was still plenty to learn, but as a physicist he was well prepared
to learn it. The ability of physicists to learn to do engineering was often a
matter for discussion and argument, generally good-natured, often humor-
ous, perhaps indicative of the wholesome jealousy Jerrold had noted. In an
essay written nearly thirty years later, Vannevar Bush, himself an engineer,
recalled: "At one time early in the war I suggested to Lee Dubridge. . . that
he had an oversupply of physicists in the team he led, and not enough
engineers. He did not agree, and of course I did not interfere. But late in
the war he said to me 'You see we did not need the engineers.' I told him,
'Hell, any self-respecting physicist can become an engineer in a year or two
if he puts his mind to it.'"[25]

Jerrold got along with the Bell Lab engineers and relished the work,
but the experience did not alter his basic views about engineers. He respected
what they knew but felt that as a group they lacked the creativity and the
ingenuity of the physicists. He liked to say that engineers had "all of the brakes
and none of the motors" (although he tended to say that about anybody who
was not ready to move as fast as he was). He was sure that if the Radiation
Lab had been run by engineers, it could never have accomplished its task
so quickly, and speed was important. "There was a war going on," he said,
"people were being killed every day, and we knew it." Nevertheless, he
conceded that at Bell Labs he learned far more than he gave, and when he
came back to Cambridge he held strongly to the engineers' opinions about
reliability and quality control. For a while he was known around the
Radiation Lab as "Bell Labs' Boston Branch."[26]

Whippany is within commuting distance of New York and Jerrold had
been able to live at home during the period he served there. By then it was
clear that he would be at the Radiation Lab for a considerable time, so when
he returned to Cambridge in early 1942, he began to look for a place to live
suitable for himself and his family. In the early summer, anticipating that
Jerrold might stay put for a bit, Leona and 11-year-old Susan moved to
Boston. Leona was expecting their second child in August. They located a
small house in Brookline, across the river from Cambridge, and took up
occupancy in June. On August 26, 1942, their daughter Johanna was born.
For a time, they were able to settle down to an almost normal life-style in
spite of Jerrold's long hours and steadily increasing responsibilities at the
Radiation Lab.

Of course it couldn't last. The following summer found Jerrold in
England, together with a Rochester physicist named Arthur Roberts, to

establish cooperative arrangements with the British on adapting *Oboe* from longer wavelengths to microwaves. *Oboe* was a crucially important method for precision blind bombing using radar beacons. Leona returned to New York. Since they still expected to live in New York after the war (Jerrold prudently renewed his leave from Hunter each year) it seemed best that they continue to have their permanent residence there, in a place they knew, where there was the support of family and friends when needed, and where, with Anna's help with the children, Leona could continue with her career.

The Radiation Laboratory, II

By the time the war ended, the Radiation Lab was a smoothly running, efficient organization, but in early 1942 it verged on chaos. It was a new kind of organization, set up in a hurry and with hardly any precedent to guide it. The original staff of the laboratory consisted mainly of nuclear and atomic physicists who had been gathered together for the completely unfamiliar task of developing microwave radar. What they lacked in specific experience about radar, however, was more than made up for by the fundamental knowledge they had as physicists and by the ingenuity they had developed in their training. They were mostly young and energetic and moved by tremendous enthusiasm. For example, Louis Turner was one of the earliest arrivals. He came from Princeton and was the author of the first comprehensive review article on nuclear fission, which appeared in January 1940. It is noteworthy that Turner's review referenced more than 100 research articles on the subject, demonstrating that many physicists, at least, understood the importance of the discovery. Nevertheless, it was not generally believed that nuclear physics was sufficiently advanced to provide any likely or substantial contribution to the war effort. In fact, the name "Radiation Laboratory" itself had been selected partly to give the impression that, like its namesake at Berkeley, the activities would be in nuclear physics rather than in microwave electronics and therefore not worth enemy surveillance.

Turner was placed in charge of the group working on microwave receivers; later he would move to the problem of designing radio beacons. He did outstanding work in both areas. Kenneth Bainbridge of Harvard was also a nuclear physicist, as well as a spectroscopist; he had been almost the first to arrive at the laboratory and had played an important role in recruiting others and in getting the laboratory going. He took on the initial responsibility for the group developing pulse-generating electronics. Rabi, of course, was a specialist in the hyperfine structure of atoms, an esoteric subject at that time even for physicists. He now took on the much more practical problem of the transmitter tubes, to see whether cavity magnetrons might be made to produce even more power than the British had been able to obtain from them. The laboratory developed as a grand intellectual free-for-all, with all the emphasis on creativity but with little time devoted to developing systematic methods and procedures. Those who worked there came to feel that they had been greatly privileged.

The scientists were inclined to set an irreverent tone. Since funds did not come from the army or the navy, but rather from special funds in the Office of the Executive, the scientists enjoyed a remarkable degree of freedom from military authority. They tended to exploit this freedom fully, often to genuine advantage when it came to solving problems in new ways. They enjoyed themselves, finding an almost mischievous delight in now and then flouting military authority and coming out ahead in what they felt was service to the greater cause. They saw their iconoclasm as a great virtue, which no doubt it was, and were inclined to be a bit boastful about it, or at least to enjoy telling about it later. In a speech given in 1977, at a Radiation Lab reunion thirty-five years after the fact, Rabi recalled:

Our arrogance was another kind of help in establishing new relations between civilian scientists and the government, particularly the military. I remember one time, a group came from the Navy who desired us to make certain black boxes that deliver certain potentials at certain outlets. And I asked the officer, the head, what this was for. And he looked me straight in the eye, and said "We prefer to discuss this in our swivel chairs in Washington." It was a very clear answer to which I didn't reply. And they came back in November of that year with the same thing. Since I was "Mr. Advanced Development," they said, "Now look, let's stop kidding. We'll bring you a man who understands radio, and a man who understands aircraft, and we'll talk about it." And they did.[27]

James Killian, a courtly South Carolinian who was then executive assistant to Karl Compton, president of MIT, served on the Radiation Lab steering committee as Compton's representative. He had his own small laboratory anecdote, which he still relished telling in 1982:

We fought for the freedom to do things that the government thought we shouldn't be free to do. . . . as an example: the Radiation Laboratory group working on the SCR-584 radar decided they needed a merry-go-round, a carousel, and the government people didn't receive this bill with immediate enthusiasm. They wanted to know why the MIT Radiation Lab people had bought a carousel—a used one. But finally we worked our way through to buy a carousel.

We had a long series of debates as to why we couldn't account for all the wrenches and screwdrivers.[28]

Knowing that they would themselves be held accountable, and perhaps not understanding well the peculiar requirements of men who frequently had their best ideas at three o'clock in the morning, it was natural that military officers would expect both orthodoxy and accountability from the scientists. It was equally natural that the scientists would become impatient with the detailed requirements of the military mind. They saw such requirements as conflicting with the more important need for improvisation and invention, to which they felt any degree of orthodoxy was inappropriate. Levels of exasperation on both sides could often become high.

Fortunately, there were those who could and did show by example that relations between the scientists and the military could be fruitful. Rabi was clearly one of those, as was DuBridge himself, widely acknowledged to be a superb director of the laboratory. On the military side, one whose contribution in this respect was important was Lloyd Berkner, whose influence on Jerrold then and later was considerable. Berkner had been trained in physics and in electrical engineering at the University of Minnesota; he had been a navy pilot in the reserve since 1927. He was an activist; a man who believed in getting things done with a minimum of red tape. In 1942 he had been called up to organize the aircraft radar section of the Naval Bureau of Aeronautics and in this capacity was a frequent visitor to the Radiation Lab.

Berkner was a large, rather orotund man who, as Albert Hill recalled him, "always sounded like he was reading from the Bible." But he was neither pompous nor opinionated, and Hill said he was a reasonable man in an argument. Louis Turner would generally offer a good-natured groan when Berkner visited, said Hill, claiming that the next several days would be spent explaining to Berkner why they could not do something he wanted immediately. Berkner could be persistent. "I always had to tell him," recalled Hill, "the trouble with you, Lloyd, is that you never listen to a damn word you say." But Berkner was a good scientist as well as a military man and understood well both the nature of and the need for improvisation. In him the scientists found a sympathetic and encouraging ally: a useful bridge between sometimes uneasy partners.

Nevertheless, the scientists might have benefited from a bit more orthodoxy, especially in the early days. Jerrold remembered that in 1942 the laboratory "was in a bit of a mess. There was no shortage of inventiveness, but components were being produced that were not compatible, You pay a price when you bring smart people together and ask them to function as a team: each person thinks he has the best idea."[29] It was a lesson he learned well, for he would have very many occasions later in his life to "bring smart people together." He came to understand clearly that teamwork among such people happens only if someone is willing to be in charge. As he liked to say in later years, "You have to let everybody say his piece, but keep the chalk in your own hand."

When Jerrold returned to Cambridge from Bell Labs in early February 1942, DuBridge asked him to deal with the issue of compatibility among the various designs. His reply was characteristic of a growing self-confidence: he said he would do it with the understanding that DuBridge would let him spend as much on the assignment as would be saved by avoiding the loss of one destroyer.

From boyhood, Jerrold had always been straightforward and blunt, once he felt he knew what was at issue—in his own terms, once he had "gotten his head straight" about something. Otherwise, he had little to say. He was not easily intimidated and could be stubborn when he felt the circumstances warranted it. These were qualities that often led others to rely on him in difficult situations, and he could be unusually effective in them. Overall, it was a style that often engendered great loyalty but sometimes caused resentment and hostility, largely because, as he himself was always quick to admit, diplomacy was never his long suit, nor was patience with those he considered to be foot-draggers or posturers. M. M. Hubbard, a Radiation Lab engineer who would subsequently become one of Jerrold's most faithful allies, remembers the resentment he felt at their first meeting, when Jerrold spoke of Hubbard's supervisor—a man whom Hubbard had known for a long time—as "not worth a damn." Hubbard thought it was a presumptuous and arrogant remark, but he concluded on the basis of subsequent events that Jerrold clearly had been right. Many years later Hubbard recalled, "He had a higher percentage of being right than anyone I ever knew, and I was a man who kept records. I liked to keep score."

As the laboratory evolved, it gradually settled down with a stable structure of research groups and committees capable of the kind of coordination necessary to such a very large enterprise. Seven research divisions came into being. Jerrold, who had earlier been director of the Radio Frequency group, became head of Division 5, the Transmitter Components Division. He thereby became a member of the laboratory's steering committee. He was now a policymaker, and he continued to be one for the rest of his life.

There were many from this period who stayed in Jerrold's life in one way or another. The linkages formed in the Radiation Lab environment remained important throughout the lives of those who were there and to a considerable extent determined the shape of the contributions they would all make to public life for decades to come.

The steering committee provided the setting for Jerrold's first interaction with James Killian, who later in the war became executive vice president of MIT and in 1948 succeeded Compton as president. Still later, Killian became the chairman of the President's Science Advisory Committee under Dwight D. Eisenhower, in the difficult period immediately following the launching of the first Soviet *Sputnik* in 1957. In that capacity, Killian continued to rely heavily on those he had known at the Radiation Laboratory, Jerrold Zacharias among them.

After the war, Lloyd Berkner, by then a naval captain, returned to his prewar position at the Carnegie Institute of Washington, where he resumed his research on wave propagation and geophysics. In 1951 he became president of Associated Universities, Inc., an organization of 13 East Coast universities formed to operate the Brookhaven National Laboratory.[30] He served in that post for nine years and then became president of the Graduate Research Center of the Southwest.

Leland Haworth was an electron physicist from the University of Wisconsin. He and Jerrold occupied adjoining offices and shared secretarial services. Haworth also was a member of the steering committee as head of Division 6, the Receiver Division. Eventually he too became associated with the Brookhaven National Laboratory, as its second director. Subsequently he became director of the National Science Foundation and a member of the Atomic Energy Commission.

Albert G. Hill received his Ph.D. in atomic physics at the University of Rochester in 1937 and was an instructor at MIT until 1941. As enrollments dropped during the war, the Institute had not renewed his appointment, and he had gone to work for the Research Corporation. He had pretty much given up hope, he remembered, for an academic career, but at the Radiation Lab he found a wholly new place to start. He too became a member of Division 5, succeeding Jerrold as its head in 1945. After the war, Hill became first the associate director and then the director of the MIT Research Laboratory of Electronics, the successor to the Radiation Lab itself. Subsequently he became director of MIT's Lincoln Laboratory and then director of the Weapons System Evaluation Group in the Department of Defense. Eventually he became MIT's vice-president for research.

Jerome B. Wiesner was an engineer from the University of Michigan who had gone to the Library of Congress in Washington after graduation to work on a project designed to preserve American folk music. Younger than Jerrold by about ten years, he became a project engineer at the Radiation Lab, part of Division 5, Transmitters, reporting to Zacharias. Before the war ended, he had become the chief engineer of Project Cadillac, the largest Radiation Lab project; its mission was to provide airborne reconnaissance and warning of kamikaze attacks. The project is often regarded as the forerunner of modern AWACS reconnaissance systems. Wiesner subsequently moved on to the instrumentation division of Los Alamos, and in 1946 joined MIT's Department of Electrical Engineering. He succeeded Hill as director of the Research Laboratory of Electronics in 1952 and in 1960 became chief science adviser to President John Kennedy. He became president of MIT in 1971.

• Edward M. Purcell, an experimental physicist from Illinois, received his Ph.D. from Harvard in 1938. At the Radiation Lab he wound up in Rabi's Division 4, working on X-band and later on K-band radar. In 1952, by then a professor of physics at Harvard, he was awarded the Nobel Prize for his work in radiofrequency spectroscopy. He was one of those on whom others always tended to call for help; Jerrold in particular liked knowing that Purcell was just a phone call away. Purcell became a famous teacher and expositor of physics; it was he who, as a member of the President's Science Advisory Committee under Killian, wrote the first public explanation of the American space program. Later he provided irreplaceable support for Jerrold's projects in educational reform.

There were many others, of course, but these examples are illustrative. The science writer Daniel S. Greenberg, writing in 1967, offered a useful metaphor, which he called a paradox: "There is no American Scientific Establishment. Yet Harvard, MIT, Caltech and the University of California are its Oxbridge. Two World War II research centers, the MIT Radiation Laboratory and the Los Alamos Scientific Laboratory, of radar and atom bomb fame, respectively, are its Eton . . . The physicists are its aristocracy."[31]

Of course, it has never been true, not now and not then, that to have been a member either of the Radiation Lab or of the atomic bomb laboratory at Los Alamos conferred special benefits by virtue of membership alone. But the associations and friendships that were formed during the war years remained important. The graduates of the great wartime laboratories did become the nation's advisors and planners, and when they needed assistance or advice themselves, they tended to call on each other. It was natural that they should do so; they had, after all, developed the kind of confidence and trust in each other's capabilities that can only result from such close personal interaction as their wartime experience had given them. As long as you knew who could do the job, why not use him?

Submarines

From the point of view of the Radiation Lab, the war was a continuing seesaw battle between opposing groups of scientists. For every measure a counter-measure might be found, given time and ingenuity; if Allied scientists were clever, so also were those of the Axis. The men at the top, like Bush, Conant,

K. T. Compton and DuBridge, had a deep respect for the quality of German science. Many of those who were actually at work at the Radiation Lab—in the front lines, so to speak, of the scientific battle—also had vivid first-hand knowledge of German capabilities. Jerrold Zacharias was certainly one of those who had learned during the thirties what German science could do.

For these men the experience of the Radiation Lab was more than learning how to fashion a better waveguide or TR box, important as that was. They were obliged to learn as well the strategy and tactics of war, and they had to learn the limitations of technology. It was a war of maneuver and surprise, of measure and countermeasure, in the laboratories as well as in the field.

After the war's end, it was found that Allied intelligence had continually overestimated German capabilities, both in radar and in nuclear physics. Allied intelligence also at first underestimated, and then may have overestimated, the suicidal fury of the Japanese as the war came closer to the home islands. But these misjudgments did not alter the lessons that Jerrold took away from his Radiation Lab experience. He often remarked later that if the Germans themselves had had better intelligence—if the German scientists had not continually underestimated their Allied counterparts and if they had not been constantly interfered with by ill-informed directives from above, including those from Hitler himself—the war might have had a different outcome. He never forgot how closely the issue had been decided or how fragile a purely technological defense can be.

The example of German submarines serves to illustrate this point. The particular relevance of this example derives from Jerrold's postwar experience, when he became an advisor to the military on countersubmarine warfare. The lessons he took from his wartime experience were essential.

During a single terrible month in 1942, shortly after the United States entered the war, eighty-two Allied merchant vessels were sunk by German submarines, most of them in the western Atlantic. In all, more tonnage of Allied shipping was sunk in 1942 than was built. The German successes were possible in those early days of the American involvement in the war because the United States was not yet prepared to provide escort protection or air cover for convoys or for coastal shipping. But the kill rate achieved by German submarines could not be tolerated for very long.

The British had already shown that the airplane could be an effective countermeasure to the submarine, provided that the submarine could be detected and attacked before it had time to submerge and escape. Radar was critical for such detection. Diesel-driven submarines had to surface to recharge their batteries, and the German submariners had usually done this under the cover of darkness. It was often possible to learn their habits—they tended to be almost ritually regular in communicating with home base at the same time every evening—and to locate them approximately by high-frequency direction finding on their own signals. Then the German submarines would become vulnerable to radar-equipped aircraft, which could locate them accurately while still some miles away and could attack with little warning. In this fashion, German submarines had been driven from the coast of England by British aircraft, and the same needed to be done in the western Atlantic. The Radiation Lab had been given the mission to design and provide the necessary equipment, and, as might have been expected, the pace at the Laboratory intensified. Jerrold wrote at the time to a colleague at Johns Hopkins University, explaining that he had no time to participate in a meeting on nuclear magnetic moments: "It seems a shame that nuclear physics, at least nuclear moments, are taking such a beating at the hands of the enemy; but we can start in again after the war and it will be peaceful compared to our present jobs."[32]

Radar acts like a searchlight. A submarine, like any other radar target, must be illuminated by radar waves in order to be seen. This could have given the German submarines a chance to use radar-wave detectors, to recognize when they had been spotted and to take evasive action. But the Germans thought that the British were using long-wave radar; not until late in the war did they realize that it was microwave radar that had been used against them so successfully. Consequently, they had not built a dependable microwave detector and did not do so until it was too late to change the outcome. Had they developed such a countermeasure, the submarine fleet might well have remained an effective force, perhaps to the point of altering the outcome of the war.

The battle for control of the sea lasted throughout the war, with radar-equipped aircraft driving the submarine packs farther and farther away from the sea-lanes, away from the staging areas for the Allied attacks on North Africa and Italy. When the invasion of Normandy took place in 1944, German submarines in the English Channel had effectively been neutralized.

Yet battles often turn on small matters, smaller even than the German failure to develop a microwave detector. Late in the war, the Germans introduced the snorkel, a retractable air tube that enabled a submarine to remain submerged while charging its batteries. Only the snorkel would be visible above the surface, and it was a very tiny target. It would have been completely invisible to long-wave radar, but even with microwaves, detection was difficult and rare. In December 1944, at the request of the navy, the Radiation Lab established Project Hawkeye to tackle the problem of the radar detection of snorkel, but with only moderate success. The Germans had discovered how to coat snorkel tubes with a paint that absorbed radiations in the microwave band, which made Schnorkel almost impossible to detect by radar. Had the war continued, this small invention might have given the submarines the upper hand. Jerrold speculated often in later years on what would have happened if the Germans had developed the snorkel sooner. He saw how difficult it would be to defend against modern submarines employing the latest in technological devices, and when he became an adviser to the military during the early 1950s, his understanding played a central role in determining the nation's policy on this problem.

1943–1945

The Radiation Laboratory was not the only project of the OSRD, nor was the Microwave Committee without parallel. Project Y was the creation of the Uranium Committee, and the site selected for its laboratory was Los Alamos, an isolated mesa in New Mexico. It had been chosen by the army on the recommendation of the laboratory's new director, Robert Oppenheimer. The purpose of Project Y was to develop an atomic bomb. Throughout the war, those who ran the project as well as those who participated in it believed that German scientists were well on the way toward achieving the same goal. There was much that gave credence to this idea; indeed, nuclear fission itself was a German discovery. Not until after the war did it become known that the Germans had in fact made little progress toward an atomic bomb.

Project Y differed from the radar project in many significant ways. Its organizers recognized almost from the first that the project would require industrial mobilization on an unprecedented scale. Whole new industries would have to be created for the separation of the fissionable isotopes of uranium and for the production of plutonium. In the view of the Uranium Committee, the organization of such large-scale activity required that it be placed under military administration. Accordingly, the project was transferred in 1942 from OSRD to the Manhattan District, an administrative unit of the army, and the project became known by the more familiar name of the Manhattan Project.

By the middle of 1943, the new Los Alamos laboratory was recruiting intensely for scientific personnel. Arrangements were made by Conant, chairman of the NDRC, for a number of Radiation Lab people to move to Los Alamos. Among those who made the move in 1943 were Bainbridge, L. W. Alvarez, who had joined the Radiation Lab from Berkeley, and

Norman Ramsey. Others from the Radiation Lab, such as H. A. Bethe and
R. F. Bacher of Cornell, had been instrumental in setting up Project Y and
had joined it at its inception. Jerrold would also make the move to Los Alamos but not until later; for the present, he was needed where he was. By
1943 recruitment of scientists for the Radiation Lab had come to a virtual
halt. In fact, the lab had difficulty holding on to those people it had because
Massachusetts draft boards, having problems in making up their quotas, were
casting covetous eyes at the laboratory staff. Draft boards were run by patriotic
civilians who, for security reasons, were not privy to the nature and
importance of the work being done. As far as they could know, the Radiation
Lab was not a military establishment and its members not entitled to any
special deferment. During the years 1943–1945 this was a constant problem
to the Laboratory personnel department.

By contrast, the Los Alamos laboratory was set up as a military
establishment from the beginning, but there was a price to pay. The official
history of the Manhattan Project describes it thus:

The normal military procedure for protecting secret information was by subdivision.
Each individual would have access only to information immediately relevant to his
work. . . . Many of the scientists had been engaged in other war research and were
convinced of the evils in obstructing the normal flow of information within a
laboratory. They were vigorously opposed to compartmentalization. However, no
alternative was acceptable that satisfied military security requirements. Those
requirements could only be met by allowing internal freedom and imposing severer
external restrictions than would otherwise seem necessary. Adoption of such a policy
necessitated an isolated location for the project. . . . The Los Alamos site and a large
surrounding area were made a military reservation. The fenced and guarded
community was made an Army post.[23]

This was a far more restrictive policy than had ever been instituted or
thought necessary at the Radiation Laboratory. It is hard to reconcile secrecy
with scientific inquiry, and, for a variety of highly complex reasons, it was
not fully successful at Los Alamos. Many of those who experienced or
witnessed these security procedures, including Rabi and Zacharias, would
take the opportunity in later years to discuss in the public literature the matter
of secrecy and national security, but it has remained a problem with no
satisfactory solution.

By the end of 1943, the entire scientific establishment of the United
States was caught up in the business of war. The mobilization of science was
complete, and, as the eventual outcome demonstrated, it was effective. By
the end of 1943, the tide of war had begun to turn, and the Allies could begin
to foresee victory. The strategic bombing of Germany's cities and industrial

bases had begun, making full use of microwave radar to make possible nighttime and foul weather bombing; the U.S. Fifth Army stood in front of German-occupied Rome, and Italy itself had dropped out of the war. Russia had regained nearly two-thirds of the territory the Germans had captured in the east. In the Pacific the hard-fought battle of the Solomon Islands was under way, following Midway, Tarawa, and Guadalcanal— stepping-stones on the long road to the Japanese homeland. France and Germany would eventually have to be invaded, and probably Japan as well, the Allies believed. Victory was still a long way off and would be reached only at the cost of many more lives and much suffering, but it was visible.

One bright episode in 1944 reminded some of the physicists of the prewar scientific occupations they had enjoyed and that perhaps they could only now appreciate properly. In a departure from normal ceremonial procedures, made necessary by the war, the Nobel Prizes for 1943 and 1944 were announced simultaneously. The 1943 award in physics went to Otto Stern for the first detection of magnetic moments; the 1944 award went to I. I. Rabi for the precise resonance measurements of magnetic moments. The awards could not be formally presented in Sweden but were instead presented at a luncheon at the Waldorf-Astoria Hotel in New York City, under the auspices of the American-Scandinavian Foundation. The prime minister of Sweden presided. The Zachariases attended the ceremony, as did all those of Rabi's Columbia associates who could manage to be in New York at the time. Millman was there, and Kellogg and Kusch; the award of the prize to Rabi was a justification of all the hours they had all spent in the lab. It strengthened Jerrold's resolve to get back to physics as soon as circumstances would allow.

The Radiation Lab now settled down as a mature, efficient, and stable organization, designing and producing the radar systems essential for obtaining the victory. Slowly, the Allied forces ground away, burning the cities and the factories, destroying the military might of Germany and Japan. The invasion of Normandy and the recapture of the Mariana Islands in 1944 made the war's outcome a virtual certainty, but it would drag on in Europe through its last agonizing months until finally, in May 1945, Germany surrendered. In the Pacific, the killing continued for several months longer. The terrible firebombing by the Twentieth and Twenty-first Bomber Commands of Tokyo, Nagoya, Osaka, and Kobe, followed by the atomic destruction of Hiroshima and Nagasaki—hammer blow after hammer blow—finally brought an end to the fighting. At last it became possible to think of what would come after.

On August 14, 1945, within hours of the emperor's historic broadcast to his people informing them of the Japanese surrender, the Radiation Lab held a victory convocation in the Great Court of MIT. There had been one earlier, in May, to celebrate the victory in Europe, but this was quite different, for three days earlier the OSRD orders to begin the closing of the Radiation Lab had gone into effect, and this convocation marked the end. It was really over now. It was time to pick up all the threads that had been dropped or broken and to begin get on with one's life.

The orders to close the Radiation Lab were part of a more general decision that had been taken at the recommendation of Vannevar Bush and agreed to by President Truman: to dissolve the OSRD itself. It would be hard to find a more remarkable, more foresighted, or more statesmanlike decision. Bush understood that the welfare of the nation required that its scientists be back at their universities, enjoying the freedom of inquiry that can never be possible in a centralized government laboratory.

Many leading scientists felt this way very strongly. The Los Alamos laboratory, for example, was not disbanded, but most of its scientists left as early as they could to return to the academic world, which was being thrust abruptly into an unprecedented period of expansion. Five years had passed with many students and faculty absent from campuses; a powerful surge back to these places was about to happen, strongly amplified by the new GI Bill, a federal program providing tuition payments and stipends for returning veterans. The colleges and universities hastened to prepare themselves.

It was inevitable that there would arise a keen competition among the universities for the best minds. Many of the men who were now sought so eagerly had risen to prominence during the war, and their prewar jobs were either no longer suitable or not so attractive that they would not be susceptible to better offers. Jerrold was one of these.

In the spring of 1945, months before the war ended. Jerrold received an offer from MIT of a full professorship in the physics department. He had dutifully renewed his leave of absence from Hunter College year after year and had only just been promoted there, in absentia, to the rank of associate professor. It was hardly satisfactory to remain at such a rank after the responsibilities that had been his at the Radiation Lab, but the authorities at Hunter had no way to be aware of what his wartime activities had been. Rabi wanted him to come to Columbia but could offer only an associate professorship. In any case, Jerrold remembered his feeling of restlessness in Rabi's lab before the war and felt strongly the need to be fully on his own. "Rabi casts a dense shadow," he said. He gave no serious consideration to a position at Columbia. Although he and Leona had always expected to return to New York after the war, it was clear now that they should follow a different path. He accepted the offer from MIT.

Tour of Duty: Los Alamos, 1945

Jerrold traveled to Los Alamos in June 1945, at the invitation of Robert Oppenheimer, to discuss joining the Manhattan Project. Oppenheimer knew Jerrold as a first-rank physicist and was familiar with Jerrold's prewar work at Columbia. He also knew that Jerrold had been highly effective for nearly five years at the Radiation Lab. He invited Jerrold to replace Admiral William S. Parsons, who was leaving for the Far East, as head of the Ordnance Engineering Division.[34] Less than a month later, Jerrold was a member of the Manhattan Project, in residence at the New Mexico site. The division was whimsically named Division Z (for Zacharias) after Jerrold took over.

The immediate task of the division was to engineer the early forms of the bomb so that they could be more easily handled by service personnel. Once again, it had become a question of distinguishing between a laboratory sort of device and a reliable field version of the same thing. The division also was charged with designing realistic simulation tests for the various auxiliary devices associated with the bomb, such as the fuse that triggered the bomb when it had fallen to a certain predetermined altitude. Problems of this sort had received insufficient attention, it was thought, and Jerrold had the right engineering and administrative skills for them by virtue of his earlier work on radar. A small test facility already existed at a nearby airfield at Sandia, New Mexico, formerly an army base, and the activities of Division Z were transferred there. (Eventually Division Z became the Sandia National Laboratories.) But events moved quickly, and the division barely had time to get organized before the bombs were dropped in August.

On August 6, Little Boy, a uranium device, exploded over Hiroshima, and on August 9, Fat Man, the plutonium bomb, was dropped on Nagasaki. On the next day the Japanese began the formal process of surrender. The war was finally over.

By the middle of September, the Los Alamos laboratory was beginning to redefine its mission and to adapt to whatever its postwar role would turn out to be. Many of its scientists were in the process of leaving or thinking hard about it. Some would return to the positions they had held before the war, but many were young men who had only come to prominence during the war itself. For them, return to their prewar appointments was neither appropriate nor attractive, and they sought new positions.

The chairman of the physics department at MIT both before and after the war was J. C. Slater, a difficult and enigmatic man known for his "microsecond smile." Slater was an able theoretical physicist who had worked with Bohr in Copenhagen in the early 1930s and had made substantial contributions to the quantum theory of the solid state. He seemed to have little interest in pursuing physics outside that field. At the Radiation Lab, he had served in Division 5, nominally reporting to Zacharias; however, colleagues found him difficult to work with, and he had been put in charge of his own very small group to deal with "special problems." Subsequently, he had worked at the Bell Laboratories on magnetron design. According to colleagues who knew them both, Jerrold got along reasonably well with Slater, but they never formed a close friendship.

At MIT before the war, Slater had not managed to make his department a leader in nuclear physics, and, as a result, the subject had been incompletely represented there. The physics department had been a good one, of course, but had followed rather traditional lines: Philip Morse was outstanding in mathematical acoustics, as were Julius Stratton in electromagnetic theory, and Slater himself in solid state physics. It was clear now, however, that nuclear physics and nuclear engineering must become central activities in the department's and the institute's postwar programs. Karl Compton, MIT's president, assigned the responsibility for building up these areas to Jerrold. Both understood that prompt and energetic recruiting for new personnel was essential. To Slater's credit, no conflict of authority developed. The structure under which nuclear science eventually emerged at MIT was a new interdisciplinary laboratory run by Jerrold, and the original suggestion for creating such a laboratory had come from Slater himself.

Jerrold's involvement at Los Alamos was opportune in bringing him into contact with many nuclear physicists just at the time when they were thinking about their postwar lives but before most of them had made any decisions. Albert Hill, who had succeeded Jerrold as head of the Transmitter Components Division at the Radiation Lab and had become one of his closest friends, believed that Jerrold's transfer to Los Alamos was probably engineered by Compton—with Rabi's help, almost certainly, since Rabi was Oppenheimer's adviser—in part so that he could do this recruiting. Jerrold was successful at it and convinced about a dozen excellent scientists, as well as a number of promising graduate students, to come to MIT. It made an enormous difference to the institute, establishing it in the first rank in nuclear science almost at once. It was an accomplishment in which Jerrold took pride, as he later admitted.

Among the scientists he recruited were V. F. Weisskopf, who had been a group leader in the Theoretical Physics Division, and Bruno Rossi, a group leader in the Experimental Physics Division. Weisskopf established a distinguished theory group at MIT, and Rossi succeeded in making MIT a world leader in the field of cosmic rays. Among the younger scientists was Martin Deutsch, who eventually did brilliant work in positron physics. Jerrold also managed to persuade a number of promising young men who did not yet have their doctoral degrees to come and finish their work at MIT. One of them, David Frisch, recalled: "Jerrold sort of blew open the psychology of the system by making grad students who didn't yet have their PhDs, offers of $3000 a year, plus tuition . . . which was absolutely stupendous, for the time . . . and it went like a dose of salts. I think Ken Bainbridge was extremely angry and I think Tolman was angry, and really it upset the system . . . and so he recruited that way easily . . . he brought, among people at that level, Herb Bridge, Matt Sands, Bob Williams . . . there were others." In a 1983 interview, Jerrold also reminisced about those times: "MIT was not alone in this kind of thing. At the end of World War II there was a kind of fraternity rushing going on at Los Alamos by Cornell, Rochester, Cal Tech, Chicago, Columbia, MIT; they were all trying to recruit some of the great people who had been assembled at Los Alamos to make atom bombs . . . [w]ell, it was a free-for-all at Los Alamos, but there were enough to go around and I brought back a first class contingent to MIT."[35]

Jerrold also wanted to build a new molecular beams lab and to get back into his own research. He persuaded Bernard Feld, a recent Ph.D. from Columbia who had been an assistant to Enrico Fermi, to come and do theory for the group he hoped to establish—to be what was commonly called the "haus theoretiker." All in all, considering the short time he was at Los Alamos, he managed to run up a remarkable score.

He was ready for his return east by the end October and, accompanied by Feld, drove cross-country once again. His green Ford which, as a harbinger of better times, had earlier been released from its wartime storage in a New Jersey garage, carried them home to new careers and new lives.

Isadore Zacharias, c. 1905.

(Facing page, top) Jerrold and his grandfather, Julius Kaufman, c. 1907.

(Facing page, bottom) Aboard the Jacksonville alligator, c. 1910.

(Above) Irma Zacharias on the occasion of a benefit concert for Beryl
Rubinstein, Asheville, North Carolina, summer, 1914.

Jerrold, soon to depart for New York . . .

. . . and soon after arrival.

(Top) Jerrold and Leona, c. 1927.

(Bottom) Jerrold in his raccoon coat, c. 1930.

(Top) Yorktown Heights, 1932. I. I. Rabi is in front. Edith Quimby, Helen Rabi, Leona, and Jerrold are in back.

(Bottom) Jerrold in the molecular beams laboratory, c. 1932.

Jerrold Zacharias

Lee A. DuBridge

Al Hill

I. I. Rabi

Jerrold and some of his colleagues from the Radiation Laboratory: Lee A. DuBridge, director; Al Hill, who replaced Jerrold as head of Division 5; I. I. Rabi, associate director; Norman Ramsey; Jerome Wiesner; Edward Purcell; J. C. Slater; M. M. Hubbard; and F. W. Loomis. From *Five Years at the Radiation Laboratory* (Cambridge: MIT 1946).

Norman Ramsey

Jerome Wiesner

Edward Purcell

J. C. Slater

M. M. Hubbard

F. W. Loomis

90

March. 14, 1956

A precise gravity meter   A absolute
method for precise (parts in 10^8 or better)
determination of local acceleration
of gravity.

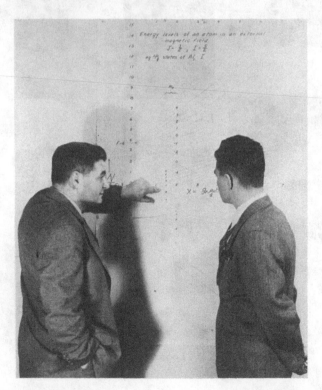

A plate is held up against
the top plate by a few
kilovolts of electrostatic
attraction. When the volts
are smartly removed the
movable plate will fall
parallel to itself in a vacuum
with the acceleration "g". We want
to time its fall thru successive
known or measured distances.

Method (a). Start a fast
cycle counter when the
plate goes thru a microwave null
and stop it when it goes thru
the next. Repeat for all nulls
within a distance of a foot &
calc. g. from the rate of increase.
of v.

Energy levels of an atom in an external
magnetic field
$J = \frac{1}{2}$, $I = \frac{3}{2}$
eg ^2S states of A, I

$X = $ Stufeld

(Facing page, top) A page from Jerrold's lab notebook, with an idea for measuring the acceleration of gravity with high precision using a microwave interferometer, one wall of which is in free fall.

(Facing page, bottom) Jerrold discussing nuclear hyperfine structure with Darragh E. Nagle, one of his first graduate students.

(This page, top) Jerrold Zacharias, c. 1960.

(This page, bottom) Reunion photo, Columbia University, May 23, 1967, on the retirement of I. I. Rabi. Left to right: Norman Ramsey, Jerrold Zacharias, Charles Townes, I. I. Rabi, Vernon W. Hughes, Julian Schwinger, Edward M. Purcell, William A. Nirenberg, Gregory Breit.

Jerrold at the PSSC studio.

4

Branching Out

The Laboratory for Nuclear Science and Engineering

Until I was thirty-five, which is when I went to the Radiation Lab, I never opened my mouth, never said a word to anybody. And I know why, I didn't have anything to say.[1]

—J. R. Zacharias

Probably no one who came to know Jerrold after the war would have recognized him from this modest self-description. The war had changed him dramatically. For five years he had been a highly successful leader of a large and complex group, under difficult and challenging circumstances, and this had given him a new confidence in himself; there would be a few who, perhaps still smarting from confrontations with him, might have preferred to call it arrogance. In any case, he discovered that he now had plenty to say and saw no reason to be hesitant. Rabi said later that it seemed to him that Jerrold had made a kind of quantum jump, had become almost a different person. Certainly he did not resemble the young and diffident researcher he had been before the war, satisfied to remain quietly in the background of someone else's laboratory. He had not really changed, of course; what was different was that he could no longer be satisfied with minor roles. He was ready to act on a larger stage.

The end of the war found the nation's scientific enterprise irrevocably altered. New attitudes and opportunities prevailed as the transition was made from wartime laboratories to peacetime enterprises. Old institutions needed to be revised and new ones created, with new procedures and new styles of research. New channels had to be found for the public support of science. It was a moment with great possibilities and, as the head of a new laboratory concerned with the most visible scientific issues of that critical time, Jerrold found himself extremely well positioned. He had become a kind of scientific entrepreneur, which he defined in the following way: "An entrepreneur is somebody who's cocky enough to believe that if he stays in charge everything will be all right. He doesn't know that his plan will work if you put somebody else in charge."[2] The place of science in the national agenda was open to further important change, and a scientist with an entrepreneurial bent was likely to make his presence felt.

The transition to a new order began abruptly with the war's end. In August 1945, the Radiation Laboratory employed nearly 4,000 people and was comparable in size to a major industrial corporation or factory. Morale was high, for the laboratory's products had been successful beyond expectations. Its research capability was unprecedented; only toward the end of the war had it even been matched by the Los Alamos laboratory. Radiation Lab activities by then covered more than fifteen acres of floor space and had required several new buildings, which had been built hastily in the stark, unadorned, barracks–like style that was somehow appropriate to a time of emergency. Yet only five months later, on December 31, 1945, size and success notwithstanding, the Radiation Laboratory officially went out of existence. Most of its people had by then departed; its hundreds of oscilloscopes, its thousands of amplifiers, oscillators, waveguide junctions, and meters lay unused and, for the moment, unclaimed. The great wartime enterprise, which had made victory possible, was simply and quickly dismantled.[3]

Four independent fragments were left behind to deal with the problems of the lab's termination. One of these was the Publication Division, with the prodigious task of overseeing the production of the twenty-seven technical volumes that would describe as much of the lab's five years of research as could then be declassified.[4] Two divisions were established to deal with contract settlement and personnel problems. A fourth fragment, the Basic Research Division, was formed on January 1, 1946, under the directorship of Julius A. Stratton, a professor of physics at MIT since the early thirties. At Jerrold's persistent urging, Albert G. Hill, his successor, colleague, and friend from the Radiation Lab, was named associate director.

The Basic Research Division was the true inheritor of all that was left of the Radiation Lab: its research aims, its traditions, and its physical inventory. The division operated independently for six months and then, on July 1, 1946, became a part of MIT as the Research Laboratory of Electronics (RLE). This was a new kind of interdisciplinary enterprise—a joint undertaking of the departments of physics and electrical engineering. According to Stratton, the idea for such an entity had been set forth by J. C. Slater in a 1944 memorandum. Slater had recommended "an Electronics Laboratory, established jointly by the two departments, not constituting a separate department, but still having an independent existence, under its own director. . . . [It] would have a budget of its own, and could have personnel of its own, such as research associates, assistants, and technicians."[5]

The model seemed to promise sufficient flexibility to adapt to rapidly changing circumstances. On December 19, 1945, President Compton announced that a second interdisciplinary laboratory was to be established: the Laboratory for Nuclear Science and Engineering (LNS&E), under the direction of Jerrold Zacharias, who had been brought to MIT principally for that purpose. The new laboratory was to be similar in administrative structure to the RLE, providing a common ground for nuclear research in physics, chemistry, biology, metallurgy, engineering, and such other fields as were appropriate.

The two laboratories started with exceptionally strong cadres of scientists: RLE drew primarily on the skills of the veterans of the Radiation Lab, while LNS&E relied largely on the talents of those whom Jerrold had managed to recruit from Los Alamos and also of those alumni of the Radiation Lab whom he persuaded to join the new venture. Of course, both laboratories also drew heavily on the existing strengths of the MIT departments. The question of funding remained to be settled, although substantial funds from the navy had been promised.[6] The challenge was to arrive at an arrangement that would guarantee the independence of research; hardly anything was more important to the scientists after five years of secret, mission-oriented work. But there was no doubt that the funds would be adequate. Times had changed since the days when Rabi had had to borrow electric batteries for the experiments at Columbia and had managed to make a grant of a few thousand dollars stretch over several years. Physics could not be done that way any more. It was now big time; it needed big-time budgets, and they would be available.

The attitude of the public toward physics and physicists in this period just after the war made big budgets possible. The public was more than grateful; its feeling verged on awe. The atomic bombs that had destroyed two Japanese cities riveted everyone's attention and provided startling testimony to the power that physicists had placed within the country's grasp. Nuclear physics itself remained a deep mystery to the public, which quickly became convinced by the popular press that the subject was incomprehensible to ordinary people. For a time, physicists looked a bit like supermen.

Before the war, scientists in general and physicists in particular had received little serious or sustained notice outside their own fields. Occasionally a comment made by someone like Lord Rutherford of England or by the American Nobelist R. A. Millikan might make the newspapers but only in the Sunday supplements, where editors have always liked to place articles speculating on one or another improbable future. The New York World's Fair of 1939–1940, in a last prewar expression of wistful optimism, had chosen for its theme The World of Tomorrow, and had featured prominently among its exhibits an entertaining and impressive prediction of a peaceful world. The world of 1960, the exhibits suggested, would be free of drudgelike labor, powered by electricity from atomic sources so cheap as not to be worth metering or fighting over. But most people had regarded that as science fiction; 1960 had seemed far off in a vague and remote future. Now, as the second half of the 1940s began, the future seemed to have arrived, sooner than anyone had thought possible, and only the physicists appeared able to understand what it might mean. The public treated them with a degree of respect that was unfamiliar to them. Bernard Feld, for example, whom Jerrold recruited to MIT, recalled:

I had at that time accepted the offer to come to MIT, and I was supposed to appear at MIT on September first, I guess—it was in 1945—and instead of appearing I applied for a six-month's leave of absence and I went to Washington, and I spent six months, or five months, whatever it was, going around from Congressional office to Congressional office trying to lobby against the bill [to put nuclear research under the control of the military]. . . .[7] It was really quite amazing how easy it was to lobby in those days. You would knock on the door, it opened, you would say, Mr. Senator, I have just come from Los Alamos and I'd like to talk to you, and of course he would welcome you in. . . . It was an Open Sesame.

Radar had attracted less public notice than had the atomic bomb. As it happened, the bombs had been exploded at just about the time the Radiation Laboratory had started to make public what it had accomplished in secrecy during five years of hard work, but this attempt to provide some kind of public accountability was almost totally eclipsed by the startling news about the atom. The story of the development of radar and its role in achieving the victory thus was made public without ever becoming well known.

There were those, however, in the Congress and, especially, the military who understood that many of the dramatic technological developments of the war had their origins in the basic research that had been done before the war, often in small, underfunded (or unfunded) nongovernment laboratories. That would now change.

In September 1945 a congressional bill (the Vinson bill) had been drafted by a group of enthusiastic young naval reserve officers for the purpose of setting up a new Office of Naval Research (ONR) under the secretary of the navy.[8] For the proposal to be acceptable, it was necessary for its sponsors to demonstrate that the universities would be willing to accept navy contracts for the support of basic research, and to make that happen, the navy had to convince the universities that it would not seek to control their research. The task of presenting the case to the universities was undertaken by Captain Robert D. Conrad, USN, a persuasive and thoughtful man with technical training, who visited a number of universities to sell the new idea. In a speech made in 1946, Conrad expressed the navy's view:

The Navy has embarked on a venture which is not only new to the Navy, but new to Government, and which is of deep significance to the national welfare as well as to the national security. This venture is the active and comprehensive support of research. . . .

Freedom of discussion and publication are essential to [research's] progress. Scientists must utilize the vast accumulation of facts inherited from their predecessors and established by their colleagues. . . . The universities and colleges are the traditional custodians of the spirit of scientific research and we must depend mainly on them to support and extend the social awareness of research as one of the foundations of civilization.

From what I have said, it should be clear that it is a contradiction to speak of directing and controlling research. An unexplored country cannot be mapped. It is proper and necessary to plan development work, but research must follow only its inner promptings. Direction by external authority defeats its own object, for neither the path nor the goal of research work can be foreseen.[9]

This was a remarkably long-sighted view, and Conrad eventually was able to get the agreement of such institutions as Harvard, MIT, Chicago, the University of California, and Cal Tech, as well as many others, to accept navy support for research. The bargain was closely driven, however, and the universities achieved what they felt they needed. The arrangement allowed a single overall contract to be drawn with a university, calling for research in broadly defined areas rather than the specification of detailed goals. Most important, the agreements with the navy specified that basic research was to be unclassified and freely publishable.

Among other advantages, this arrangement allowed universities for the first time to provide financial support to a substantial number of graduate students who were sufficiently advanced to be classified as research assistants. As Albert Hill pointed out, a whole generation of graduate students in the basic sciences had disappeared because of the war:

It was that organization [ONR] which saved American science—save is too strong a word, but certainly they advanced it by two or three years, which is a lot of time. . . . During the war, they [the graduate students] were all in the trenches somewhere, or being reserve officers, or what have you, and we lost an academic generation of scientists. The ONR . . . recognized this. Now, the Navy has no charter to do education, but they have a charter to do research, and to educate graduate students you do research [and] pay them out of research funds . . . and this helped tremendously.[10]

The availability of this support more than any other contemporary measure ensured the continuing preeminence of the United States in postwar science.

These were the circumstances prevailing in the nation at the birth of LNS&E, and it was principally up to Jerrold to fit the new laboratory into this scheme of things. Until well into the 1950s, the major sources for the support of basic research were the Department of Defense (the ONR had been joined in supporting such research by the Army Signal Corps and the Army Air Corps Communication Command), the Public Health Service, and the Atomic Energy Commission. Even after the establishment of the National Science Foundation (NSF) in 1950, well over half the support for university research continued to come from the military and most of the rest from the Atomic Energy Commission.[11] The continuing failure of the Congress to provide substantial civilian-based support for basic research resulted from a contentious democratic process: agreement could not be reached on questions of political accountability and presidential control of the new NSF. In the meantime, the gap in support was bridged principally by the ONR, which took a generally foresighted and liberal view, setting procedures and policies of contracting, funding, and peer review that served so well that they would be closely followed by the Atomic Energy Commission and by the NSF itself once it was established.

Thus it happened that the new Laboratory of Nuclear Science and Engineering was able to begin operations in its first year with an annual budget of $360,000, provided by ONR. In its second year, the budget jumped to about $1.4 million and stayed at that level for the next seven or eight years. These are large sums. Taking inflation into account, they would be equivalent to amounts eight to ten times larger today: the $1.4 million of 1947 thus translates into about $13 to $14 million in 1990 dollars. The LNS&E was clearly a substantial operation.

Settling Down Again

Jerrold was now forty years old, going on forty-one, and not the least bit disposed to slow down just because the war was over. What had excited and pleased him most about the wartime experience was the pace at which he and his colleagues could make things happen, and he saw no reason not to match or exceed that pace in peacetime. A fast tempo suited him. Having seen what could be done in a hurry, provided only that people wanted to do it, he was impatient with the more careful and deliberative procedures that he generally encountered in peacetime. Many years later, he continued to express this view: "It's very tempting to look at the restrictions put on moving fast . . . a nuclear power plant now takes twelve years—Jiminy Crickets! Read the Smyth Report! Look at the time scale! You do it fast, and you have to know when to stop driving for the extra decimal point that's not going to get you anything."[12]

Jerrold's list of things to do immediately was imposing. First, he had to get a brand new laboratory going, with some eight or ten new or revitalized research programs, and he intended to join fully in the effort to win for MIT a significant place in the postwar scientific world. This was terribly important to him and remained so for the rest of his life. He had developed an absolute and lasting loyalty to MIT during the days of the Radiation Lab.

He also wanted to get his own atomic beams lab started. He thought it unwise and inevitably unfair to set up his own research within LNS&E, where he would be competing for funds with his subordinates. Instead, he arranged to establish his lab within the Research Laboratory of Electronics. He knew that Stratton and Hill would never hesitate to say no to him if the circumstances warranted, in spite of their close friendship to him, and for the most part, other people recognized that, too. The arrangement eliminated any possible conflict of interest.

He and Leona also needed to reorganize their lives to accommodate the new turn his career had taken. Their home would be in the Boston area from now on, and they needed to settle on living arrangements. As apartment-dwelling New Yorkers, they had never owned a home and had no equity with which to buy one. Jerrold had to borrow money even to make the down payment on the comfortable house he found in suburban Belmont. It was a lucky find; when he located it, after much searching, he reported to Leona that not only was it ample, but there was something pleasant to look at out of every window. Belmont had long been home to many members of the faculties of Harvard, MIT, and the other nearby colleges and universities, and Leona and Jerrold found it congenial. Leona wanted to reestablish herself professionally, and after she moved up from New York with the children that summer, she was able to obtain an appointment as a research associate in ophthalmology at the Harvard Medical School. She enrolled Susan in Belmont High School, found someone to look after Johanna, and went to work. Gradually their lives settled into new patterns.

The Laboratory of Nuclear Science and Engineering began formal operation on the MIT campus in April 1946, sharing space and some facilities, such as machine shops, with the Research Laboratory of Electronics, in the buildings that had so recently housed the Radiation Lab. The new laboratory was rapidly organized into separate research groups, each with its own budget and its own group leader reporting to the director. There was a steering committee for the laboratory, chaired by Jerrold, the members of which were the heads of the relevant academic departments: physics, chemistry, electrical engineering, and so forth. There was an Advisory Committee for Science, consisting of the senior group leaders, which also met under Jerrold's chairmanship. The arrangements were put together with the kind of speed and efficiency that Jerrold admired so much; within three months of the start of operations, the basic structure of the laboratory was in place, and running smoothly. There were fifty two staff scientists at work in eight research groups, supported by some fifteen nonstaff group members (some were beginning graduate students) and twenty four administrative and technical workers, including machinists, glassblowers, and the like.

Jerrold's transition to MIT was not frictionless, of course. He was in many respects still the new boy on the block, in spite of his five years of service at the Radiation Lab. He had been put in charge of this new enterprise rather abruptly, and there had been little consultation with others who were affected. Robley Evans, for example, a member of the physics department, had been in charge of the MIT cyclotron before the war, making radioactive isotopes for medical purposes.[13] He had been one of those with whom Captain Conrad of the ONR had conferred on his visit to MIT, and Evans was preparing his own proposal to ONR in response. He subsequently recalled talking to George Harrison, the dean of science, about it in 1945:

A Navy captain, Bob Conrad, came to my office . . . and said that the Navy was committed to set up an Office of Naval Research . . . and they had scanned the country and they regarded my laboratory as a national asset—how much money did I want? . . . I think this is related to the founding of the Laboratory for Nuclear Science and Engineering, which turned out to be headed by Zacharias. . . .

So we talked about that, and there were several discussions. . . . [Someone] said that . . . he thought there was an instrument fellow at Los Alamos called Jerrold Zacharias who was very good on instruments and would be helpful . . . in the electromagnetic propagation and things of that type, and wouldn't it be a good idea to have him come, invite him to the Institute?

And he came, and was here. Then George [Harrison] came up to my lab, and cornered a couple of us . . . "How about having Zacharias be head of the Laboratory of Nuclear Science and Engineering?"—in a way in which the only answer possible was, "Sure, George."

So they folded in my visit from Captain Bob Conrad.[14]

It was all a bit high-handed, perhaps, but much was at stake. Under such circumstances it would have been surprising if there had been no resentment at all, however transitory. But nothing of that sort ever interfered with the operation of the laboratory. Evans was named a group leader at the beginning and was able to pursue his research; in the end, he said, that is what mattered to him.

Some early shifting of groups and personnel took place in the first year or so. By 1947, fifteen months after the laboratory had begun, the roster included 97 staff scientists in ten groups; the number of nonstaff, administrative, and technical personnel had increased to 106. The groups and their leaders had settled down to the following:

Cosmic Rays	B. Rossi
High Voltage Research	J. G. Trump
Nuclear Cross Sections	R. J. Van de Graaff
Radioactivity and Cyclotron	R. D. Evans
Synchrotron	I. A. Getting
Theory	V. F. Weisskopf
Fission Element Chemistry	D. N. Hume
Inorganic Nuclear Chemistry	C. D. Coryell
Organic Nuclear Chemistry	J. D. Roberts
Neutron and Gamma Ray Shielding	C. Goodman

E. R. Gilliland, of the chemical engineering department, became the associate director, and Mac Hubbard, Jerrold's lieutenant from Radiation Lab days, became the assistant director. The organization had something of the flavor of the Radiation Lab setup, although there were differences, reflecting the differences in scale and in purpose. But overall it was an arrangement that seemed familiar and that Jerrold felt he knew how to manage. With Hubbard to look after the day-to-day operations, Jerrold could turn his attention to other matters.

Making Policy

Once the Laboratory was up and running, Jerrold was ready to get his own research program started. These activities, together with supervising graduate students and teaching MIT's bright undergraduates—they were far more sophisticated, said Jerrold, than those he had taught at Hunter—constituted a full and busy program. But policy issues now also claimed a significant part of his time, and they continued to do so for most of the next decade. He was drawn into nearly every consultation and every committee at MIT that had anything at all to do with nuclear science, with questions of MIT-government relations, or simply with the institute's research policy. He was frequently the representative of MIT in matters of national concern, invited to participate partly in his own right, partly in the name of the institute.

These activities were not all equally important, to be sure. The question of where MIT's synchrotron should be built, for example, was important to MIT, but it did not rank with the question of where a major national laboratory should be built. Nevertheless, all such issues demanded their share of time, and Jerrold tended to give them all equally careful attention. Two examples will serve to illustrate.

A National Nuclear Science Laboratory

Brookhaven National Laboratory is located about two-thirds of the way out on Long Island, approximately 75 miles east of New York City. If Jerrold had had his way, it would have been located at Fort Devens, Massachusetts, 30 miles northwest of Cambridge. He had proposed that as part of his campaign to secure maximum national influence for MIT but lost this particular battle fairly early. When he thought about it later, he realized that in order to have put up a better case, he would have had to be prepared to involve himself in the running of the new laboratory—perhaps not to be its director but to be deeply involved, nevertheless—and that was more than he had wanted to do.

Jerrold's involvement with Brookhaven began in an interesting way. He could not recall, later, why he, DuBridge, and Rabi happened to be visiting Ernest Lawrence in Berkeley in early 1946, but the conversation turned to the question of how those in the eastern United States could get going in high-energy nuclear physics, which until then had been largely a California specialty. High-energy physics would require a new generation of big accelerators, machines capable of producing intense streams of very energetic particles. The energies would have to be comparable to those until then found only one or a few at a time in the cosmic rays. Jerrold recalled:

Berkeley was making the big machines; they were inventing them, building them, funding them, and finding people to exploit them, and Lawrence was very candid. He said, "You can't do this in the East, you haven't got what it takes." . . . We made up our minds, the three of us, that something had to be done. We had to bring to bear the powers of [the] universities . . . So we started stirring up the old guard. Now the old guard had been, until recently, assembled one way or another at the Radiation Lab of MIT. The Radar Lab. This included Bob Bacher who was then I think at Cornell, Lee DuBridge who was at Rochester, Rabi at Columbia and me at MIT. I won't try to pull together the group names. But it shouldn't surprise anybody that Rabi, Ramsey, Bacher, DuBridge and I joined forces to promote a laboratory where we could do what we called nuclear engineering . . . [15]

Stirring up the old guard brought rapid results. On March 23, 1946, a first meeting of representatives of nine universities took place at Columbia University. Rabi and Norman Ramsey attended, as did DuBridge; John Slater and Jerrold Zacharias represented MIT. Henry D. Smyth of Princeton, George B. Pegram of Columbia, and Jerrold were able to report at the meeting that they had already managed to meet with Colonel Kenneth D. Nichols of the Manhattan District and that they had received assurances of substantial support from that quarter.[17] Smyth spoke in terms of an outlay of $5 million to $20 million over the next two or three years; Jerrold said that $30 million to $40 million would be more like it, since he wanted the new facility to have both nuclear reactors and what were then referred to as "electronuclear machines"—in other words, accelerators.

The meeting quickly established itself as the Initiatory Advisory Group, elected DuBridge its chairman, and Ramsey its secretary, and appointed a Planning Committee consisting of one representative from each institution. Jerrold was the MIT representative.

One week later, the Planning Committee met in Rabi's Columbia office and established a number of subcommittees. Jerrold headed the one on the so-called electronuclear machines and served on another with Ramsey, Bacher, and Smyth to recommend a site. At its first meeting the previous week, the Initiatory Group had passed a sense-of-the-meeting resolution: "that the lab should be located for maximum accessibility as near as possible to New York City." As far as Jerrold was concerned, a location 30 miles west-northwest of Cambridge, Massachusetts, met that criterion, and a few others besides:

Now we knew that we needed space, we needed land, and we didn't need—as a primary need—easy access. We wanted to go where people weren't living. If we were going to build power reactors we knew they would be dangerous. We didn't know how dangerous. I didn't even want to work with them, but I wanted some place where MIT could be involved in nuclear engineering of power reactors. Well, I was all for putting it out at Fort Devens . . . and Bacher's reaction was "absolutely not. It's only forty miles away [sic] from MIT and MIT will take it over." Now this was more a comment about me than about MIT, but Bacher didn't want Fort Devens. It would have been charming. People could have lived in Concord and Lexington and it would have been wonderful.[18]

The others were much more concerned with accessibility than was Jerrold, although they all shared concern about the dangers of situating the laboratory near a population center. They did not, however, take into account that the laboratory itself would attract a considerable population of workers to its vicinity. Reflecting on this long afterward, Jerrold felt that they had missed an important opportunity to deal with important safety issues. They should have recognized, he thought, that it would be impossible in the long run to keep people away from reactors and accelerators: "As soon as you establish a laboratory, there's something for people to live near to, and so, . . . [Brookhaven]'s got too damned many people around it already. We wanted people away from them . . . Somebody should have said that we have to go after a reactor which is inexpensive, reliable, safe, cheap— that has all the appropriate qualities that you can manufacture. I think it's fair to say that the first team of professional scientists did not get into the nuclear power business. We were glad to let somebody else get into it. And that's no way to run a railroad."[19]

Like many others, Jerrold was endowed in his later years with keen and unforgiving hindsight, which he tended to focus on all the missed opportunities of his earlier years. He did not spare himself or those who had been with him. He believed all his life that once you could see a problem clearly, you could pull together the right group of people to solve it. He never doubted, for example, that a safe nuclear reactor could have been developed in the late 1940s or early 1950s if only some of what he called the "first team" had gotten to work on it, just as he came to think that the "first team" had missed the boat on the control of nuclear weapons in the same period. He always thought that Rabi probably knew what was needed. "You know, it's amazing," he said, "but wherever you turn, there's Rabi. And he says what's got to be said. People don't always listen to him. And he doesn't necessarily always follow through with what he believes. Sometimes he's more modest than I think is appropriate. Rabi's not really an entrepreneur in some way."[20] It was certainly essential to see the problems, but without what he called entrepreneurism, nothing much would happen. The problems needed more entrepreneurial management than they got. "Our error was omission," he said. "We didn't pay attention."

The nine universities in due course formed themselves into Associated Universities, Inc. (AUI), and selected the former Camp Upton, near the town of Yaphank on Long Island, for the site of Brookhaven National Laboratory. At about that time, DuBridge was appointed president of the California Institute of Technology and resigned from the group; Philip Morse, of the MIT physics department, was selected as the first director of the laboratory. He was succeeded two years later by Leland Haworth, Jerrold's close colleague at the Radiation Lab, with whom he had shared offices. Jerrold had sponsored both Morse's and Haworth's appointments.

Jerrold continued on the board of trustees of Associated Universities, Inc. for many years as the representative of MIT.

A Proton Accelerator for the Navy?

From 1947 on, the Laboratory for Nuclear Science and Engineering was one of the major recipients of ONR support, and it was natural that it became something of a showpiece for the navy. The scope of its activities was impressive, and its director was enthusiastic and personable. Captains and admirals trooped through Cambridge expecting to have the newest developments explained to them and they generally went away satisfied. Entertaining visiting brass was an important activity of the laboratory as the 1940s drew to a close. Not surprisingly, Jerrold was frequently called on for advice and counsel on scientific matters.

In the spring of 1949, the chief scientist of the Office of Naval Research was Alan T. Waterman, a highly respected expert in the electrical properties of solids. He was faced with a serious dilemma. A proposal had come to him from a group of research people at the Naval Research Laboratory (NRL) for the construction at NRL of a proton accelerator capable of reaching 3 billion electron-volts of energy.[21] His research people had indicated to him privately that they were unhappy at not being able (as they saw it) to participate in mainstream high-energy nuclear physics, and Waterman feared that they might well leave NRL if their proposal was refused. Nevertheless, the proposal represented a sharp departure from navy policies, which were designed to support basic research in civilian laboratories. He therefore appointed a special Nuclear Science Advisory Committee and placed the matter before it for advice.

Jerrold, by this time well known in navy circles, was asked to serve on the committee. His co-committee members were S. K. Allison, Chicago; L. A. DuBridge, CalTech; C. C. Lauritsen, CalTech; W. V. Houston, Rice; A. H. Compton, Washington University.; W. A. Noyes, Rochester; and J. R. Oppenheimer, Institute for Advanced Study.

The committee's response, drafted by Jerrold, was negative, on technical grounds. The navy's proper business, said the committee, was nuclear power, with particular emphasis on nuclear propulsion; everything of interest to that concern was to be found in the energy region below 20 million electron-volts. A 3-BeV proton-synchrotron could not be considered part of a nuclear power program. Furthermore, the small staff of nuclear physicists at NRL would be swallowed up by the project; instead of nuclear physics, they would find themselves doing the electrical engineering that machine design and construction required. On practical grounds, therefore, the committee concluded that the project should not be pursued.

While the committee's response was straightforward, dealing with the question on strictly technical grounds, the full issue was not resolved so easily.

By September, the NRL scientists had rethought their idea and were now proposing something more modest: to make a copy of the quarter-scale model of the Berkeley 6-BeV machine. The Berkeley model had already been operated successfully at slightly more than 1 BeV. Since the design had been completed, it would not make the same demands on the NRL staff, and Waterman stated that it could be done within the existing NRL budget. The question Waterman posed to the committee was this: "Is this a proper thing for NRL to do? . . . If it is granted that there should be a service type of organization on research for nuclear physics, I think you would agree that it is necessary to give them [the ten divisions of NRL] facilities for basic research in that field too. Otherwise, it would not be possible to keep a competent staff. But it is an important question. Here is a large item. . . . It is, after all, a policy question, you see."[22]

The advisory committee was therefore obliged to consider questions of policy as well as purely technical matters. The members were still highly doubtful about whether there was a valid scientific justification compatible with the navy's mission. They continued to question whether the operation of such an accelerator had more than a remote relation to the problems of nuclear power, which dealt mostly with "engineering and high temperature, and things of that sort, but not so much the basic nuclear physics."[23] They questioned whether the NRL function of providing scientific advice and training to the navy might not in this instance be carried out at the Brookhaven or Berkeley installations.

Many memoranda and letters passed among the committee members. Arthur Compton, president of Washington University in St. Louis and brother of Karl, put his finger on an essential point. In the end, he said, the issue was this: "that by setting [a bevatron] up in the Naval Research Laboratory it would be a step in the direction of concentrating naval research, the research supported by the Navy, in that one laboratory to the exclusion of the universities. That is the danger we have to look out for."[24] DuBridge also wrote to the same point. "It seems to me," he said, "that the military establishments are primarily designed to serve the military. This means that their primary function is to do general development work aimed toward military problems . . . Making it possible for an outstanding scientist together with one or two assistants to undertake as a sideline an interesting basic investigation is a very different thing from devoting a very large fraction of the resources of a division of NRL to a basic research problem whose connection with Navy interests may be remote."[25]

This was the same question about government laboratories that Vannevar Bush had confronted in 1945 when he had recommended the dismantling of the Radiation Lab, and it was the same question that would engage Jerrold—and others—again and again as MIT strove in subsequent years to deal with the question of military-oriented research on its campus. But on this occasion, the committee could speak clearly and unequivocally. Basic research belonged in the civil domain; it was an enterprise for government to support but not to control. Research, if it were to be basic and fundamental, must be free; it could not be undertaken with one eye on the navy's mission. Grateful as everyone was for the support of the ONR, this line was not to be crossed.

The navy's proton accelerator was not built.

Back to Research

American physics had languished for five long years. The great achievements of the Radiation Lab and Los Alamos, and of all the other specialized laboratories that had been part of the war effort, represented great advances in technology, but these were not advances in fundamental physics. Little new basic knowledge had come out of these efforts. However, new skills and new techniques had been developed that would prove immensely valuable, and new minds had been sharpened for the attack on the problems that awaited. Physics was about to experience another extraordinary burst of creativity and accomplishment, comparable to the developments of the 1930s.

There were two lines of research that Jerrold knew he wanted to explore. The last experiment he had done before the war had been to measure the spin and magnetic moment of K^{40}, a long-lived, low-abundance radioactive nucleus. There were many other nuclei of a similar nature, the spins and moments of which remained to be measured. Having made one such measurement successfully, he had a small head start, although he knew that such head starts do not usually last long in fields that become very competitive. He had in mind a long-range systematic program for measuring as many of these spins and moments as he could.

The other line of inquiry brought him back to hydrogen again. Well before the war had interrupted physics research, physicists had generally recognized that there were some serious problems with the theory of electrons and their electromagnetic interactions. The theory of these phenomena—quantum electrodynamics, or simply QED—provided a tantalizing puzzle. QED gave many answers that agreed extremely well with experiment, but every attempt to use QED to calculate certain particular properties of the electron led to infinite, and therefore unacceptable, results.

In spite of continued efforts, this remained a source of persistent frustration to theorists. Weisskopf, looking back on those interesting days, referred to the prewar situation as the "fight against infinities."[26] Now, in 1945, as the physicists got back to work, it was time to resume this fight. Suggestions once again began to appear about how to deal with these difficulties.

On the experimental side, there had been some modest and sketchy prewar results suggesting that even theoretical predictions that seemed to agree with the experimental results might not be completely correct. A very small discrepancy had been noted in the hydrogen spectrum, for example. It was not certain; it was very tiny, and it might have been due to some subtle instrumental error.[27] But if it was real, it should be explored.

Large discrepancies, naturally, are almost always caught right away. But sometimes it is the small discrepancies, not quickly apparent, that lead to interesting physics. This particular discrepancy was only a small fraction of a percent in the value of the electron magnetic moment, but that was significant, assuming it was real.[28] In any case, there was now new postwar technology available, most of it the result of microwave radar development at the Radiation Lab. Measurements could now be made using microwave frequencies, for example, that would have been impossible before the war. It was time to make accurate tests of all the theoretical predictions. Jerrold decided to measure the so-called electron g-factor—that is, to make a new determination of the hyperfine structure of the hydrogen atom. This would be equivalent, from the point of view of theory, to re-measuring the electron magnetic moment in the hydrogen atom.

The measurement required the use of a beams apparatus such as those he had built in Rabi's Columbia laboratory years earlier. He asked Rabi to provide him with one of the Columbia setups, but Rabi had revived his own research program by this time, and declined. (His refusal illustrates the kind of protocol that had grown up over the years in American research laboratories and perhaps elsewhere as well: by custom, the equipment belongs to the laboratory, regardless of who builds it. On occasion, a piece of apparatus may move with an experimenter but usually only as a courtesy, when there is no further use for it in the original laboratory.) Rabi was as eager as Jerrold to get his research going again. He had lost many of the people who had worked with him before the war, and Columbia had not partici-pated vigorously in the recruitment rush at the war's end. Of course, Columbia had succeeded in getting Rabi himself to return, no small thing, and there were others who had never left, having spent the war years at the Columbia Radiation Laboratory working on magnetrons. Rabi was begin-ning now to reinvigorate his laboratory, finding new young people and getting new experiments going. Jerrold had no choice but to start building his apparatus from scratch.

And so he did, almost, but not quite, managing to catch up with Rabi in the process. Together with two graduate students, Darragh Nagle and Rene Julian, whom he had brought to MIT from Los Alamos, he started building a new apparatus for the g-factor measurement, designed to be similar to the one Jerrold had built for the K^{40} experiment. They would use the same flop-in method that Jerrold had successfully devised six years earlier. The apparatus could then continue to be useful later on for nuclear magnetic moment measurements on other radioactive nuclei.

A heavy cylindrical brass casting, some 8 or 9 inches in diameter and a meter or so long, was produced to house the experiment. In Jerrold's lab, these castings quickly became known as torpedo tubes, which they rather resembled. The usual and expected vacuum troubles began, almost at once. Cast brass is rather porous, for example, and the interior of the casting had to be tinned, or coated with a thin layer of solder. Leaks would nevertheless occur, as they always had, and had to be tracked down; by now Jerrold had an almost instinctive ability to diagnose such problems quickly, but such nuisances manage to take time anyway. All of the auxiliary equipment—the detector, the hydrogen source, the gauges, and the magnets—had to be fabricated, put into working order, and integrated into the system.

If he had known that Rabi was working on the same problem, per-haps he might have worked harder or faster; he might have abandoned some committee work or cut back on his administrative obligations. But it is doubtful, in any case, that spending more time in the lab at that point would have availed him of very much. Even during the war, when many projects could be moved forward with impressive speed, there were still matters that could not be accelerated. Some work takes a certain amount of time; to do it faster is to do it badly. An experiment develops a rhythm, a cadence, determined by the pace in the machine shop, the time needed for a casting to be delivered, and so forth. The timing arises out of the need for a number of people to work usefully together. An experiment can be hurried some but not much.

In April 1947 a brief note appeared in the *Physical Review*, the principal publication of the American Physical Society, bearing the names of J. E. Nafe, E. B. Nelson, and I. I. Rabi, all of Columbia, and it gave the result that Jerrold and his group had been working to obtain: the measured value of the electron magnetic moment in hydrogen was off by about a tenth of a percent from the theoretical value.[29] Nagle, Julian and Zacharias did not finish their measurement until the middle of the summer; their publication did not appear in the *Physical Review* until the November 15 issue.[30]

There is not much joy in being the second to publish a result. By November, the principal interest in the Zacharias measurement was that it agreed with Rabi's earlier one. The Zacharias group could take some satisfaction, in knowing how close they had come, starting from nothing, to getting there first. But they were disappointed perhaps especially Jerrold as he sought to escape what he called Rabi's shadow. Jerrold said later: "It wasn't a race; we were simply working as hard and fast as we could. For years now, I must admit that I've quietly resented being put in second place by my own apparatus. The lesson in all of this, which I've learned over and over, is don't compete with Rabi, the man is incredible, inevitably right, and always comes out on top."[31]

If Jerrold was unhappy at the outcome of his competition with Rabi, however, he never showed it. His close friends and associates from Radiation Lab days, Albert Hill and Jerome Wiesner, who at that time saw Jerrold nearly every day, were unaware of anything like that; they simply saw Jerrold cheerfully hard at work.[32]

As it turned out, it was neither the Rabi experiment nor the Zacharias experiment that attracted the most attention. Instead, a related experiment, carried out at Columbia by Willis Lamb and Robert Retherford, caught everyone's interest. Lamb was a prewar California discovery of Rabi who, with the latter's sponsorship, had been appointed an instructor at Columbia in 1938. He had spent the war years at the Columbia Radiation Laboratory working on magnetron design theory. Retherford was a graduate student who had worked for a time before the war with J. M. B. Kellogg, Jerrold's earlier collaborator. Retherford was a talented experimenter; Lamb was primarily a theoretician, but the experiment was nevertheless his design.

Lamb and Retherford used a modified atomic beams apparatus with a microwave oscillator to measure the hydrogen fine structure. There were two states of the atom that theory predicted should be characterized by precisely the same energies; Lamb and Retherford found that each of these energy levels was shifted by a tiny amount, but the shifts were not precisely the same; the two states did not in fact have identical energies. The energy difference became known as the Lamb shift, and its measurement won the Nobel Prize for Willis Lamb in 1955.[33]

Between the publication of the Rabi group and the publication a few months later by the Zacharias group, there took place a meeting that turned out to be of major significance to the history of physics: the first Shelter Island Conference on Theoretical Physics.[34] This was a select gathering of twenty-three of the nation's most active and promising theoreticians, who had been invited to meet for a few days in early June in a quiet and sequestered place to discuss the problems of quantum electrodynamics. Rabi was invited to discuss the current state of experimental knowledge, and he arranged for Willis Lamb to be invited as well.

The announcement at the conference of the Lamb shift, following so closely on the publication of the anomalous electron g-factor measurement by Rabi's group, was the stimulus for the astonishing burst of theoretical problem solving that quickly followed. It may be that the state of theoretical physics was such that the outstanding problems of quantum electrodynamics were ready to be solved; perhaps it required a Bethe, a Schwinger, a Feynman, returning to these problems after the war, to do it.[35] Some techniques for solving these problems had already been available in the late 1930s, but without the new experimental results as a spur, the problems had not been tackled properly.

By this time, Jerrold's second line of experiments was beginning to pay off. He and his students, with theoretical support for a time from B. T. Feld, had begun a program to measure the spins and magnetic moments of all the radioactive nuclei that could be handled in a molecular beams apparatus, and the work was going well.[36] Jerrold, remembering the difficulties he had had with K^{40} due to its low abundance, now built a mass spectrometer using new and simple magnet technology that had been developed during the war. This enabled the low-abundance isotopes to be separated from the others, and it simplified the experiments considerably, as well as making them more accurate.

There were literally hundreds of isotopes that could be measured, and they could not hope to have the field to themselves indefinitely. Nevertheless, their work in this area put the laboratory on the map and established MIT as an important center of molecular beams research. A steady stream of graduate students passed through Jerrold's lab. The measurement of each new magnetic moment required techniques sufficiently different from one another and each sufficiently novel so that the experiments provided the substance of a whole sequence of Ph.D. theses.

For Jerrold, however, this line of research became less interesting as time went on. Each experiment was only a variant, after all, of the ones that preceded it. The new problems that came with each measurement were not new in kind. He began to spend less time in the laboratory, and by 1949 or 1950 it was essentially being run by two of his more advanced graduate students, Vincent Jaccarino and John King. Jerrold's attention was elsewhere; he would eventually return to the lab with new ideas, but for the time being, his attention became diverted to military matters and to matters of public policy.

5

Scientists and the Cold War

Disarmament and Rearmament

The withdrawal of scientists from military matters, however desirable that may have seemed, proved to be only temporary for a number of them, Jerrold included. Some scientists would be drawn back only on a casual, occasional basis; Jerrold, who could never do things by halves, would eventually find himself giving a considerable portion of his time to such matters and, in fact, playing a significant role in some of them. The world situation was becoming increasingly difficult for the United States as the cold war intensified, and Jerrold believed that it was important for him to contribute what he could. But it is also clear that military matters had a fascination for him that he was hard put to resist.

By the end of the 1940s, the cold war had begun to impose substantial new demands on the U.S. military establishment, which found itself obliged to turn once again to the scientific community. The American armed forces had been considerably reduced in size and potency as a result of a rapid postwar demobilization, accompanied by a sharp reduction in military budgets. As late as 1948, despite the worsening situation in Europe and elsewhere, President Truman was still insisting on a limit of less than $15 billion for military spending in the 1950 fiscal year budget.[1] Undoubtedly, the U.S. monopoly in nuclear weapons made it seem reasonable to place economic aid to Europe and Japan ahead of preserving or increasing military strength. However, the sense of security provided by the nuclear monopoly turned out to be much more temporary than anyone had expected. Leslie Groves, for example, who had served as the commanding general of the Manhattan Project, had predicted in testimony before a congressional committee in 1946 that it would take fifteen to twenty years for the Soviet Union to produce a bomb. Bernard Brodie of Yale University, who in most respects was an uncommonly prescient military analyst, described in 1947 the War Department's estimate of Soviet nuclear capabilities: "For a number of years, perhaps as many as eight to fifteen, only the U.S. will possess atomic bombs in significant quantities."[2] But only two years later, in August 1949, the Soviet Union exploded its first atomic weapon, nicknamed "Joe I." America would not feel secure again for nearly half a century.

As a result of America's wholesale disarmament, military planners had placed the atomic bomb at the center of American military policy. Having dispersed its powerful ground forces, leaving itself without the ability to wage and sustain a long war of any kind, America appeared to have no alternative. In this early and relatively primitive stage of nuclear weapons planning, the bomb was considered solely as a means of inflicting overwhelming destruction on an enemy—the more overwhelming, the better. The idea of tactical nuclear weapons was not considered a serious option, and most planners did not believe that tactical weapons would even be possible. Brodie stated at the time: "The atomic bomb is not an all-purpose weapon; in fact, it is rather limited in its employment due to its great destructive power (which is not significantly reducible at present), and its relatively high cost as a single weapon."[3]

Brodie's assessment was wrong, in respect to both bomb design and cost, but in 1947, when he wrote, many of the simple facts about nuclear weapons were still kept secret from even the most respected and influential analysts. Ideas about nuclear deterrence were still primitive, defense against atomic weapons was thought not possible.[4] A nation might seek to disperse its industry and its populations, but it could do that only at a huge and highly uncertain cost; but there would surely be no way to avoid or escape atomic bombs. For many, this was a compelling argument for building up America's first-strike capability: building up the Strategic Air Command, developing larger and more destructive bombs, and—partly because budgets continued to be limited, but also as a matter of philosophical choice—minimizing expenditures for a defense that they believed could not work.

The issues were not entirely technological, of course. The role of the atomic bomb was a central one, for example, in the bitter interservice rivalry that characterized the early postwar years, but overall, that rivalry was more political than not. Nevertheless, every aspect of the military debate had a strong technological component, and it was only a matter of time until the scientific community was drawn in.

The sharp competitive struggle among the military services had begun even before the war, had worsened during it, and had deteriorated still further following the passage of the National Security Act of 1947 (frequently referred to as the Unification Act). Relations between the navy and the other two services already showed serious evidence of strain by the war's end.[5] In April 1945, for example, Vice-Admiral John S. McCain, a ranking naval aviator, had written to Secretary James Forrestal to complain about what was then still the army air force:

My Dear Mr. Forrestal,

General LeMay's statement that B-29's have rendered carriers obsolescent is the first overt act in the coming battle for funds.

No matter how fair the words, or beguiling the phraseology, and regardless of intent, a unified command, a single service or department of national defense, will of necessity be an instrument for an extra-constitutional and an interested division of funds prior to submission to the disinterested Budget (Office of the President) and a presumably disinterested Congress. The Army banks on controlling the individual who will head this single unit, and historically, they will be correct in that assumption.

There will be little planes, as well as big planes, that will sink all kinds of ships and perhaps amphibious tanks can be built up into that role for public consumption. This will appeal to the grand American illusion that wars can be fought cheaply.

It is beginning to look to me that the war after the war will be more bitter than the actual war.

Which is, of course, a shame.[6]

McCain predicted well. The rivalry between the navy and the air force affected seriously every area of operations.

Project Lexington

It was inevitable that the new technologies developed during the war would provoke many new proposals for their further development and use. Few people had any real sense of what was possible or practical; the scientists with military experience, who might have been expected to be able to analyze and advise, were no longer readily available, and government and military agencies naturally turned to the universities. Starting in 1948, the practice developed to commission short ad hoc studies on the technical or scientific issues of the cold war. Because these studies were often carried out in the summers to conform to the schedules of the academic year, analyses of this type soon became known as summer studies.[7]

The first important summer study was Project Lexington, commissioned by the Atomic Energy Commission in the summer of 1948 and operated under the auspices of MIT. Its purpose was to examine the possibility of nuclear-powered flight. MIT had earlier acquired as surplus the Lexington Field Station, an emergency command post for an army antiaircraft battalion during the war, and this structure, largely an underground bunker, became both the locus for the study and the origin of its code name.

The idea of a nuclear-powered airplane had arisen almost immediately after the war and had at once been given a high priority by the air force. When the air force formally achieved parity with the army and navy under the Unification Act, one of its first actions was to set up an atomic energy division, with "atomic propulsion of aircraft as its principal objective."[8]

The decision to try to develop a nuclear-powered airplane was strongly influenced by the interservice rivalry that seemed to affect every aspect of military thinking. If the atomic bomb were indeed to be the central feature of U.S. military power, then whichever branch had responsibility for delivering it would quickly become dominant, with all that that would imply regarding budget share and influence, in both the newly created Department of Defense and in Congress itself. At a time when there was not enough money to go around, the outcome of that competition seemed vital.

In 1948, the exercise of American military force in most parts of the world would have required the use of naval carriers as advanced, mobile airfields. The navy had under development a new flush-deck supercarrier, the *United States*, which it hoped would serve as the prototype of a fleet of advanced air platforms from which atomic attack could be launched, not only on the coastal ports and bases of an enemy but deep into his heartland as well. The air force competed for its share of the nuclear pie (a term coined by Albert Hill) by accelerating its program for testing the long-range B-36 bomber and coupled that with a vigorous public relations effort.[9] The competition was still undecided; it was not yet possible to know which way Congress or the president would lean.[10]

A nuclear-powered airplane would change the nature of the debate. Since such an aircraft would not need to refuel at all, its range would be unlimited, and the nation's nuclear weapons could be delivered anywhere without the need for any bases, mobile or otherwise, in forward areas. Project Lexington might provide an edge in the interservice competition; it was of great political as well as strategic importance.

The project was developed under the chairmanship of Walter G. Whitman, the respected head of MIT's Department of Chemical Engineering. Whitman had had a distinguished wartime career as chairman of the basic chemistry division of the War Production Board and had served on the National Advisory Committee for Aeronautics. He would go on in subsequent years to become a member of the influential General Advisory Committee of the Atomic Energy Commission during its critical years and would hold a number of other important posts.

Jerrold was named associate director of the project and, as he later recalled, was a frustrated and unhappy participant in it. He sensed early on that the decision to develop a nuclear-powered aircraft had already been made before the project began but that it was a wrong-headed decision, made for political rather than sound practical reasons. And he felt unable to modify what was going on; as he put it, "As associate director, I was free to associate, but not to direct."

According to Jerrold, Whitman interpreted the charge to the project quite literally: it was to determine only the possibility of nuclear-powered flight, not to consider either its practicality or its desirability. Among the project's physicists, aircraft designers, metallurgists, and others, there had been a forward-looking group, said Jerrold, who felt that they should try to predict the future of chemically powered vehicles: airplanes and rockets, all the alternatives to nuclear-powered flight. But that was not done. The only question that was allowed was "Can it be done?" The equally important question of "do you want to do it?" was not examined. Speaking of it many years later, Jerrold told an interviewer:

I wanted to look at planes powered by fossil fuels and see if there was an alternative— to see if you could get long-range planes without having reactors flying over our heads that could crash. We could have looked at things like in-flight refueling, but the director wouldn't allow it. He wouldn't assign any staff to it. He wanted us to stick to the technical questions about a nuclear plane.

We had a perfect setup to do a good study, but it turned out lousy. The report was too confusing—we admitted that you could make a nuclear plane, but we didn't say that you'd never want to do it. It taught me one of the most important things about a study. You've got to allow people to argue the general issues, not just look at technological ones.[11]

The technological problems that needed to be solved to make nuclear flight feasible were immense. Foremost was the need to shield the crew of the aircraft from the nuclear reactor power source. Shielding is a straight-forward matter of interposing absorbing material between the reactor and the individuals being protected; there is no substitute for providing sufficient shielding mass. Making the reactor small would provide a geometrical advantage, so that the total shielding mass might have been brought within reasonable limits but that would have implied a reactor operating at very high temperature, and the refractory materials that would be necessary did not then exist. Thus there was no prospect for an early version of an airplane with a nuclear power plant. Nevertheless, the strictly limited scope of the project required the conclusion that, yes, nuclear-powered flight was possible.

This unfortunate conclusion of Project Lexington served air force goals very well. The program for the development of a nuclear airplane continued for more than a decade; as late as 1959, with the nation already well into the missile age, Chief of Staff of the Air Force General Thomas D. White testified before Congress: "While certain scientific advisers do not feel that we are ready to start building a nuclear powered aircraft, I consider that our aircraft nuclear powered program (ANP) is sufficiently advanced so that we should proceed with construction of the prototype development airframe and should accelerate the propulsion phase of the program."[12] The air force continued to use the Lexington report as justification whenever the concept of nuclear flight was challenged on technical grounds; more than $1 billion was spent on it before the program was finally cancelled by President Kennedy in 1961.

A Desperate Year for the Navy

The issue was bomb delivery: if you delivered the bomb, you got the big piece of the pie. There was a tremendous hassle over the B-36 bomber versus carrier-based delivery systems. It was a real horrendous battle. Because of it, when Truman had to pick a Chief of Naval Operations, he went down a hundred numbers to get a guy who hadn't been involved in it.

—Albert G. Hill

Jerrold would soon be asked to direct a summer study himself on behalf of the navy. That study, known as Project Hartwell, had to do with Soviet submarines and was hugely successful; it became a model and the standard by which other summer studies were subsequently judged. In order to understand that success, it is useful to look at more of the background, for it continued to be the interservice rivalry that lent the summer studies so much importance.

By the latter part of 1948 the struggle between the admirals of the navy and the generals of the air force had become public, a matter of almost daily newspaper comment. The dispute centered on the roles of the aircraft carrier versus the long-range bomber in American military planning, although the issues ran much deeper. Underlying this battle was the "interested division of funds" that McCain had foreseen and that could now be plainly seen to depend on who would have responsibility for delivering the atomic bomb. America had not begun to rearm, although both the State Department and the National Security Council were developing plans for rearmament. In most respects American military strength remained in the diminished state in which it had been left by postwar demobilization.[13]

The rivalry among the services had already reached the point where the secretary of defense felt it necessary to intervene. In March 1948, at a high-level meeting in Key West, Florida, an agreement—really a sort of peace treaty—had been reached among the army, the air force, and the navy, under the supervision of Secretary James Forrestal, on the roles and missions assigned to each branch. The agreement outlined as the navy's mission "the seeking out and destroying of enemy fleets, maintaining sea supremacy and conducting anti-submarine warfare." The agreement further specified the navy's principal air function: "to conduct air operations as necessary for the accomplishment of objectives in a naval campaign." Virtually all other air combat operations were assigned to the air force.[14] The agreement was probably as fair and reasonable as could have been expected: the Navy agreed not to develop its own strategic air force, and the air force recognized the right of the navy to participate in any all-out air campaign. The navy was not to be denied use of the atomic bomb.[15] But the navy admirals did not trust the air force to live up to the agreement, and, indeed, the air force generals continued to insist that since strategic bombing was the key to American military policy, the nation's security could be found only in the striking power of the Strategic Air Command (SAC).

In spite of the agreement, the navy's political barometer continued to fall in the ensuing months, reaching a low enough level to make 1949, in the words of one naval historian, "a desperate year."[16] Certainly there was much to distress the naval high command. Most of the admirals had not come to terms with the creation of a single Department of Defense, and they continued to oppose many aspects of unification. Although they had won on the question of whether there would be a single, all-powerful chief of staff (there would not be), and in spite of the Key West agreements, the central questions of the navy's mission remained contentious. Secretary of Defense Forrestal, ill and ineffective, resigned his post on March 3, 1949; he was succeeded by Louis Johnson, a West Virginia lawyer. Although Forrestal had been less than decisive in protecting the interests of the Navy, he at least had had some sympathy with the navy point of view. Johnson, it soon became clear, did not.

In April 1949, soon after taking office, Johnson ordered the cancellation of the supercarrier United States. The action was taken within short weeks of laying the ship's keel; incredibly, the decision had been made and announced without consultation with the Secretary of the Navy, John L. Sullivan, who promptly resigned. Hardly any other action at that time could have seemed more significant; the cancellation of the supercarrier came as a terrible shock to the senior naval command. In clear and devastating contrast, the almost simultaneous acceleration of the B-36 testing program by the air force met with loud congressional approval. It looked very much as though the principal offensive role in any future war would go to the air force, and mastery of the sea would be considered of only secondary importance.

Naval morale sank to its lowest after October 13, 1949, the date on which Admiral Louis E. Denfeld, the chief of naval operations, together with other high-ranking naval officers, testified before the House Armed Services Committee on the progress of unification. Denfeld and his fellow officers had been promised by the committee chairman that they might testify freely, "without fear of reprisals." Denfeld, a vigorous opponent of both unification and what he believed to be excessive reliance on strategic bombing, bitterly complained that the navy had not been taken into full partnership in the unified defense team and that it was being "starved" by the emphasis on the B-36. Promises notwithstanding, he was promptly fired from his job by a furious President Truman, who later recorded in his memoirs that he had removed Denfeld in order to restore discipline; the situation, he wrote, had taken on "the aspects of a revolt of the entire Navy."[17] Certainly the event caused dismay and consternation in the navy, especially as Denfeld's dismissal was widely and publicly regarded as "a victory for the Air Force in its furtherance of strategic bombing as the keystone of American striking power."[18]

In all, the navy had inevitably come to feel that it had few friends or allies in high places, and fewest, perhaps, in the White House. Franklin Roosevelt had been a navy partisan, having once served as assistant secretary of the navy (1913–1920). The navy had enjoyed a generally benign atmosphere in the Oval Office all during Roosevelt's long administration. Harry Truman, by contrast, had been an artillery captain, and most navy men strongly believed that a land-oriented former soldier could not understand the complicated business of war at sea.

But Truman evidently did understand men, and he understood the nature of command. He now reached well down the seniority list to find a man to replace Denfeld, selecting Forrest P. Sherman, a naval aviator, and promoting him over the heads of eight officers senior to him. Commentators at the time took note of the surprising nature of the appointment. The military reporter of the *New York Times*, Hanson W. Baldwin, wrote: "He [Sherman] is controversial, not in an inter-service way, but in an intra-service way. For Admiral Sherman is probably disliked by a majority of ranking Naval officers and definitely liked by a minority. His appointment, in a sense, splits the Navy, but brings to the post a man who, during inter-service negotiations that led to the Unification Act, proved capable of 'getting along' with the Air Force."[19]

It had been Forrest Sherman, in fact, who had worked together with General Lauris Norstad of the Army Air Force to produce the original 1947 agreements on unification, without which, most people agreed, nothing at all would have been accomplished. It was undoubtedly his performance in those negotiations that had earned him Truman's trust and respect. Sherman brought to his assignment not only good diplomatic skills but good judgment as well. He sensed that some quick action was needed to resolve a struggle that perhaps never should have begun and certainly should not continue. The navy was in disarray; it was time to put an end to the admirals' revolt, for nothing would be gained by continuing to fight lost battles. As a member of the Joint Chiefs of Staff, Sherman must have been keenly aware of the significance of the document then being drawn up in the State Department, later to be known as NSC-66, that would soon lead to substantial American rearmament. Sherman set about refocusing the navy's attention on the mission that had been defined for it.

The Problem of Soviet Submarines

As Jerrold had uneasily foreseen in the closing days of the Radiation Lab, the unsolved problem of snorkeling submarines remained to plague the navy. This was an immediate and substantial military problem to which the Joint Chiefs of Staff now directed Admiral Sherman's attention: how to counteract the increasing size and probable effectiveness of the Soviet submarine fleet, by then considered by the Joint Chiefs to have become a substantial threat to American security.

Intelligence reports suggested that at the end of the war the Soviet Union had obtained perhaps as many as twenty of the German Type XXI submarines, as well as the services of the German technicians who had built them; the Soviets were consequently believed to be capable of producing many, perhaps hundreds, more.[20] These submarines were equipped with snorkel tubes, that is, underwater breathing tubes. By using the snorkel tube, a submarine could remain submerged while charging its batteries, the time when it had previously been most vulnerable. The snorkeling submarine would be practically undetectable by radar from any appreciable distance. Furthermore, modern technological developments might soon enable submarines to move fast and far under water more quietly than ever before. Somewhere in the future lay the prospect of nuclear-powered submarines capable of virtually unlimited periods without surfacing at all; but even without such advances, the present near-invisibility of Soviet submarines made them a serious threat.

If the Soviets succeeded in deploying improved propulsion systems that would require their submarines to come to snorkel depth much less often, U.S. mastery of the seas could not be taken for granted. Admiral Raymond A. Spruance, commander-in-chief of the Pacific Fleet, speaking at the War College in February 1948, explained the difficulty: "The new submarine with high submerged speed and great underwater endurance is probably the greatest threat that exists today to the safe use of the sea. Until a solution is reached to the problem of how to destroy this submarine, and until the forces are made available for this work, we shall be in a poor position to operate our armed forces overseas against an enemy who has a large fleet of them and knows how to use them efficiently."[21]

The Initiation of Project Hartwell

When I went to Washington in 1960 [as science adviser to President Kennedy], I knew more about how to get things done than the Admirals and the Generals knew, because I had had sort of a post-graduate course in that from Jerrold, on Hartwell and the other summer studies, things like that. He taught us all how to go about it.

—J. B. Wiesner

Zach was the head man. He was with a bunch of smart people . . . but he had to sew it together and keep people working, and he did a wonderful job . . . This [Project Hartwell] has many times been called . . . the most successful ad hoc study ever made, and I think there's no question it is.

—A. G. Hill

Summer studies, some are not.

—J. R. Zacharias[22]

In March 1950, following the recommendation of its own Committee on Undersea Warfare, the navy contracted with MIT for a new, intensive summer study on antisubmarine warfare. The initiative came originally from Forrest Sherman. Sherman's principal technical adviser was Mervin J. Kelly, executive president of the Bell Telephone Laboratories and widely regarded as the nation's foremost communications engineer. Sherman discussed with Kelly the problem of the Soviet submarine program and sought his advice. Al Hill remembered the story as he heard it: "This may sound like 1940s thinking, but remember, in 1950 the nuclear threat was not as immense as it is now . . . so Kelly said to Sherman, 'Look, you've got an immense number of problems here and what I would do would be to go to some place like MIT, since they have a lot of screwball scientists who will work on anything, and get a short study made, see what they think of it.'"

An approach was consequently made by naval representatives to President Killian at MIT, who readily accepted the assignment on behalf of the institute. After consulting with Julius Stratton, by then MIT Provost, Killian invited Jerrold to take charge of the study.

Initially, the navy sought help only with the limited problem of detecting submarines, but Jerrold, having learned an important lesson on Project Lexington, would not settle for that. His response had the self-assurance that many took for arrogance and was reminiscent of the response he had once given Lee DuBridge in the early days of the Radiation Lab: he would do it provided certain conditions were met. This time, however, he was not making conditions to another physicist whom he knew well and who would be inclined to accept that sort of impertinence in the good-natured way it was meant. Instead, it was the chief of naval operations himself who was expected to agree. Admirals are generally not accustomed to bargaining about conditions.

To his credit, however, Forrest Sherman sought to understand what Jerrold wanted. After Jerrold flew to Washington to meet with Sherman, he told Al Hill what had been said. Hill remembered it this way:

Jerrold said he told Sherman, "If I am going to do this study in three months' time, it's going to take an awful lot of brain power, and if I go to the people that I want to go to and say, Will you work three months inventing a weapon to be used against submarines, they'll laugh me out of court. They'll say, You can't order up an invention in a short length of time. All you can do is order up some thought."

It ended with Zach saying "What you want is a study on the security of overseas transport. You don't want a study on how to shoot up submarines. You want to move goods across the ocean." And Sherman accepted it immediately.

Jerrold later recalled the conversation more specifically: "The great Admiral Forrest Sherman was very sympathetic . . . he said, "I don't care if you try to tell us how to reorganize the Navy, so long as you tell us something about how to detect submarines."[23]

By the time the MIT study began, its scope had been broadened to include all aspects of overseas transport, including, but not restricted to, antisubmarine measures. The study was formally established as Project Hartwell after a visit to MIT on April 17, 1950, by Admiral T. A. Solberg, chief of naval operations under Admiral Sherman, and by Solberg's deputy, Admiral C. B. Momsen, assistant chief of naval operations.[24] The work was to begin on June 5, to continue until August 26, culminating in a two-day oral briefing for a group of about thirty people, including an under secretary of the navy. The final report was delivered on time on September 21, 1950 and bore the title, "A Report on the Security of Overseas Transport." Everything was to be classified Secret, even the contract that arranged for the study.[25] When the 500-page report was completed and laboriously hand assembled, the navy sent an armored truck to cart seven hundred copies of it back to Washington.

The timetable was short; Jerrold got to work at once. The first order of business clearly had to be recruitment, and he began with those who were closest to him: Al Hill, Mac Hubbard, and Jerry Wiesner. He persuaded E. R. Piore, Lloyd Berkner, and J. B. Fisk to help with the recruitment and organization. Jerrold had met Piore, a solid state physicist, years earlier at Columbia and had renewed the acquaintance while Piore was at MIT in 1948–1949 on a temporary research leave. Berkner was an expert in radio-wave propagation; he was a high-ranking officer in the naval reserve and at that time was at the Carnegie Institution of Washington. (He became the president of Associated Universities, Inc. in the following year.) James Fisk was deputy to Mervin Kelly at Bell Labs and eventually succeeded him there as president. Berkner became a working member of the Hartwell group; neither Piore nor Fisk was free to do so, but they remained members of the project steering committee. Piore, who was also an active reserve officer in the navy, did manage to devote several weeks of his annual leave to the project.

The annual spring meeting of the American Physical Society was scheduled to take place in Washington in the last week of April, and many of those whom Jerrold wanted for the project would be there. Hill recalled that Jerrold called together a preliminary session of Project Hartwell at the National Academy of Science in Washington to get things organized and to begin hiring. In spite of that early start, not everyone was in place until the first of June.

In all, thirty-three scientists took part in the study. Ten of them were alumni of the Radiation Lab or of Los Alamos, or both: L. Alvarez, R. H. Dicke, I. Getting, A. G. Hill, M. M. Hubbard, C. C. Lauritsen, J. R. Pierce, E. M. Purcell, J. B. Wiesner, and Jerrold himself. Another nine were members of the faculties of either Harvard or MIT. Three were from Bell Labs and two from the Carnegie Institution. The list was altogether remarkable. Jerrold stated that he looked for people who were so accomplished in their own fields that they felt comfortable outside them: "We only wanted people who were too busy to participate. Anyone who could devote two or three months to a project without sacrifice, conflict or guilt had too much time on his hands and was not the kind of person we were interested in . . . [We] could tap the alumni rolls of the Radiation Lab and Los Alamos, and the rosters of MIT, Harvard, the Carnegie Institution and Bell Labs, for openers."[26]

The list of participants was a tribute to Jerrold's growing ability to persuade remarkable people to join him in whatever project he had under way. He succeeded, in part at least, because he understood very well what such people required. He had long known that in a group of outstanding people, each person thinks that he has the best ideas himself; Jerrold realized that such attitudes must be accepted as part of the price paid for independent people with independent ways of thinking. "I have always been willing to tolerate prima donnas," he said, "as long as they can sing." And besides, using his metaphor from Radiation Lab days, he meant to "keep the chalk in his own hand." He did, but he had enough sense to use it sparingly.

The project began on time, opening with what Jerrold called a "three-week road show": a briefing at the Pentagon, briefings in Key West, Florida, and then a return to New England for further briefings at the New London Naval Submarine Base in Connecticut, and finally back to Cambridge. The actual work of the group took place in Lexington in the underground bunker that had housed the less-than-satisfactory Project Lexington twenty-four months earlier. The climate was different this time, however: after Jerrold got people working in small groups on their specific tasks, he knew enough to let them alone. Again, Hill's memory serves: "We were admirers of Jerrold and we were perfectly willing for him to be our peerless leader. But we wouldn't have stood still if he had written us a list of how? when? where? why? on the board more than once. If he'd repeated it, we would have thrown him out, you know. We were his contemporaries."

Hartwell's Recommendations

No doubt Hartwell reminded its participants of their wartime assignments, in at least three respects besides the austerity of its setting: first, in having collected, on a matter vital to the national interest, scientists of extraordinary quality; second, in adopting an intense, around-the-clock method of working; and third, in making clear from the outset who the adversary was. The participants did not doubt it was the Soviet Union; the first sentence in their final report named the enemy. The common understanding on Project Hartwell was that recent Soviet military developments seriously threatened American national security. The underlying assumption was that the cold war implied an urgency comparable to what they had experienced during World War II.

What made Project Hartwell important was the impact of its recommendations on the navy. Hartwell demonstrated, in a way that Project Lexington had not, that real and immediate problems of military technology could be addressed usefully and effectively by civilian scientists; it showed that this important resource was still available to the military. Future summer studies were strongly influenced by the Hartwell success. The two senior navy scientists who evaluated the entire series of summer studies in 1966 described Hartwell as "exemplary for the studies of the next ten years. The Navy's improved understanding due to Hartwell, of one of its [then] pivotal problems had an electrifying influence on several segments of its research and development program."[27]

Such an impressive and influential outcome would have been hard to anticipate. How could a group consisting largely of civilians without substantial naval experience have so great an impact in so short a time? What did they do for the navy that the navy could not do for itself?

By opening up questions about the whole of the navy's mission, the project was able to take a systems approach to the navy's problems. That had not been done for a very long time, and the quarrel that had raged between the navy and the air force was evidence of the defense establishment's inability to think things through. There were, of course, many barriers besides the obvious competitiveness: for example, it had been the long-established practice in the navy to compartmentalize the thinking and the planning that formed the basis of policy, and to restrict internal communication on to a need-to-know basis. A section of the Hartwell report was devoted to criticizing the policy of "secrecy is security." The report pointed out that because of this policy, there was no bureau within the navy concerned with entire systems, of which ships and planes are components. The project's insistence on a fresh look at navy systems provided the opportunity to break through to a more reasoned approach.

Another barrier was the lack of adequate technical background of naval officers. The project members discovered that the officers charged with the navy's atomic weapons program believed that since the Hiroshima bomb had been 50 inches in diameter, all atomic weapons would have to be that size and therefore could not be designed for 30-inch torpedo tubes, much less for 18-inch tubes. But the 50-inch dimension had been selected only because that was the size of the bomb-bay doors on a B-29 aircraft. Even before the final report was drafted, strong representations were made to Admiral Sherman about technical training, and as a result, the training programs of naval officers in atomic weaponry were altered before the summer was over.

Still another example of what the project contributed was its conclusion that the safety of overseas transport required the development of fast merchant vessels. Again, the idea was acted on even before the final report was delivered. Vice-Admiral E. L. Cochrane, USN (ret), a distinguished career officer who was a member of the project and had been the head of the department of naval architecture and marine engineering at MIT, went on leave before the project ended to become chairman of the U.S. Maritime Commission. In his pocket when he took up his new task were sketches developed by Hartwell for a new class of high-speed merchant vessels, later named the Mariner class. With navy support, more than half a billion dollars was eventually committed to building a new, fast merchant fleet. Clearly such a recommendation could not have emerged if the project had been restricted to looking for antisubmarine weapons.

In addition to the recommendations that could be implemented immediately, there were others, equally valuable, on such matters as the modernization of ports and cargo handling facilities (to reduce substantially the vulnerable period when a ship was immobilized) and much excellent technical advice on the development of new approaches to communication, new forms of sonar and airborne paramagnetic detection methods. The navy had surely gotten more out of Hartwell than it had expected; but the project's great success had as much to do with the receptiveness of the navy high command as with the talent of the Hartwell group. In particular, the ready response of Admiral Sherman ensured the project's success, for Sherman made clear that the 500-page report was henceforth to be the "bible of underseas warfare." Subsequent summer studies did not always find such an enthusiastic reception, however excellent their advice may have been.

Ten Thousand Atomic Weapons

Project Hartwell also called for the development of a large number of what were essentially tactical nuclear weapons.[28] This may well have been the first serious proposal for such a development; the proposal certainly ran counter to the conventional thinking of the time. The Hartwell report stated, "In our hands, atomic weapons can probably provide an effective anti-submarine weapon; they can also be effective against enemy port installations and submarine bases... The production of atomic weapons should be materially increased in order to obtain an adequate number for the many possible tactical uses. The Hartwell Group believes that our goal should be about 10,000 atomic weapons."

This was startling advice for a number of reasons, of which two should certainly be mentioned. First, due both to the requirements of the Atomic

Energy Act and to the navy's need-to-know policies, few senior naval officers knew enough about atomic weapons to realize that such weapons could be manufactured in the small 1- to 5-kiloton size that would be useful in antisubmarine warfare. Second, the number was staggering. At the time this recommendation was made, the entire U.S. nuclear arsenal contained probably fewer than 300 nuclear devices, and it was certainly not evident that enough fissile material could be produced in any reasonable time.[29]

In fact, most of the Hartwell participants had also been relatively ignorant about atomic weapons. The only members of the project who during the war had had Q clearance, which gave them access to nuclear secrets, were Luis Alvarez of Berkeley, Francis Friedman of MIT, Charles C. Lauritsen of CalTech, and the project director, Jerrold Zacharias. But the need-to-know philosophy was quickly dropped within the project, and no effort was made to have different levels of secrecy and clearance. Consequently, calculations could be provided for all to know on such matters as the effective underwater range of nuclear depth charges (about a mile for a depth charge of a few kilotons).

The idea of developing small nuclear weapons suitable primarily for localized use by the navy almost certainly did not originate with Hartwell. The Hartwell final report itself refers briefly to an earlier study, code-named Project Maid, which may have examined the proposition first. Rosenberg has described "a small group of naval officers . . . [who] proposed as an alternative to strategic atomic bombing that atomic weapons be used primarily against tactical military targets, such as armies, airfields, oil supplies and submarine pens, which would have to be destroyed to prevent the Soviet Union from taking Western Europe."

But until Hartwell, the idea of developing tactical nuclear weapons had not taken root and did not appear anywhere in the public debate. American policy continued to emphasize strategic atomic bombing.

The Hartwell group believed that the Soviet Union would not be deterred from a limited war by American weapons that were too enormous to be used in anything but an all-out war. After all, the Soviet blockade of Berlin in April 1948 had been attempted in spite of the American monopoly in nuclear weapons, and the United States had been obliged to respond with conventional forces. The Soviets had not been deterred from invading Czechoslovakia in 1949, and only a month after the start of Hartwell, the North Koreans had invaded South Korea; again, the threat of large nuclear weapons had been ineffective.

Hartwell therefore adopted the view that nuclear devices of tactical size, because they were usable, would be effective where the big weapons would not. Unfortunately, in recommending the development of small nuclear devices, the Hartwell group failed to recognize that a great psychological gap separates using even the largest nonnuclear weapons from using even the smallest nuclear bombs. Hartwell failed to note the extreme risk involved in reducing atomic weapons to usable size. They apparently did not realize, or in any case did not discuss, how unlikely it was that the use of such weapons could be restricted in any way, once they were made small. The Hartwell recommendations did not, attempt to look past the immediate advantage of tactical weapons.

The Hartwell report began its discussion of atomic weapons by listing eight widely held ideas that were to be regarded as misconceptions:

1. Atomic bombs will probably never be used in large numbers. (By large numbers in this connection we mean a thousand or more.)
2. They are necessarily 5 to 6 feet in diameter.
3. They necessarily weigh several tons.
4. Safety to plane and pilot requires that they be delivered only by large bombers and from high altitudes, 30,000 to 40,000 feet.
5. They are useful only against large cities.
6. The U.S.S.R. cannot produce them in significant numbers in the near future.
7. They are extremely expensive.
8. Secrecy still spells security in the atomic energy field.

Next to each of these statements was printed "<u>False</u>." The list itself was followed by this statement:

<u>These ideas are simply not true, and their acceptance can have the gravest consequence. The adoption of these misconceptions by our military experts constitutes another of the relevant facts of the atomic energy picture of 1950.</u>

Four general conclusions then followed:

1. Sufficient nuclear explosives can be made available for tactical uses.
2. Planning and development of tactical atomic weapons for the Navy should have high priority.
3. The tactical uses of the atomic weapons must be considered from two points of view: (1) how we may use such weapons; (2) how such weapons may be used against us and how we can counter the threat.
4. The secrecy policy as applied under the Atomic Energy Act of 1946 is endangering national security by enforcing an inadequately low level of work on the development of atomic weapons for tactical uses. Several changes must be made in this policy if an adequate level of thought, research and development on tactical atomic weapons is to be achieved.

These ideas were radical at the time—and welcome to the Navy. As the summer of 1950 ended the Hartwell proposals offered the navy new roles and functions in the atomic arena. The rules of the game of defense were changing; President Truman decided to endorse NSC-66 with its plan to rearm the nation. Substantial new funds would soon be allocated, and totally new war plans would have to be drawn.

The Hartwell recommendation for the navy to obtain large numbers of usable atomic weapons eventually became deeply troubling to Jerrold, as did the nuclear arms race itself. He did not see how Hartwell could have honestly recommended otherwise, but he came to believe that the growing dependence on nuclear weapons meant that a great opportunity had been missed in the days just after the end of World War II. While it was still early enough for the scientists to have played a significant role in controlling nuclear weapons, he said, they had not managed to do it:

We didn't formulate a clear statement of how to think about atomic bombs and how to contend with their existence. We didn't get together and work our way through the political and military issues. When I think of all the collected military and scientific know-how at Los Alamos, it is a shame that we didn't even think to pick up that ball, never mind run with it. We may not have had all the answers, but we at least could have taken a start at some of the questions.

Why didn't the military and political issues command the same stubborn and relentless energy as the technical problems?[31]

Part of the reason, he believed, was that the entire Los Alamos project had been conceived in the same narrow spirit as the Lexington Project in 1948. The purpose of the Manhattan Project had been strictly defined: to produce a gadget, a device, nothing more. The Radiation Lab had been less restricted. Its purpose had been to take the existing devices (magnetrons and so on) and to consider how to use them, to work out how they would be integrated into existing systems. This is what he meant when he referred to systems engineering and to himself as a systems engineer. Hartwell was a success story, he felt, because it was able to look beyond the gadgets to the entire system and examine what overall purposes were to be served.

By contrast, at Los Alamos they had considered only a single question: can you make a bomb? At the very least, Jerrold eventually came to believe, they should have concerned themselves with how to use the bomb; the question of whether to use it at all would have followed very quickly. But such concerns had been virtually impossible to pursue at Los Alamos, and a good part of the reason for that was the military insistence on secrecy. General Groves, after the war, wrote: "Compartmentalization of knowledge, to me, was the very heart of security. My rule was simple and not capable of misinterpretation—each man should know everything he needed to know to do his job, and nothing else."[32]

Even Harry Truman had not known about the atomic bomb project when he succeeded to the presidency in 1945, and he had had to be briefed too quickly and perhaps too superficially before meeting with Stalin and Churchill at Yalta. Speaking of it in later years, Jerrold commented somewhat wryly:

Even if those of us at Los Alamos had decided to or been allowed to wrestle with the policy issues, there is no guarantee that anyone would have had the necessary clearance to listen.

It's a pity that there was this confusion about secrecy and security and that the weapons were isolated from the policy process because there was plenty of time to get it straight. Our stockpile of atomic weapons grew at a snail's pace.[33]

The scientists had attempted to involve themselves in arms control, but, Jerrold thought, the efforts had been somewhat misdirected. Some had worked very hard on political matters immediately after the war, but the issue on which they concentrated had been whether the control of atomic energy should be in civilian or military hands.[34] Important as that undoubtedly was, Jerrold felt it distracted from the issue of control itself; the debate should have been more about *how* to control atomic energy and less about *who* should control it. When that particular hard-fought and exhausting battle about *who* was decided in 1946, in favor of a civilian Atomic Energy Commission, most of the scientists withdrew, believing that they had won.[35] Most of them had already contributed five years to the war effort; they had spent long years away from their studies and their students and in many cases away from families as well. The scientists could hardly be faulted for wanting to regain normal lives.

Jerrold was not alone in later perceiving a missed opportunity; Rabi, for example, had similar feelings. He thought that the scientists might have had enormous influence on programs to control atomic weapons in those years, but they had failed to capitalize on it fully. In a recorded conversation with Jerrold many years later, Rabi commented: "The atomic bomb spoke with a very loud voice . . . All they [the scientists] had were these duplicating machines, and they went out and they spoke throughout the country, and they defeated that [the May-Johnson bill], and the politicians saw that there was a power there . . . of course, when the Fall came, they all went back to their universities . . . and we more or less disappeared, the scientists disappeared from public notice."[36]

There were, of course, many other reasons why the arms race got out of hand. Perhaps the scientists might have done more at Los Alamos and after; but whether one regarded the scientists' movement as a success for what it accomplished or a disappointment for not going further, most of the crucial choices were actually made elsewhere. They were made, Jerrold said, in Yalta and Potsdam, in Washington and in Moscow. They were decided by the

Congress in the Atomic Energy Act of 1946, which made it illegal for the United States to share atomic secrets with any foreign power, including its allies. That act, Jerrold felt, made any efforts at international control illusory. The issues were decided by pressure from the army and the navy, and especially from "a young, but cocky Air Force" (Jerrold's words), which had determined to push hard for massive nuclear armaments. Equally, the issues were decided in response to pressures from the Soviet Union. Josef Stalin was resolved that his nation must be the military equal of the United States, and his generals could no more be ignored than could our own.

Could the scientists have made a bigger difference to arms control? Might the course of events have been deflected in some way, early on when the arms buildup was still small? On balance, Jerrold thought probably not, but he always regretted that he, and the others, had not gotten it clearer, had not tried harder. He thought he more or less knew how they should have begun, and how they still might begin: with a group of people "who don't see their way through to the end, but who know that somehow or other, if you've got an idea, you can make it happen . . . you start in the upper left-hand corner of the blackboard, and you write it down. Then what are the next steps, and the next steps, and the next steps?"[37] He was seventy-nine years old when he described that way of beginning. But whatever opportunity might have existed in the early days, it had not been seized—not for want of effort or desire, Jerrold thought, but for want of leadership in the political world. It simply hadn't been there when it was needed: not then, he thought, and not since.

Project Charles

By 1951, the United States had fallen into a general mood of suspicion and fear of the Soviet bloc; there was a general perception that the communist threat was immediate. Soviet MiG-15 aircraft had been reported in action in Korea as that war ground on without resolution, and the communist Chinese had begun to engage UN forces. There was little reason for optimism. One of the signs of the dark mood of the country was that purges, perjury trials, blacklists, and hearings before congressional Un-American Activities Committees increased in frequency. The United States moved into the ever more paranoid era labeled with the name of Senator Joe McCarthy.

The debate over how much emphasis to give to strategic nuclear bombing in American defense planning grew even more acrimonious as the United States began to rearm and new military choices began to be available. Many in the air force high command continued to believe deeply that the only appropriate American reaction to the Soviet threat was to develop quickly and to maintain a first-strike capability. As the B-36–supercarrier debate had clearly illustrated, this meant developing long-range superbombers capable of penetrating the Soviet Union and, in order to justify such superbombers, it came to mean developing the superbomb as well. Within the Strategic Air Command, the offensive striking arm of the air force, this principle had been elevated to something like dogma.

These ambitious and aggressive attitudes were fostered in part by the competition for funds, prestige, and power, but it was also due, Jerrold believed, to a mostly unwarranted, self-aggrandizing view of itself by the Air Force:

Out of World War II came a young, but cocky Air Force, wedded to the myth of "Victory Through Airpower." Following the war, the Air Force was convinced and set out to convince the world, that it had won the war with the bombing of Hamburg, Dresden, Berlin, Tokyo . . . but . . . the bombing of an exhausted foe . . . did not contribute to the winning of anything of value; it only produced indiscriminate destruction.

Nonetheless, the Air Force started to campaign for an atomic weapons program that would produce bigger and more destructive weapons deliverable from Air Force bombers, building on the "Victory Through Airpower" model. The Air Force proposed and lobbied actively for the country to put all of its eggs in one basket, to concentrate the resources available for the country's defense in atomic bombs, in the hands of the Air Force.[38]

The positions taken in the defense-offense debate were not always realistic. Jerome Wiesner pointed out that one of the most characteristic features of the arms race has always been to ascribe to the other side the capacity to do that which one hopes to do oneself:

It turns out that at the time [about 1949–50] we were running an arms race with ourselves, because U.S. intelligence estimates were so poor. Our estimates of Soviet capabilities were estimates of where we were going to be, with a year or two delay . . .

Given that time, which is hard to remember . . . we had a hysteria and we all engaged in it, because I believed, as did most other people, that the United States was in mortal danger, and we had to do what we could, technically, to protect it . . . We all believed it. That was before we had access high enough up in the government, and learned to be suspicious of such things.[39]

Since the prevailing mood in the air force was so very much in favor of developing offensive power, the proponents of that point of view tended to regard any allocation of resources to other problems, such as defense, as diminishing what was available for the items at the top of the priority list: superbombs, superbombers and nuclear-powered aircraft. Nevertheless, a 1948 directive of the Joint Chiefs of Staff had made the air defense of the

United States almost entirely an air force responsibility, and the Joint Chiefs were not ready to abandon defense completely in favor of offense.

Against this background, the summer studies continued. Project Charles was a "summer" study of the air defense of the North American continent. It began in February 1951 and ran for nearly eight months; it had its beginnings in contention, ran its course through intramural quarrels and disputes, and wound up with a report that Jerrold felt left much of its job unfinished. On the other hand, it led to the creation of an important new national laboratory, remarkable for being under civilian control, capable of pursuing the job further. Personality conflicts arose throughout Charles and there were troublesome undercurrents that might have foretold, had anyone been prescient enough to interpret them, what would erupt two years later in the infamous Oppenheimer hearings. As with Project Lexington, Jerrold felt that Project Charles had again been an effort to choose a solution without knowing fully what the problem was, although in this case a more thoughtful analysis had sometimes won out.

The story of Project Charles illustrates as well a remark that Jerrold made some years later, commenting on the small number of scientists on whom the government relied for advice: "We all know each other . . . People always think that because the U.S. has a population of 170 million and there are a lot of people in the Pentagon, it all has to be very impersonal. Science isn't. It's just us boys."[40]

In fact, many of those involved in starting Project Charles knew each other from the Radiation Lab. The project itself, once it was underway, involved others as well, but at the beginning nearly all the participants carried baggage left over from their earlier interactions with each other. Sometimes, as old animosities reawakened, it got in the way.

Jerrold had not particularly wanted to be part of Charles. The project arose at a time when he was just getting back to work in the lab after Hartwell, and he had a full schedule. He did not get on particularly well with some of those who were involved; he described one as "oppressive and overbearing" and another as "a zero without a rim—the complete negative." But there were others he thought well of, and he judged there was at least a reasonable chance of success. Jerrold was like many other practical, hands-on types, who believe deep down that if you want a thing done right, you had better be involved in the doing of it. He couldn't have stayed out of it, anyway, once Jim Killian asked for his help. Jerrold had come to believe, along with Killian and Stratton, who jointly constituted the administrative leadership of MIT, that there was an important role for universities to play in the defense of the nation. In a speech made some years later, Jerrold said,

We didn't get into the underseas warfare business [Hartwell] because we thought "let somebody else do something." Our friends up the street—Mr. Conant at Harvard—said Harvard should not engage in classified research, it should not get into the military province . . . and Jim Killian, one of the real clear heads at the time, said "If we don't, who will?"

. . . The students of today [1971] don't understand that. It's not proper to say that the safety of the country should be sloughed off on somebody else because it's dirty work—it's not our business.

There is a need for an institution such as this one to come to the aid of its country when it's in trouble. And the truth is that the country was in trouble, on a military basis. The atom bomb—atom bombs are so attractive, [people think that] the way to protect the country is with those great big beasts of bombs. But it's too big a weapon. It's as if the only defense you have is to kill yourself. That's what atom bombs are like. . . .

MIT was willing to go along on the unpopular subject of air defense [which] in the eyes of the control of the Air Force was extremely unpopular . . . MIT knew it needed to be done and MIT was prepared to back it, and MIT shouldn't back away from it.[41]

The current air defense system was clearly inadequate, and something had to be done. Not willing to raise defense to the same level as offense, the air force sought a primarily technological solution, what Jerrold described as a "quick fix." Anything that went beyond that would be seen as both a diversion of resources from offense and a distraction from the immediate, restricted military problem.

The current plan of defense, if it could be called that, was a relic of the war. The country had been divided into separate independent defense areas, with little or no communication among them. Major gaps existed in radar coverage, especially with respect to low-altitude penetration by enemy aircraft; there was no provision at all for early warning, and the existing radar systems would have been easily saturated by multiple targets. The tasks of identifying large numbers of targets and directing interceptor fighter aircraft were well beyond the capacity of what was then in place.[42] Writing two years later, in 1953, Killian and Hill described the continuing inadequacy in air defense: "If there should be a surprise air attack on the United States next week, next month, or even within a year, our defending forces would be able to intercept and destroy only a small percentage of the invading planes. . . . This may not be adequate to ensure the continuity of our government. . . . This may not be adequate to ensure the continuity of the United States."[43]

On the advice of its Science Advisory Board (SAB), the air force had established a technologically oriented committee to look into the state of air defense and to recommend appropriate changes. This group, known as the

Air Defense System Engineering Committee (ADSEC), began work in December 1949.[44] It was chaired by George E. Valley, an associate professor in the MIT physics department and a research associate in Bruno Rossi's cosmic ray group at LNS&E.

Apparently the committee worked in a somewhat free-wheeling way. The official history of the Air Force Science Advisory Board notes: "As Dr. Valley described their operations, they functioned as informally as possible, making most of their recommendations verbally to the Air Force officials who sat with them."[46] Within a few months, ADSEC had sketched out a new and highly innovative scheme for a network of air defense stations. It included the important recommendation that the new system make use of the Whirlwind computer then being designed at MIT. This was to be the world's first high-speed general-purpose digital computer, which was expected to be capable of solving air-intercept problems in real time—that is, fast enough to be useful in actual air-intercept situations. Whirlwind was originally a project supported by the Office of Naval Research, and it was in some difficulty both because it was turning out to be very expensive and because there did not seem to be any problem at hand to which it might be applied. The navy had decided to close down the project. Jerome Wiesner then happened to point out to Valley in an informal exchange that Whirlwind might be useful to ADSEC in the air defense problem. Valley had followed up the suggestion, which turned out to provide the salvation of Whirlwind. The project was soon transferred to the air force, from which it received a considerable infusion of money.

But if ADSEC, through George Valley's intervention, saved Whirlwind, so did Whirlwind redeem Valley's ideas at ADSEC. In principle, Whirlwind could sort out multiple radar inputs and do the calculations quickly enough to direct interceptors to their targets. No other calculating device of that time could come close to doing that. In order to test the proposed integrated system of interlocking radars, with Whirlwind as the centerpiece, ADSEC had recommended to the Air Force that a pilot system be built.

Perhaps it is not surprising since five of the seven members of the committee were staff members of MIT, and another was an alumnus of the Radiation Laboratory, that General Hoyt Vandenberg, air force chief of staff, decided to invite MIT to build a laboratory to develop the pilot system. Adequate funds would be made available. James Killian recalled:

In the view of Vandenberg and the Air Force, MIT was uniquely qualified to do this job. MIT . . . had a great interest in this field, but we were not excited about taking on another big laboratory . . . so we played hard to get.

I remember sitting on a park bench in Lafayette Square with Louis Ridenour, who at that time was chief science advisor to the Air Force. I asked Louis while we were sitting there . . . why MIT should undertake a Laboratory of this kind. "It will make MIT a world center in the field of electronics," [he said], and that has stuck in my mind ever since.

[Secretary of the Air Force Thomas K.] Finletter got upset at our reluctance to take on the Laboratory. He came up [to Cambridge], sat himself down on the sofa in my office, and we had a couple of hours talk. . . .

I had been well advised by Zach, by Jerry Wiesner and Al Hill, that we should—if we undertake an air defense program, then it ought to be tripartite. It ought to involve all three services, not just the Air Force, and that in addition we ought to have some outside group have a hard look at this.[47]

Finletter and Killian did some hard negotiating in their couple of hours. It was Killian who prevailed: the advice of Zacharias, Wiesner, and Hill, in which Mac Hubbard had joined, was followed.[48] Although the air force provided most of the funding, the army and the navy were brought in, and the project did become tripartite, as recommended.

The advice jointly given by Zacharias, Wiesner, Hubbard, and Hill had arisen initially out of their dismay that so serious a matter should be decided upon with so little study. ADSEC had not even provided a written report; it had all been informal. No serious consideration had been given to alternatives, nor had any study been made of how to integrate an air defense system into national military policy. Al Hill said later:

What stuck in my craw was that I'd been through Hartwell, and here was a major project to be based on a very small study that hadn't been reviewed by anybody except the people who wanted it. . . . I wasn't against it. I just didn't want to buy it like that. . . . We said [to Killian] "You have to have a study, a major study, and it can't be limited to the ADSEC findings, it has to look at the whole air defense problem." And that was the origin of Charles.

They [the air force] knew what system they wanted, and they thought all we had to do was build it. And we, in the end, said "Nuts, we'll look at the whole problem." Now, the ADSEC recommendations weren't bad, but you just know that no small group of part-time people working for less than a year was going to solve that problem.

The air defense of the United States required far more, in their opinion, than shooting down enemy bombers with air force fighter airplanes. The army's radars, its antiaircraft weapons, and its Nike missiles, as well as the navy's carrier fleet and sea patrols, had important roles to play. Jerrold urged that they think as well about the civilian sector, particularly about the complex issues of passive defense, including such matters as the dispersal of industry, the effects on transportation systems and electric power networks. He was concerned that these immense problems had not been sufficiently considered. He later recalled: "What hit me hard in the big bomb business was the difference big bombs made in radioactive fallout. Our group [in Project Charles] knew that radioactive fallout had a dimension which had not been studied hard enough. . . . I remember going to the blackboard once and drawing a diagram: part of it was a few bombs on target, part was a dozen bombs on target, and as soon as you got above a hundred, it was called 'the horrors' . . . I remember using the word 'horrors.'"[49]

The problem had an international character as well, not yet considered. Albert Hill commented: "I was the first one to suggest the study, I'm pretty sure, and Jerrold was the one who got the Canadians and the English involved, and I think that was smart, especially the Canadians, because we'd have to do it [the interception] over their land, over their territory."

The recommendation for a tripartite summer study rather than a quick implementation for the air force alone did not sit well with either the ADSEC group or with other interested officials in the air force, all of whom had expected to see their proposals promptly accepted. Again, Hill commented:

At that time, Louis Ridenour was the chief scientist of the Air Force and Ivan Getting was a civilian in a job that normally had a major general . . . in Air Force Headquarters. Getting and Ridenour were very, very high on Valley's stuff, as was Dave Griggs, who came along replacing Ridenour later. They had all known each other at the Radiation Lab, and they were all very partial to the Air Force, which was this new, growing service. They were running into a lot of trouble because the Air Force was squabbling with the Navy and, having divorced themselves from the Army, they were squabbling with the Army, and vice versa. The scientists tended to line themselves up as pro-Air Force or anti-Air Force. It was idiotic.[50]

Considering the extraordinary success he had achieved with Hartwell, Jerrold was the natural and obvious choice to lead the new project, but under the circumstances that choice was impossible. Valley was one of those with whom he had never gotten along, even in Radiation Lab days, and, in addition, Jerrold's views on what he saw as the arrogance of the air force were becoming well known. Killian therefore made a diplomatic move: he invited Wheeler Loomis, the Illinois physicist who had been associate director of the Radiation Lab and who was well liked in all quarters, to direct the project. Jerrold was named associate director, a job he probably would not have accepted under anyone except Loomis. That had been his title on Project Lexington, and it had galled. He hoped matters might turn out differently this time, but he was disappointed, not because of any constraints imposed by Loomis but because he came to feel that Loomis had lost his wartime edge; his leadership, said Jerrold, lacked strength and initiative.

In spite of the difficulties and differences in viewpoint, Project Charles got underway promptly in February 1951 and completed its work about eight months later. The basic ADSEC recommendations were accepted, and a proposal was adopted for the design and construction of a pilot "Cape Cod system": "An experimental air control and warning network . . . should be established in Eastern Massachusetts by connecting between 10 and 15 radars and height finders to the WHIRLWIND digital computer at MIT."[51] Valley's ideas were judged sound, as far as they went, and were to be implemented. Personality conflicts had not, in the end, gotten in the way of judgment. Years later, Jerrold remarked on some of those conflicts: "I think it's important, and I would like to say it myself, that the moving spirit behind all of this was George Valley, and the fact that some of us didn't get on well with George from time to time—sometimes we did, and sometimes we didn't—should not obscure the fact that George had ideas, and was pressing in the Air Force for something that the Air Force really did not want."[52]

Jerrold nevertheless remained dissatisfied and unhappy. He knew that the vulnerability of the United States—the problem that had been pointed to by Killian and Hill—could not be removed by a set of localized radar systems, although such systems might be necessary in a larger scheme. He felt that the Charles report was something of a whitewash. Shortly after the report was issued, Jerrold wrote a long private letter about it to Julius Stratton, using that term, saying that he felt that Loomis's direction had been uninspired and that he objected to continuing ADSEC's working style. It was a style, he said, that favored the quick fix, relied on liaison officers to interpret the findings, and discouraged discussion and thorough analysis. He felt that Loomis and Valley had been unwilling to consider anything but the limited Cape Cod–type system, which he felt was far too small in scope.

A reading of the Charles report will show what I mean. On quick fixes . . . if anything were to happen today, there is a guarantee that our air defense would be zero—no more, no less—zero. This is not to be found in the report. On long range defense . . . the report is a whitewash. . . .

I know that it is possible to [undertake] such considerations by a large, enthusiastic, agreeable group. Subgroups dominated by liaison officers with no time for discussions does not do. On Project Charles our director insisted on: first, write it; then, see if you believe it. Since writing is such a time consuming thing and can really only be done by one person at a time—there is no way to make group decisions by writing first. . . . This point was a sore one all during Project Charles.[53]

Project Charles recommended that its work continue and that a special research facility be created for that purpose. The recommendation was accepted by the project sponsors, and a new permanent national research installation, the Lincoln Laboratory, came into existence. It was to concentrate on air defense and was to be operated by MIT on the same tripartite basis that had prevailed on Project Charles. Wheeler Loomis became the first director of the Lincoln Laboratory, but remained there for less than a year; he was succeeded by Albert Hill. Jerrold served as a consultant through the following summer and then returned to his own research lab in RLE, now headed by Hill's successor, Jerome Wiesner. Valley became a division head within Lincoln Laboratory; in 1958 he became chief scientist of the air force while on leave from MIT.

The Charles study was not limited to the ADSEC recommendations. The project report contained discussions of many other issues of air defense: early warning from remote stations, suggestions for new kinds of air-to-air weapons, and new ways of using existing defensive weapons such as those of the army. It made suggestions for improving naval communications, improving ground radar equipment, controlling multiple interceptors, and using a ground observer corps.

The Project Charles report also included preliminary discussion of some of the difficulties of passive defense: the dispersal of populations and of industry, stockpiling, conservation, and reconstruction. Even today, these enormous problems remain unsolved, and constitute a strongly sobering influence. There had been considerable resistance to including them in the study, but Jerrold had insisted. The air defense of the continent could not depend on gadgets alone. Like any other truly complex problem, it was a problem of systems engineering.

The Summer Study of 1952

Harry Truman had never been partial to advice from scientists; when he set up the first President's Science Advisory Committee in April 1951, he established it in the Office of Defense Mobilization (SAC-ODM) rather than in the White House, thereby managing to remain well insulated from it. Oliver Buckley, retired director of Bell Telephone Laboratories, was the committee's chairman. The committee membership was impressive, but it lacked access to the president and could not be effective.[54]

Buckley stepped down from the chairmanship in May 1952 and was succeeded by Lee DuBridge, but these were the waning days of the Truman administration, and not even DuBridge could accomplish much. The committee continued in this curious way into the Eisenhower period, meeting in a somewhat desultory fashion not more often than once a month. DuBridge struggled to find a role for the committee, strengthening it by adding colleagues he knew and respected. Jerrold became a member of SAC-ODM in 1952, as did Rabi, James Fisk of Bell Labs, and Charles Lauritsen of CalTech, among others. But the committee still had nothing to do; there was no agenda. Killian recalled that in November 1952,

some of us who were accustomed to action began to question whether we should take the initiative to disband.

To face this question squarely, it was decided to hold a meeting outside of Washington, where we could take our hair down quietly. Consequently, arrangements were made for a three-day session in early November, 1952, at the Institute for Advanced Study, Princeton, Dr. Oppenheimer having offered to be host there.

At this crucial meeting we examined the need for more science and innovative technology to improve our defense and how the resources of these fields could be brought to bear on policy-making at the level of the National Security Council.[55]

They decided to continue the committee, as Killian recalled it, hoping that Eisenhower would want it, and eventually he did. But not until the summer of 1954, two and a half years after Eisenhower's election, did the committee gradually begin to be upgraded. Finally, in 1957, responding to the severe pressure brought to bear by *Sputnik*, the first Soviet satellite, Eisenhower moved the committee out of ODM and into the White House. It was then given direct access to both the president and his staff. As PSAC, the President's Science Advisory Committee, it became highly influential, but in the years before that upgrading began, committee membership had been almost useless. Such advice as a reluctant government received, at least in the military-scientific domain, came far more often, and rather more irregularly, from summer studies like Hartwell and Charles, and then, generally filtered through the military.

Project Vista was still another summer study, run at the California Institute of Technology starting in the spring of 1951, running more or less at the same time as Project Charles. Its purpose was "to conduct a broad study of the problems of ground and air tactical warfare with especial reference to the defense of Western Europe. The object was to examine the tactics, techniques and equipment available for use in combat."[56] Specifically, Project Vista recommended the development of tactical nuclear weapons for possible use in the European theater. It "argued for the tactical use of nuclear weapons as initially a more effective way to stop the Soviet armies than strategic bombing. With this recommendation, it thus challenged directly the prevalent ideas of the Strategic Air Command (SAC). The Air Force sought to modify and then suppress the report."[57]

Project Vista had among its participants some of those who were serving or would soon serve as members of SAC-ODM and who perhaps felt somewhat keenly the frustrations of that service. Vista was headed by William A. Fowler, a noted CalTech nuclear physicist (subsequently a Nobelist in

1983), who had been a student and protégé of Charles Lauritsen, and included Robert Oppenheimer and I. I. Rabi, as well as Lauritsen, among its participants. Some of the others involved were Robert Bacher, R. B. Brode, Robert Christy, Dale Corson, Leland Haworth, William Higinbotham, and Norman Ramsey, all highly respected as physicists and as original thinkers and all with substantial experience in the wartime laboratories. Although Jerrold was listed among the participants, he came only as a visitor, as did Al Hill, each for a few days. Fowler nevertheless recalled that Jerrold's influence was felt.

The recommendations of Vista were consistent with those of Hartwell, although they went much further (and were correspondingly more controversial). Certainly the Vista membership did not need any outside stimulus to conclude that tactical nuclear weapons ought to be developed; it is not likely that they needed the example of Hartwell, although Lauritsen, at least, had served on the earlier project. The idea of tactical uses for nuclear weapons was bound to arise. Just as Hartwell could not have ignored the possibility, neither could Vista. Nevertheless, in spite of logic, the Vista recommendations were bitterly received by the air force; incredibly, in the loyalty hearings that Robert Oppenheimer would soon face, his participation in Project Vista would be offered as a reason why he was not to be trusted.

Approximately at the same time as Vista, yet another summer study, known as Project East River, was conducted in New York City, under the auspices of Associated Universities, Inc. It was directed by AUI's president, Lloyd Berkner. Its mandate was to consider the question of civilian defense in the event of a surprise action by a hostile force.

East River pointed out that U.S. defense plans contained no provision for early warning of an enemy attack. This glaring deficiency had indeed been one of the serious concerns of SAC-ODM, one that its members felt had almost certainly been inadequately communicated to the president. The East River group reported straightforwardly that without early warning, there could be no defense at all; the SAGE system would be worthless. Evidently similar observations had been made by Vista participants. Oppenheimer, Rabi, and Lauritsen now felt that a special study should be made of that problem, and they suggested that the study would be appropriate work for Lincoln Laboratory. The three Vista scientists convened a small meeting at the Statler Hotel in Boston some months later, which both Jerrold and Al Hill attended, and there it was proposed that Jerrold should direct such a study, if one could be gotten started. At Jerrold's insistence, the three Vista members present agreed to be listed as senior consultants on the new study. Jerrold was quite sure that they would need the prestige of such distinguished names, first to help convince the air force of the importance of the problem, then to assist in the recruitment of others for the study, and, finally, to add weight to whatever recommendations the study might eventually make.

It was exactly the kind of activity that Lincoln Lab had been created to carry out, and Al Hill remembered that as director of Lincoln, he "bought it in a hurry." Project Charles evidently had not solved the problem of the air defense of the North American continent, nor had it ended Jerrold's involvement in the problem. He supported the proposal to carry the studies further, especially as it offered the opportunity to carry out some of the work that he thought Project Charles had left undone.

The discussions and recommendations of Hartwell, Vista, and East River had confirmed Jerrold's belief that it was a mistake to rely only on offense; big bombs and big bombers were not enough. He continued to favor constucting a practical defense. He wanted to go beyond the achievements of Project Charles and eventually succeeded, but it took all of his powers of persuasion to convince a reluctant air force to authorize yet another summer study on defense and, later, to implement its findings.[58]

The new study of continental air defense was begun at Lincoln Laboratory in the summer of 1952, under Jerrold's direction.[59] As it turned out, the three who had originally proposed the study, Oppenheimer, Rabi, and Lauritsen, played only minor roles in it as consultants and had little to do with its recommendations. Oppenheimer and Rabi were deeply involved at that time in the complex decisions concerning the development of the hydrogen bomb. The question of whether to launch a crash program for its development had engendered very strong feelings, and neither scientist had time to give to this new study once it was launched. Lauritsen, a member of countless government and military committees, could not manage more than a part-time consultancy to the project.

The 1952 study led to the creation of the DEW (distant early warning) line, which consisted of a string of radar stations, established every hundred miles or so, on the barren tundra in the far northern regions of the Canadian archipelago. The chain of stations was reminiscent of the British Home Chain of World War II, but the technology was now far more sophisticated. The DEW line stretched across the top of the continent, north of the Arctic Circle, with airborne radar networks over the oceans on either end of the line. The idea was to obtain the earliest possible warning of an air attack crossing the North Pole. "It is perfectly obvious," said Jerrold in a speech made nine years later, "that if the United States is subjected to bombing threat by aircraft, our own forces and our people should have every warning they can get. But in 1952 this was not universally obvious. I can't tell you why;

I didn't understand it then; I don't understand it now. I can only assure you that we had great trouble convincing many people in high places of the correctness of this assumption."[60]

The engineering difficulties of the project were formidable, since during most of the year, access to the stations was possible only by air. An effective warning system requires a 24-hour radar watch, and communication from radar sites to Air Defense Headquarters must not only be instantaneous but absolutely reliable. Fortunately, new techniques of automatic radar surveillance were then being developed, and new methods of reliable long-distance communication by ionospheric and tropospheric scattering were just becoming practical.[61] But automation was still fairly rudimentary in 1952; radar stations still had to be manned, and the problems of support for human operators at remote stations might have turned out to be even more severe than the purely technical problems had the project not taken them into consideration. It was a good example of the way that the summer studies, when carried out à la Zacharias, tended to lead problem solvers into unexpected and uncharted domains, which often turned out to be critical. As Jerrold pointed out, "One does not take a course, preliminary to a Ph.D. in physics, on the human problems of the Arctic." He went on to recall what was involved:

Many military installations required cooks for the motor pool and a motor pool for the cooks . . . but that simply would not work in the Arctic. . . . By simple mechanisms we had to make it possible for a man to be able to cook, sleep, tend his chores, tend communications, man the radars, and in general act like the general purpose human being that any camper must be. In the Arctic there isn't much brush to cut, but in some places there are plenty of mosquitoes to fight. . . . we had then to make sure that the human problem of that dreary mosquito-bitten existence was possible.[62]

Just as it had been difficult to get the study started, it was difficult to get these ideas accepted. Jerrold remembered that following the summer study there had been a series of committees: "They were all earnest, patriotic, understanding; but first a committee was set up to evaluate our proposals, and another committee to evaluate that committee in turn, and so on and so on, up to five committees; but we won out. It takes courage and nerves as well as hard work to stay through this committee-izing process, and I don't recommend it as a steady diet."[63]

He sought diligently for every argument and every device that would demonstrate the practicality of the ideas developed in the summer study. One of these ideas was that a 24-hour radar ought to provide a sound signal for the operator as well as a blip on a radar screen. Borrowing his own childhood expression, he told the committee that radar observers "have eyelids, but they don't have earlids." The observers could not always be on the watch for a visible indication, but they would hear a sharp sound no matter what else they were doing.

He hit on a dramatic demonstration of the principle. Frank O'Brien, his head machinist at LNS&E, recalled how the idea came about:

Zach saw on my car seat a toy I bought for my daughter. It was a small dog house that contained a small plastic dog which would leap out instantly when you shouted his name "Sparky." After playing with it for a few minutes, Zach said "By Jiminy, I've got to have that. You buy your daughter another one." He was so enthusiastic about it that I said, "Certainly you can have it; but what on earth are *you* going to do with it?" "I need it, I need it," said he, "to dramatically convince some people at a certain conference that a particular thing we have in mind must be done!" I asked, "Which people? What thing must be done?" He replied, "Never mind who or what. I'll tell you in about five or six years." [64]

What Jerrold did was both simple and bold. He arranged a test without warning his audience ahead of time of what would happen: an aircraft would take off from Bedford Airport, near Lincoln Laboratory, and would fly over a certain place at a prearranged time. It would be detected as it flew past by a radar equipped with the new automatic range gate circuits that were being proposed for the Arctic stations; the resulting radar display would be accompanied by a sharp "ping," and Sparky would pop out of his kennel.

It worked. It worked again. It worked every time the generals and cabinet officials asked to see it. Sparky made the point in a way that could not be misunderstood or ignored. Certainly nobody who saw the demonstration ever forgot it, and it became quite famous for a time; regrettably, however, its fame was restricted to those with sufficiently high security clearance to be allowed to hear about it.

L. D. Smullin, a professor of electrical engineering at MIT and a member of the 1952 summer study, described the outcome: "In October of 1953 it was agreed to set up three experimental stations at Point Barrow, Barter Island and Skull Cliff [stations located well north of the Arctic Circle] during the following summer. To accomplish this, ships had to be loaded by March. This left four months to decide what was needed, to invent it and build it, buy it and pack it . . . not only radars and electronics but houses, heating units, fire fighting, cooking and living equipment, etc., etc. When the last boat left a northern site in September, the radar crew left behind would have to be self-sufficient for the next nine months." Smullin went on to recognize the difficulties that had been overcome by Jerrold's enthusiasm and hard work: "It all came to pass! . . . It is clear that it was Zach's drive, his unblushing use of 'Sparky' . . . and his commitment to the importance of warning in air defense that made this program go."[65]

Once again, the air force had been persuaded to implement a system of defense in spite of the SAC's philosophical commitment to offense. That philosophy remained unchanged, but the continued emphasis on defense, with its implications for a limited kind of warfare, began to disturb to those in the air force high command who were determined to have superbombs.

These issues, which had absorbed Oppenheimer, Rabi, and many others as well, which had run like a troublesome underground current through all the summer studies, eventually burst into public view in the Oppenheimer loyalty case. Jerrold was part observer and part participant in that, but it was deeply upsetting to him, as it was, and indeed still is, to nearly the entire physics community, and beyond. It became all the more distressing to Jerrold as the case developed, and the 1952 summer study appeared to become one more item in the load with which he felt Oppenheimer was unfairly burdened. Never, he thought, had an honest effort toward the defense of the nation been so meanly rewarded.

The Oppenheimer Case

On December 3, 1953, a discussion on a sensitive and delicate matter took place in the Oval Office of the White House, presided over by President Eisenhower.[66] Present were Herbert Brownell, attorney general of the United States; Robert Cutler, chairman of the Planning Board of the National Security Council; Allen Dulles, director of the Central Intelligence Agency; Charles E. Wilson, secretary of defense; and Admiral Lewis Strauss, chairman of the Atomic Energy Commission. The subject was J. Robert Oppenheimer.

The atomic bomb had made Oppenheimer into a national popular hero of almost unprecedented dimension. He was an unlikely candidate for such a role; he was hardly a man of the people. He was brilliant and erudite and made few concessions to those who were less so. He was mercurial, moody, and sometimes opaque. He could be, and often was, arrogant. Nevertheless, scientists and the public alike believed that the development of the atomic bomb during the war years could not have happened without his leadership. The success of the Manhattan Project had been a personal triumph as well as a national one. Oppenheimer's fellow physicists were, by and large, extremely proud of him. The popular press tended to portray him on an intellectual plane beyond the reach of ordinary mortals.

In the years since the war, he had served inter alia as chairman of the General Advisory Committee (GAC) of the Atomic Energy Commission.

It had fallen to him to preside over the intense secret debate about whether to mount a crash program to develop the hydrogen bomb. He had managed to alienate many of those who were eager for an all-out effort, including Admiral Lewis Strauss, present that December day in Eisenhower's office. On at least one earlier occasion Strauss had felt himself publicly humiliated by Oppenheimer's condescending tone, and he was not inclined to be forgiving. On the question of the hydrogen bomb, Strauss saw Oppenheimer's deliberate and measured approach as foot dragging, and he was impatient and irritated with him on that score as well. It is likely that none of those present at Eisenhower's meeting had much patience or tolerance for the moralistic tone that had characterized the advice of the GAC concerning the hydrogen bomb. There was a general tendency to attribute this tone to Oppenheimer himself.

The meeting in Eisenhower's office had been called to discuss a sensational and bizarre charge laid against Oppenheimer by a man named William L. Borden in a letter to J. Edgar Hoover, director of the Federal Bureau of Investigation. Borden had been the executive director of the Congressional Joint Committee on Atomic Energy and had recently left the government for a job in an industrial corporation. He wrote at length, concluding: "My own exhaustively considered opinion, based upon years of study of the available classified evidence, [is] that more probably than not, J. Robert Oppenheimer is an agent of the Soviet Union."

Brownell had brought the letter, together with the FBI's file on Oppenheimer, to Eisenhower's attention. Borden's letter came at a time when the president and his administration were already feeling plagued by Senator Joe McCarthy's reckless allegations of communist infiltration in government. Eisenhower seemed helpless to contain McCarthy and curiously hesitant; time after time he declined to stake his own political popularity against the senator's demagogic appeal. The accusations against Oppenheimer somehow increased the president's sense of political vulnerability and danger. The meeting in the Oval Office was therefore to be an exercise in damage control. Rather than taking on McCarthy directly, Eisenhower had determined to move preemptively to keep McCarthy and his crew at arm's length. Stating that he was "not in any way prejudging the matter," he directed that a "blank wall" be placed between Oppenheimer and any sensitive or classified material. Eisenhower said that he had seen the Borden letter and thought that there was "no evidence that implies disloyalty on the part of Dr. Oppenheimer." But he added, continuing to play it safe, "this does not mean that he might not be a security risk." He appointed a three-man committee to investigate the charges, and Oppenheimer's security clearance was suspended. The matter was to be held fully confidential; McCarthy was thus blocked. When later McCarthy charged that the development of the hydrogen bomb had been held up for eighteen months by "Reds in the government," the Oppenheimer matter was well beyond his reach.

On December 23, 1953, Kenneth D. Nichols, general manager of the AEC, wrote to Oppenheimer informing him that the commission "has developed considerable question whether your continued employment on Atomic Energy Commission work will endanger the common defense and security, and whether such continued employment is clearly consistent with the interests of national security." This was followed by a lengthy bill of particulars, and then, "you have the privilege of appearing before an Atomic Energy Commission personnel security board."

Oppenheimer responded at equal length, requesting the hearing that apparently was considered a "privilege"; he published his dramatic reply in the *New York Times*, which succeeded in riveting public attention to the matter. The hearing took place from April 12 through May 6, 1954, and it was a wonder of contradictions, conflicts, and high drama. It was conducted in secrecy, with all participants, including the thirty-nine witnesses, enjoined from any public revelation of the proceedings. It was not a trial except in the court of public opinion. Courtroom rules of evidence were not followed; for example, Oppenheimer's attorneys were not given the right to review classified material. Nevertheless, judgment was pronounced. Confidentiality notwithstanding, when the proceedings were ended, the entire transcript of nearly a thousand pages was abruptly—and surprisingly—made public.

Jerrold's role in the Oppenheimer hearings was not a central one, but it was illustrative of the Alice-in-Wonderland character of the proceedings. The testimony itself can hardly be improved on for conveying the flavor of what transpired.

In The Matter of J. Robert Oppenheimer: Extract from the transcript of the Hearing before the Personnel Security Board of the Atomic Energy Commission, April 29, 1954[68] *The witness is David T. Griggs:*[69]

Q: There has been some mention of a group called ZORC. Was there any such group as that that you knew about?
A: ZORC are the letters applied by a member of this group to the four people, Z is for Zacharias, O for Oppenheimer, R for Rabi, and C for Charlie Lauritsen.
Q: Which member of the group applied it?
A: I heard it applied by Dr. Zacharias.
Q: When and under what circumstances?
A: It was in the fall of 1952 at a meeting of the Scientific Advisory Board in Boston—in Cambridge—at a time when Dr. Zacharias was presenting parts of a summary of the Lincoln Summer Study.

Q: In what way did he mention these letters? What were the mechanics of it?
A: The mechanics of it were that he wrote these . . . letters on the board—
Q: . . . Wrote them on what board, a blackboard?
A: Yes.
Q: And explained what?
A: And explained that Z was Zacharias, O was Oppenheimer, R was Rabi, and C was Charlie Lauritsen.
Q: How many people were present?
A: This was a session of the Scientific Advisory Board, and there must have been between 50 and 100 people in the room.

In The Matter of J. Robert Oppenheimer: Extract from the transcript of the Hearing before the Personnel Security Board of the Atomic Energy Commission, May 4, 1954. The witness is Jerrold Zacharias:

Q: Did you or did you not, Dr. Zacharias, on the occasion of the 1952 Scientific Advisory Committee meeting, in the fall of that year in Cambridge, write on the blackboard in the course of that meeting the term "ZORC" and explain it?
A: To the best of my knowledge and belief, I did not write on the board the letters "ZORC." . . . I have even gone so far as to check the memory of a few other people on this very point, and none who has been questioned remembers any such thing.

The witness is Albert G. Hill:

Q: Do you recall any incident occurring during . . . the meeting of the Scientific Advisory Board in which the word "ZORC" or anything like that figured?
A: I cannot recall any such thing. The statement was made in Griggs' testimony that Zacharias wrote this on the blackboard. I cannot believe that, because it would have been a cute trick in a very public and formal meeting, and I know Zacharias well enough to know that I would have been quite angry with him had he done it. I am convinced he did not do it. . . .
Q: I think I should tell you, Dr. Hill, that I am very much concerned, as are my colleagues on the board, about the fact that there is testimony before this board which indicates very clearly that some one or more witnesses have not told the truth to this board. There has now developed in this proceeding a real question in some cases of veracity . . .
A: May I ask you a question, sir?
Q: Yes, sir.
A: Were you referring to the "ZORC" incident?
Q: Among others, yes . . .

Hill always regretted that he had been prevented at that point by an over-zealous attorney from adding some remark reminding the members of the board that their duty was to determine who was telling the truth.[70] He was sure it was not Griggs. It is unlikely that it would have made any difference to the overall result, however, had he managed to say anything more; the board would almost certainly have let the matter drop at that point in any case. The hint of conspiratorial doings had been injected into the proceedings. Proof was not required; the damage was done as soon as the idea had been planted.

The effect of introducing ZORC into the testimony had been to suggest the existence of a cabal of some sort surrounding Oppenheimer—to suggest a plot of some devious nature, intended to weaken the air force. The term had originated, as far as is known, in an article in *Fortune* magazine, in the spring of 1953, wherein ZORC was described as a real, well-defined entity.[71] The summer study of 1952, the so-called Lincoln Project, was offered as an example of the work of this cabal. The *Fortune* article claimed that the fictional conspiracy had, through the agency of the Lincoln Project, suggested replacing the Strategic Air Command, the pride of the air force, with a "jet-propelled, electronically hedged Maginot line"—at a price of $50 billion to $150 billion. In fact, the recommendations of the summer study had been nothing like that, in either substance or in price. The DEW line was intended for early warning and had not been intended to be a line of defense. Neither had it been proposed as an alternative to the Strategic Air Command, which would always be essential nor had the cost of the DEW line even approximated the figures cited in the article. Outrageous figures of similar magnitude kept appearing, however, without apparent need for authentication, in magazine and newspaper articles.[72]

Jerrold gave his view of the *Fortune* article in his testimony: "I felt it was a journalistic trick to bring into focus the kind of scurrilous charges that were being made in the article. . . . Whether all journalistic tricks are dirty, I don't know. I rather feel that this one was."

Elsewhere in his testimony, he made clear what the interaction actually had been among himself, Oppenheimer, Rabi, and Lauritsen:

One evening Dr. Lauritsen and I had a discussion about air defense and the participation of Lincoln and how it would be possible to make an air defense in the face of a growing threat. . . . Dr. Lauritsen and I thought it would be a good idea to set up a study group to investigate the question of defense of the North American continent.

I got in touch with Dr. Hill . . . we decided that it would be a good thing to do, that it would help air defense if we did it, and it would likely help the Lincoln Laboratory's growth.

We had a discussion about this with Dr. Lauritsen, Dr. Oppenheimer and Dr. Rabi. I remember that it was in a room in the Hotel Statler. Five of us, as I remember it, certainly Dr. Hill was there.

For Griggs, such a meeting was sufficient evidence of conspiracy. Unwilling to recognize differences of opinion as either healthy or normal and encouraged by rumors and hearsay, he himself had developed grave doubts about Oppenheimer's loyalty. Griggs told how these doubts had arisen:

Shortly after I came to Washington [as chief scientist of the air force] . . . I was told by Mr. Finletter that he had serious question as to the loyalty of Dr. Oppenheimer. I don't know the basis for his fears. I didn't ask. I do know he had access to the FBI files on Dr. Oppenheimer. . . .

However, it was clear to me that this was not an irresponsible charge on the part of Mr. Finletter or on the part of General Vandenberg [air force chief of staff], and accordingly I had to take it into consideration in all our discussions and actions which had to do with the activities of Dr. Oppenheimer during that year [1951–1952].

With the origin of his suspicions thus established, Griggs proceeded to testify about Oppenheimer's behavior, as he perceived it through the screen of those suspicions:

It became apparent to us—by that I mean to Mr. Finletter [and his associates], that there was a pattern of activities all of which involved Dr. Oppenheimer. Of these one was the Vista project—I mean was his activity in the Vista project . . . We were told that in the late fall of, I believe, 1951, Oppenheimer and two other colleagues formed an informal committee of three to work for world peace or some such purpose, as they saw it. We were also told that in this effort they considered that many things were more important than the development of the thermonuclear weapon, specifically the air defense of the continental United States, which was the subject of the Lincoln Summer Study. No one could agree more than I that air defense is a vital problem and was at that time and worthy of all the scientific ingenuity and effort that could be put on it. We were, however, disturbed by the way in which this project was started. . . .

It was further told to me by people who were approached to join the summer study that in order to achieve world peace—this is only a loose account, but I think it preserves the sense—it was necessary not only to strengthen the Air Defense of the continental United States, but also to give up something, and the thing that was recommended that we give up was the Strategic Air Command. . . . I should say the strategic part of our total air power, which includes more than the Strategic Air Command.

David Griggs's position on defense was already a matter of record. He had not only opposed the 1952 summer study, seeing in it a threat to air force offensive capability, but he had also tried to interfere with the study's beginning, by attempting to influence James Killian and Julius Stratton against it. Jerrold reacted forthrightly. He testified, "I would not want to bring up Dr. Griggs on charges of being disloyal in his effort to sabotage an effort in which I was the major promoter. However, let me say that it is a bit of a pity that dueling has gone out of style. This is a very definite method of settling differences of opinion between people than to try to bring out all detail in a hearing."

Griggs had sounded the prevailing theme of the hearing: he did not feel, he said, that the General Advisory Committee (GAC) of the Atomic Energy Commission, under Oppenheimer's chairmanship, "had been doing anywhere near as much as it could do to further the development of the thermonuclear weapon." Griggs claimed to see "a pattern of activities" in Oppenheimer's behavior on the GAC.

It was this allegation of opposition to the hydrogen bomb that was really at the heart of the attack on Oppenheimer's loyalty. Although the Oppenheimer case had been triggered by a reckless and unsupportable charge of disloyalty and treason, the real issues in the case had to do with the nation's military posture and with all of the problems of rivalry and ambition connected to it. The crucial question, as his attackers saw it, was whether to rely on a massive offensive capability or on an enhanced defense. The possibility that there might be a rational balance that included both seems not to have been conceded, at least not by the proponents of the superbomb. Almost five years earlier, on October 30, 1949, the GAC had advised against a crash program for the development of the super—advice, it must be emphasized, that it was duty-bound to give. Oppenheimer had reported the committee's view: "No member of the Committee was willing to endorse this proposal." Remarkably, the GAC had chosen to add an ethical dimension to its predominantly technical advice:

We have been asked by the Commission whether or not they should immediately initiate an "all-out" effort to develop a weapon whose energy release is 100 to 1000 times greater and whose destructive power in terms of area of damage is 20 to 100 times greater than those of the present atomic bomb. We recommend strongly against such action. . . .

We believe a super bomb should never be produced. Mankind would be far better off not to have a demonstration of the feasibility of such a weapon, until the present climate of world opinion changes. . . .

In determining not to proceed to develop the super bomb, we see a unique opportunity of providing by example some limitations on the totality of war and thus of limiting the fear and arousing the hopes of mankind.[73]

Two of the GAC members, E. Fermi and I. I. Rabi, sought to include an even stronger moral element in their opposition to the development of the super. They added to the report an opinion:

Necessarily such a weapon goes far beyond any military objective and enters the range of very great natural catastrophes. By its very nature it cannot be confined to a military objective but becomes a weapon which in practical effect is almost one of genocide.

It is clear that the use of such a weapon cannot be justified on any ethical ground which gives a human being a certain individuality and dignity . . .

It is necessarily an evil thing considered in any light.

For those who favored the rapid development of the super as the nation's response to the first Soviet atomic explosion, these views of the GAC were unacceptable. The addition of what probably was read as pious, moralistic advice must have been particularly galling to those with an inclination to find the language of the fellow traveler in whatever they disagreed with. Edward Teller, generally acknowledged to be an extraordinarily capable physicist, disagreed with the GAC on both technical and policy grounds; in addition, he completely rejected the idea that there was an ethical dimension to the question. He could not bring himself to charge Oppenheimer directly with disloyalty, but his testimony before the Personnel Security Board had the same effect:

Q: . . . Do you or do you not believe that Dr. Oppenheimer is a security risk?
A: . . . I feel that I would like to see the vital interests of this country in hands which I understand better, and therefore trust more.

Teller had agitated single-mindedly and obsessively for the development of the super since 1942—he referred to his "almost desperate interest in the undertaking"—and had now managed to obtain air force backing for his ideas. Secretary of the Air Force Finletter was particularly insistent that a crash program be started; if the AEC would not initiate such a program at Los Alamos or would not establish a second weapons laboratory for that purpose, Finletter threatened that the air force would establish and fund such a laboratory itself and build it around Teller and his proposals. The air force position was strongly supported by the Joint Committee on Atomic Energy, one of the most powerful committees of the Congress.

Thus, for a brief time the military policy of the United States hung in a precarious balance. Which way would it tip? The answer was not long in coming: the advice of the GAC did not prevail. On January 31, 1950, just four months after the first Soviet atomic explosion and three months after the GAC gave its opinion, President Harry Truman ended the debate: "I have directed the Atomic Energy Commission to continue its work on all forms of atomic weapons, including the so-called hydrogen or superbomb. Like all other work in the field of atomic weapons, it is being and will be carried forward on a basis consistent with the overall objectives of our program for peace and security."[74]

One year later, again against the advice of the GAC, Teller got what he wanted: the second laboratory was created at Livermore, California, not far from the Berkeley campus. But for many of those who, like Teller, continued to press hard for the super, it was not enough. Oppenheimer continued to be associated with defense-oriented activities such as Vista and Lincoln. Although he did not play a central role in them, he lent these projects the prestige of his name and reputation, and they flourished. His opponents were not appeased by their victory in the debate over the crash program; they wanted more. Oppenheimer's influence persisted, they felt, and would have to be reduced or eliminated.

Years later, Jerrold pointed out that someone like Borden, the writer of the initial accusatory letter to the FBI, had almost certainly been warped by the general paranoia that had seized the country at the time. Borden's behavior was unforgivable, but he was not the genuine villain of the piece. Such people saw spies everywhere; they did not believe that the Russians could have succeeded in manufacturing an atomic bomb without some tremendous disloyalty within our own ranks. They did not understand, for example, that the only genuine atomic secret was the fact that a bomb could be built. Once we had built and used one, there was no important secret left. The Bordens of this world were vicious enough, but they probably did not really understand what was going on.

But there were others, Jerrold said, who knew exactly what they were doing. They were out to destroy Oppenheimer's influence in any way they could, and if that meant destroying Oppenheimer as well, they would not balk at that. Jerrold remained unrelenting toward these people for the rest of his life. He abominated Lewis Strauss, whom he viewed as duplicitous; he despised Kenneth Nichols, who, as general manager of the Atomic Energy Commission, had drawn up and presented the set of charges on which the hearing was based. It was a document that succeeded outrageously in mixing Oppenheimer's opinions with questions about his loyalty. It was gratuitously harmful, Jerrold believed, a shameful exercise in character assassination. After the Oppenheimer case, Jerrold never referred to Nichols except as "the man who murdered Robert." Jerrold's view of Edward Teller was equally severe. He saw Teller as the personification of all that was dishonest and disloyal, a man who had betrayed a friend. Jerrold would not attend a meeting and would not even stay in a room if Teller was present. He made clear that he would have nothing to do with him.

In The Matter of J. Robert Oppenheimer: Extract from the transcript of the Hearing before the Personnel Security Board of the Atomic Energy Commission, April 21, 1954. The witness is I. I. Rabi:

[There is no call for] this kind of proceeding against a man who had accomplished what Dr. Oppenheimer has accomplished. There is a real positive record . . . We have an A-bomb and a whole series of it, . . . and what more do you want, mermaids? This is just a tremendous achievement. If the end of that road is this kind of hearing . . . I thought it was a pretty bad show. I still think so.

In The Matter of J. Robert Oppenheimer: Extract from the transcript of the Hearing before the Personnel Security Board of the Atomic Energy Commission, April 23, 1954. The witness is Vannevar Bush:

The National Academy of Science meets this next week and the American Physical Society meets, and I hope that they will do nothing foolish. But they are deeply stirred. The reason that they are stirred is that they feel that a professional man who rendered great service to his country, rendered service beyond almost any other man, is now being pilloried and put through an ordeal because he had the temerity to express his honest opinions. . . .

I think this board or no board should ever sit on a question in this country of whether a man should serve his country or not because he expressed strong opinions. If you want to try that case, you can try me. I have expressed strong opinions many times, and I intend to do so. They have been unpopular opinions at times. When a man is pilloried for doing that, this country is in a severe state. . . .

Excuse me, gentlemen, if I become stirred, but I am.

On April 6, 1954, just before the hearing began, Jerrold wrote to Oppenheimer about his own arrangements to testify. He added:

. . . But more important,—just give them the standard JRO . . . you have nothing personal to fear—really not—and your stand is so important for the nation. I guess all I mean is—give 'em hell. We're all with you. Love to Kitty.

Jerrold

On June 28, 1954, Jerrold wrote to Oppenheimer in a similar spirit, but tempered by the events of the preceding months:

Dear Robert—

I have been reading my eyes out on that 990 pages [the transcript of the hearing] of mixed blessings. This is just to assure you that you have real cause for gratitude—you have been paid the finest tribute a man can get—a sincere appreciation by the friends he cherishes, and only mild sniping by his enemies. Take comfort in it—it's a lot of what really matters. Love to Kitty—

Jerrold

On the very next day, the Atomic Energy Commission, acting on the recommendation of a majority of the Personnel Security Board, announced that by a vote of 4 to 1 it had decided that Dr. Oppenheimer should be denied access to restricted data. The "mild sniping" had not been so mild after all; it had been cruel and effective.

Yet Jerrold's judgment of it turned out to be close to the verdict of history. What has ultimately counted in that verdict was indeed the sincere appreciation of Oppenheimer's colleagues and friends.

Indeed, action which in some cases may seem to be a denial of the freedoms which our security barriers are erected to protect, may rather be a fulfillment of those freedoms.

—Findings and Recommendations of the Personnel Security Board in the Matter of J. Robert Oppenheimer[75]

Entrepreneurship

A Time to Measure

However absorbing the summer studies were, they did not completely interrupt Jerrold's life. Remarkably, he managed to make room for them while keeping on with all his regular activities: running a major laboratory, teaching, and doing research.

Jerrold was hardly unique in those busy days in having many claims on his attention. By the 1950s, scientists found it almost normal to be seriously occupied with matters outside their laboratories, and even the Oppenheimer case, upsetting as it had been, did not alter that pattern. Any airline flight between Washington and Boston was likely to have aboard several members of what Jerrold liked to call the Greater Cambridge physics community, engaged either in advising some government committee or seeking federal funds for a new research project. Briefcases that contained six- or seven-digit budget estimates and a list of telephone numbers of admirals and generals in the Pentagon might also carry the notes for a professor's next-day lecture. The physicists led complicated but somehow manageable lives, and Jerrold, like others, seemed able to keep a sensible balance among his many activities.

Jerrold took his teaching seriously and worked hard at it in those years. One form of undergraduate teaching that particularly appealed to him was the seminar, which, he said, should consist of seven or eight students, although any number up to about twenty would be all right. The seminar should include at least two "grownups": people who were familiar with the subject to some extent but secure enough to admit when they did not know something. Some good give-and-take discussions could arise that way, he said, and that kind of exchange was essential to the teaching process: "When you're sitting in a room listening to somebody lecturing, and you're letting it pass, just letting the words flow over, and saying, Yes, I do agree, or No,

I don't agree, then that's one thing. But if you're trying to formulate what you really think, and what you're going to hit that guy with as soon as you can interrupt, then that's another. Part of the learning-teaching business is to get someone to come back at *you*."[1]

He first organized such a seminar in 1950, for a group of a dozen bright seniors, to discuss the important modern experiments in physics. Jerrold was bothered by what he saw as an artificial and unnecessary separation of theory from experiment in physics education. A few years later, he would tackle this problem in a fundamental and comprehensive way, but for now, he simply knew it was important to talk with students and to make sure they knew about the experiments as well as the theory. He wanted to discuss with them the breakthroughs in quantum electrodynamics; the Lamb shift experiment, the hyperfine structure of atomic hydrogen, and the recent discovery of positronium by his colleague Martin Deutsch were examples.[2] In other words, he wanted to talk about the things in physics that involved precision, ingenuity and resourcefulness. These were what he loved the most.

The seminar met once a week, from seven in the evening until around midnight. Colleagues occasionally joined in; sometimes Al Hill came, or Martin Deutsch to discuss his own work; sometimes Francis Friedman, a younger member of the department, attended. Friedman was a nuclear theorist who had spent the war years in the Manhattan District Plutonium Project at the University of Chicago; he could steer the seminar through the dense theoretical thickets it would sometimes encounter.

John King, now a faculty member at MIT, had been a member of the seminar group as a senior and remembered it fondly. "Jerrold was expansive," he recalled, "full of ideas, not formal . . . I sort of knew him as a person, not as some remote professor." Later King worked as a graduate student in Jerrold's laboratory, but he still speaks nostalgically of his undergraduate experience:

In those days [1949–1950] we had a senior lab with Sanborn Brown in charge. I went to Zacharias, he seemed much more magnetic and interesting. . . . We tried to detect hydrogen atoms with high efficiency, and we tried to improve on the Pirani gauge by making H⁻ ions in a mass spectrometer. . . .

I'd been neglecting all my course work to work in the lab, so I had bad grades. Zach said "You should be a graduate student here," so I said "All right," and so I was, and I've been here ever since.

One evening in 1951 or 1952, Jerrold remembered, the seminar began to discuss experiments in molecular beam magnetic resonance, so closely bound up with Jerrold's scientific life. The group talked about the prewar experiments in Rabi's lab, the postwar work on the hyperfine structure, and the use of the mass spectrometer. Jerrold remembered they were talking about the theory of relativity, and he brought up the possibility of using a resonance beam apparatus to make an atomic clock; one so accurate that it could test directly the predictions of the theory. It would be a clock more stable and more precise than any that had been built before.

The idea of a clock based on the frequency of an atomic transition was not new, nor was the related idea of using the atoms in a magnetic resonance beam to establish a frequency standard.[3] Rabi had suggested it in a 1945 speech, and others were thinking about such schemes. Norman Ramsey, who had been Jerrold's academic neighbor at Harvard since 1947, had a great interest in the possibility of such a clock, and the National Bureau of Standards was also working out various approaches to the problem with the help of another of Jerrold's former Columbia colleagues, Polykarp Kusch. But thinking about a clock and making one are two very different things. As Rabi himself had said so many years earlier, speaking of the resonance method itself: "There was nothing really new. The real important thing was to *do* it, and discover the precision which you get out of it."[4] Rabi's words applied equally well to making an atomic timekeeper. The trick was certainly to do it, and Jerrold decided to try.

It is worth pausing to see why the idea was so exciting, why an atomic clock was likely to be such a giant step forward. For a clock to keep good time, it needs to depend on some very stable repetitive action, some event that occurs with as constant a frequency as possible. The problem of making a clock is the same as the problem of obtaining a stable frequency standard. In other words, a clock needs a frequency regulator. An old-fashioned grandfather clock uses a pendulum, which is a fairly good time-keeper for ordinary purposes; provided it is not moved from place to place, provided that the temperature and the atmospheric pressure remain constant, and provided there is no significant wear and tear on the pendulum support. There are better and certainly more useful regulators than a pendulum: the vibration of a wafer of quartz, such as is used to regulate a modern wristwatch or clock, is normally stable to within a second or two per month and with care can be made accurate to a few seconds in a year.

The rotating earth, which is the basis for astronomical time, is a more stable regulator than a quartz crystal. But even the earth's rotation is not uniform; it includes many tiny wobbles. For the most part the wobbles are predictable, and suitable corrections can be made for them, but when the demands of precision become high enough, small, unpredictable deviations become important. The limits of the astronomer's clock are determined, therefore, by many uncontrollable variables: how precisely the earth retains its shape, as well as how accurately the moment can be judged when a star crosses a fine line in a telescope or transits behind the edge of the moon. Regulated by the rotating earth, the astronomical clock can keep time accurately to about a tenth of a second in a year, that is, about one part in several hundred million. That is high precision, to be sure, and for many scientific purposes it is high enough. But the rotating-earth clock is not stable enough to permit testing of the theory of relativity, for example, nor can it be used to examine the stability of atomic transition frequencies. What would a more accurate clock reveal about such things?

Anybody familiar with the problems of timekeeping would have recognized that improved accuracy would also be important outside the domain of pure science; a better clock would increase the accuracy of navigation, for example, beyond the 5- or 10-mile accuracy then possible with ships' chronometers. No doubt there would be other applications as well—in Jerrold's words, "a better mousetrap always finds a mouse." It was frustrating that as late as 1951, the astronomer's clock was still the best timekeeper available. But if a clock could be regulated by the frequency of some particularly stable atomic transition, far higher accuracy might be achieved. An improvement of a factor of one thousand ought to be readily achievable, Jerrold thought, although it ought to be possible to go much further; he said later, "Surely we could try an intermediate device (10^{11}), something not too difficult that was surely possible to do, given the strengths of the Research Laboratory of Electronics and the years of our research with molecular beams . . . Furthermore, we needed our own intermediate clock in working toward a clock of super precision.[5]

Thus, two clocks were contemplated. The standard molecular beams technique would be adapted to produce the intermediate clock—the "small clock," as it came to be called. But a new idea would be required to go beyond the small clock's accuracy. Unexpectedly, this new idea came out of the discussion in Jerrold's undergraduate seminar.

One of the advantages of the seminar method is that new ideas sometimes pop up. The discussion becomes animated, and it is not always clear where the ideas originate; Jerrold was never certain who came up with the idea of using gravity to slow down the atoms in a beams apparatus: "Somebody, I can't recall whether it was Hill, Friedman, a student, or even me, put forward the idea of building a vertical apparatus ten, twenty or thirty feet in length. . . . Why not build an apparatus that would take advantage of gravity? Point the beam up, let the atoms spend a full second moving through the flopping field, a half-second rising up and a half-second falling back. . . . It was an idea rich with potential."[6]

The stability of an atomic frequency depends directly on how long the atom vibrates undisturbed in the region of observation. If the atom could be made to spend as much as a full second in the flopping region of the beams apparatus, the arithmetic suggested they might achieve an accuracy almost a million times better than the astronomer's clock. The idea was simple, as exciting ideas often are: just shoot the atoms upward out of an oven, and let the faster ones escape while the very slowest ones fall back down to a detector. They would be like so many water droplets from a garden hose, each droplet falling back under the influence of gravity: a fountain of atoms, in fact, where only the slowest ones would be captured. Of course, the idea would not be easy to implement—it was clear, for example, that a vertical configuration for the apparatus would bring about many new and formidable problems—but it was a challenge and an opportunity that an experimenter like Jerrold found irresistible.

He began by discussing these ideas with Jerry Wiesner, now the director of RLE, who would have to authorize the funding if it were to be done. He discussed the fountain idea with Norman Ramsey, who was delighted with it; Ramsey said later that it was so elegant an idea that he wished he had thought of it. Al Hill encouraged Jerrold, and other colleagues and friends were equally positive. By the middle of 1953, funding from RLE was in hand, and work could begin.

The project would be difficult, but the more difficult it seemed, the more it looked like fun and the more attractive it became. Jerrold was becoming restless; he felt that he was marking time in his research, just keeping things going. He was no longer challenged by magnetic moment measurements; the freshness and the excitement of the earlier days seemed to him to have gone, and the work was becoming routine. He had allowed the lab to be run for the last few years largely by his junior colleagues, almost without his help or often even his presence.

It was true that there were hundreds of magnetic moments yet to be measured and difficult techniques still remained to be developed. Someone might spend a useful lifetime filling in a table with those hard-won numbers, one by one, but Jerrold felt he did not want to be the one to do it. Breaking new ground early had always seemed to him far more exciting than harvesting the crop.

And so the clock project began. Building the small clock was the first order of business. The requirements were severe enough to give anybody pause. In addition to being as accurate as possible, the clock would have to be reliable, stable, and reproducible. Like the grandfather clock of folksong, it must "number life's seconds" without ever stopping. Jerrold made a bold decision; recalling his experiences in the Radiation Lab, he described again the essential difference between a laboratory device and a production model: "We decided to build a molecular beam apparatus six feet long, designed to run day and night without attention, an unusual feature for such an apparatus at that time. In fact, there were those who considered us bold to try to make a clock which would run continuously even in the laboratory where we would be present to tend it. But we decided to go even further: to try to design such a device so that it could be put into commercial production, and therefore be generally available."[7]

It was certainly not obvious that a commercial device was possible. Neutral cesium atoms were to be the working substance, and Jerrold pointed out that never before had a vacuum tube been sold or even built commercially that used neutral molecules as the working fluid. He remembered later that even molecular beams experts had had doubts about it: "Polykarp Kusch and I made a bet. He said, 'You're crazy.' I said, 'OK, you want to bet? The standard bet is a magnum of champagne.' You see, a vacuum tube has electrons in the beam, but cesium atoms? A molecular beam apparatus? That only a few people in the world can make work?"[8]

The project drew in new people. A graduate student, R. D. Haun, was set to work designing new, highly efficient oven-and-slit arrangements for the cesium source. J. G. Yates, a visiting lecturer in electrical engineering from Cambridge University, went to work on the electronics.[9] The supporting technical staff of RLE—especially Frank O'Brien and Fred Rosebury, two highly skilled technicians—went to work fabricating the new sorts of devices the researchers dreamed up. John King had earned his Ph.D. by that time and had won a place on the MIT faculty; he now devoted a summer to setting up the experimental work on the small clock. An undergraduate senior, Rainer Weiss, came around one day looking for an opportunity to do something useful in a research lab. Was there something he could do on this new project? He reckoned he could learn to run a lathe or a drill press, and before he came to MIT, he had earned some money wiring up hi-fi systems, so he knew a little electronics—"very little," by his own recollection. Jerrold took him on, warning him that while the small clock was a virtual certainty, the big clock was at least a three-year experiment with a 75 percent chance of failure. The first assignment Weiss received, John King recalls, was to make an insulated box in which to store dry ice: a pedestrian beginning, perhaps, to what would become a bright and productive career, for Weiss, like King, eventually became a permanent member of the MIT faculty. Weiss became Jerrold's right-hand man on the big clock project, much as John King had been on the small cesium device, and stayed with the project right to the end—until, as Jerrold said, the odds caught up with them.

They were spared the need for manufacturing long, ultraprecise magnets due to an invention of Norman Ramsey. Ramsey had realized that the all-important spin flips in the atomic beam occur as the atoms either enter or leave the magnetic field region; this enabled him to borrow the idea of interferometry from the field of optics and to design a magnetic resonance apparatus using two geometrically short magnetic field regions at either end instead of a single very long one. It was a great simplification.[10]

The small clock was highly successful, and a slightly modified version of it, the Atomichron, was put into commercial production by the National Company of Malden, Massachusetts, in 1956. The far more ambitious big clock experiment could not be made to work. Jerrold and Ray Weiss went at it hard for two years, day and night, but were never able to detect the slow atoms that theory predicted and that they expected to find. Eventually they were forced to conclude that virtually all the slow atoms were scattered and lost due to collisions with other cesium atoms. Cesium atoms turn out to be unusually large, although the experimenters had not known that when they started; the atoms collide with each other frequently and unavoidably, and the slow atoms are both speeded up and scattered into wrong directions, away from the detector.

Jerrold had had some warning of what might happen in a paper published by Immanuel Estermann, Otto Stern, and their co-workers at the Carnegie Institute of Technology in Pittsburgh.[4] The Carnegie group had also found a deficiency in the number of slow atoms in a cesium beam and had attributed the lack to atomic collisions. But Estermann and Stern's equipment had not even approached the detection capabilities of the big clock. Jerrold, Ray Weiss, and their co-workers had constructed larger ovens producing more intense beams; they had developed better detectors and were able to achieve much higher vacuum conditions. Jerrold had continued to believe that success was possible and that they would eventually detect the slow cesium atoms. They calculated that there would be about 10 million slow cesium atoms produced by the oven each second; their equipment was so sensitive that they would have detected as few as two or three atoms per second. It was difficult to believe there would not be at least a few. Jerrold summed it up in a 1971 lecture: "You can just tackle the wrong experiment. . . . Ray and I, in a way we did tackle the wrong experiment, but a lot of stuff came out of it. We tackled an experiment that was so hard it couldn't be done, by us, anyway. We were working very hard, and Ray is very skillful, so it couldn't be done. What came out of it . . . was a little auxiliary cesium clock which is now the standard of time for the world."[12]

The Atomichron, as commercially manufactured by the National Company, made use of much new design generated by Jerrold and his group.

The mass spectrometer that Jerrold had developed for the measurements of the magnetic moments of radioactive nuclei was adapted for the commercial clock; in addition to incorporating the Ramsey split-field idea, new vacuum techniques were incorporated, and a new form of titanium ion pump was developed. There were no more cast brass "torpedo tubes," with their problems of porous walls; vacuum technology had moved on to stainless steel. The clock was so well designed that it would never have to be opened to replenish the cesium; the clock used less than a millionth of a gram per day. The clock, essentially maintenance free, was a tour de force of production design.

The Atomichron immediately became the frequency standard in laboratories around the world. It proved to be so stable, reliable, and reproducible that in 1967 the cesium atom replaced the rotating earth as the fundamental keeper of time. By formal international agreement, the second was officially defined to be the duration of 9,192,631,770 vibrations of the cesium atom.

The development of the cesium clock had significant effects on Jerrold's life and on the lives of those who were involved in it with him. John King continued in molecular and atomic beams work and eventually became Jerrold's closest associate in the laboratory. They did important experimental work together, including measuring the hyperfine structure of stable bromine isotopes (the subject of King's Ph.D. thesis), and they initiated a painstaking and difficult program, in which King finally succeeded on his own to measure nuclear octupole moments.[13] King ultimately became Jerrold's scientific heir, to whatever extent such a concept holds: he took over the entire beams laboratory operation when Jerrold finally withdrew from it. Jerrold liked to refer to him as the "snappiest guy in the business." King currently holds the distinguished Francis Friedman Professorship in physics at MIT.

Ray Weiss went on to earn a doctorate in Jerrold's laboratory, as John King had done before him. Weiss and Jerrold developed a particularly close and lasting relationship, which quickly expanded beyond the laboratory to include their families. Weiss is now a full professor of physics at MIT, where he is engaged in ambitious experiments to test the general theory of relativity. In their scope and originality, as well as in their intent, these experiments are reminiscent of the most exciting aspects of the big clock experiment; clearly Weiss also has received an important intellectual legacy from Jerrold.

The National Company, which had helped to develop the production model of the clock, found considerable profit in it. The company had previously been a small electronics firm producing amateur radio equipment and specialized items for the military. Its involvement in the clock came about through the intervention of Mac Hubbard and Jerry Wiesner. Hubbard, through acquaintances made back in Radiation Lab days, had become a director of the company and in turn had nominated Wiesner for a similar membership on the company's board. Wiesner, impressed by the potential of the clock, subsequently suggested to Jerrold that he approach the company to engage its interest in the project. An arrangement was made in 1954, when work on the small clock was still at an early stage, under which Jerrold became a consultant to the company with royalty rights. National also agreed to provide some of the development funds for the small clock, and invested about $50,000 in it at an early stage. In December 1954, Jerrold became a National Company director.

Such arrangements—faculty members as directors, consultants, and strong participants in commercial ventures—serve well to demonstrate the entrepreneurial attitudes that MIT encouraged and fostered in its faculty and staff. The technological flowering of Massachusetts industry during the 1950s, if not totally due to MIT policies, was surely aided significantly by those policies.

The Atomichron, which was introduced to the public at a press conference in the Overseas Press Club in 1956, quickly became a commercial as well as technological success. Jerrold recalled: "We put them [the clocks] into commercial production, and there are about a hundred of them around the world. That was sort of a by-product. Incidentally, I made about two hundred thousand dollars out of that: royalties, fees, one thing and another."[14]

The company's success in the clock business was unfortunately not repeated; National never developed any other comparably successful product. For a time, however, it did well, and so did those involved in it. (Mac Hubbard, incidentally, did not benefit from the company's financial success. He had accepted appointment in 1954 as associate director of the Lincoln Laboratory in 1954, serving under Albert Hill. Anticipating that Lincoln Laboratory would become a major customer of the National Company, Hubbard had resigned his directorship to avoid any possible appearance of conflict of interest.)

Something to Fall Back On

Hycon Eastern was a small electronics firm located on "Research Row," at what is now the lower end of Memorial Drive in Cambridge, just a stone's throw from MIT. The company was established in May, 1955, but Jerrold and Jerry Wiesner had planned it several years earlier, in a time of some stress.

Wiesner described how it came about:

Starting Hycon Eastern was almost a direct response to our interaction with Joe
McCarthy. We got into some pretty rough tangles. . . . Cohn and Schine came
around, looking to get into the personnel files of Lincoln Lab and RLE, and they
got turned down flat.[15] Senator Saltonstall got worried about MIT tangling with
McCarthy, and actually called Jim Killian . . . and said, "I think we should patch
up this relationship." There was a meeting . . . called by Killian.

I said to my wife when I left in the morning, "I'll probably come home without
a job," because I thought Killian would look for some compromise. The meeting
started with Jim [Killian] saying "What's going on?" Saltonstall said, "This is a very
dangerous man, you people are having a scrap with him and he's going to do a lot
of damage. I think I should get you two together, see if you can patch it up." Killian
took a deep breath and said, "I've been following all this. I don't think we have
anything to patch up with him."

That was essentially the end of the conversation. So I didn't feel I had to resign.

At the same time, Jerrold and I had been talking, and this made us say, "Gee,
maybe we ought to have some outside thing to fall back on." He and I used to consult
for Raytheon, and National, but we didn't have anything of our own. Lots of our
friends had started companies and had done very well, so we decided we should
do something.[16]

He and Jerrold had already engaged in joint ventures for some time,
as he indicated. Jerrold also recalled how things had been for them in the
beginning when they had first arrived in Cambridge:

I was loaded with too many things to do—because professional salaries were so bad.
At MIT my salary in 1946 was $7000 a year and the house I bought cost $22,000,
so I had to find a way of earning a living. . . .

So Jerry Wiesner and I—we joined forces for quite a while—we talked to
Lawrence Marshall who was then president of Raytheon and he offered to pay us
$5000 a year as consultants—each—and that was pretty good. And one day we were
out there, we had dragged Ed Purcell along, don't ask me why, and they were trying
to make something called a microwave oven, at 12.5 cm. wavelength. And they
kept getting the chicken toasted too much in one place and not enough in another.
The nodes and loops of the microwave oven, the "Q" of the oven was too high.
So Purcell—it wasn't either Jerry or me but we were there—said, "Why don't you
put in a little fan that just sort of rotates slowly?" So they did. And it rotated the
nodes . . . Nodes dope, Purcell. So that paid for our [down payment].

It was natural that they would involve their friends in establishing a new
company. Rabi was brought in by Jerrold and became part of the venture,
and Ed Purcell (who, in addition to solving the "chicken-toasting" problem
had won the Nobel Prize in 1952 for experiments in nuclear magnetic
resonance) also joined in.[18] By the time the company was established, a
technical advisory board had been set up that, in addition to Jerrold, Wiesner,

Rabi, and Purcell, included R. H. Dicke of Princeton; E. A. Guillemin of MIT, R. V. Pound, of Harvard; and H. J. Zimmerman of MIT.[19] The idea of having technical advisory boards in industry was not common at the time, although it became widely used later. They had put together a stellar lineup for their board, and it was extremely effective.

Al Hill was not then available; he had just resigned as director of Lincoln Lab and was in the process of transferring to Washington, to become director of research of the Weapons Systems Evaluation Group. But Mac Hubbard, who had stubbornly declined to continue as assistant director under Hill's successor, was available; he was called on to run the new company as its president. All that was needed to get started was capital.

Jerrold was often in Washington in those days, as was Wiesner. On one of these trips, Jerrold had met Trevor Gardner, then special assistant for research and development in the Office of the Secretary of the Air Force. Gardner was a graduate engineer who had gone into business management in California after his wartime service as a development engineer in OSRD. He had been very successful at the "management of matters technical," as he put it, and had become president of a Pasadena company, Hycon, Inc., specializing in high technology and employing about a thousand people.[20] Gardner shared many of Jerrold's views about an increased emphasis on electronics for defense. He tended to be strongly in favor of any new technological development, so much so, in fact, that James Killian was moved to describe him in his memoirs as "technologically evangelical."[24] Gardner was thus cordial and willing to be helpful: he put Jerrold in touch with his friends and ex-partners at Hycon.

As Hubbard described it later:

[Hycon] had made fixed-fire rockets for the Navy during World War II. . . . they had made millions, multi-millions, but they didn't have any program when nobody wanted to buy rockets any more. . . . So they wanted an East Coast affiliate. [They] put up about two-thirds of the money we needed [in exchange for 56 percent of the stock]. . . . It was to carry us for two and a half years, but we were in the black within eighteen months."

Hubbard also recalled how valuable it had been to have a technical advisory board as distinguished as theirs: "any proposal we made to anybody in the military . . . got read immediately."

The aim of Hycon Eastern was to "exploit novel developments in communications, digital data processing and nuclear electronics." Like many other companies formed around that time, they intended to contract with the military, typically on a cost-plus-fixed-fee basis, to produce the pilot model of some device that the military thought it wanted; the development costs would thus be subsidized. If the military chose to contract for further production, well and good; otherwise, they planned to market the device commercially, which was the way to important profits.

Wiesner pointed out that there were both advantages and pitfalls in the way they had organized their talents: "Zach and I, Purcell and Rabi, we were all Board members. Our trouble was that we knew too much for a little company. We kept pushing projects that would have been great for an outfit the size of Raytheon. . . .

[For example], we made high precision crystal oscillator clocks for Avco, part of its B-70 program. Purcell had an idea for a satellite navigation system that we could develop. But it was a very much bigger project than we could handle."

Nevertheless, the company did well. It sold high-frequency crystal lattice filters at a cost of about one-thousandth of previous similar devices and highly portable digital counting clocks with a long-term accuracy of about one second in thirty years and a short-term accuracy perhaps ten times better. They made digital-readout oscillators, and Hubbard recalled that "they sold like hotcakes."

The circumstances and the times were highly favorable to the growth of such an enterprise. The system of contracting initially with the military allowed the exceptional talent in the universities to be brought together with the nation's investment resources, which often could be available only through the military. In the absence of a central agency in the United States analogous to Japan's Ministry of International Trade and Industry (MITI), this system became one of this country's principal modes of investment in new enterprise. It worked exceedingly well, to the nation's advantage as well as to that of the entrepreneurs. By April, 1957, the U.S. electronics industry had "become the fifth biggest U.S. industry, with 4200 companies, a work force of 1,500,000 and sales of $11.5 billion annually . . . From coast to coast the speed of the new giant's growth is staggering . . . New England's electronics expansion has changed the name of Route 128 near Boston to "electronics highway"; Massachusetts alone has some 500 electronics plants . . . In Los Angeles, where a new electronics plant is built every fortnight, there are already 470 companies . . . "[22]

The members of the board of Hycon Eastern were paid a flat retainer fee of about $5,000 a year, as well as consultant fees when appropriate. Hubbard recalled: "It was a bargain rate . . . they didn't ever charge for their extra time. They were all given stock options of varying amounts. They all had actual stock in a short length of time, and they were very glad for it, because the stock was purchased by them, I think they purchased the stock at ten cents a share, and most of them realized either twelve or fifteen dollars a share for it."

After some five or six years, the company was absorbed by Itek, Inc., a larger high-tech Boston-area corporation. Rabi, Wiesner, Zacharias, and the others did well in the undertaking. They had invested their specialized knowledge, their time, and their energy, and they had emerged at the beginning of the 1960s with the financial security they had sought. Jerry Wiesner felt it changed his options and his future: "I left in 1961. I sold my stock when I became Kennedy's Science Advisor. . . . It [the proceeds of the sale] made the difference, made it possible for me to go to Washington, for example, and live comfortably and then come back here [to Cambridge]. So for me it was very important. I always assumed Jerrold did about as well as I did, and that it was important for him in the same way, it opened things up for him to do things he wanted to do."

Jerrold had already embarked on a new chapter in his life, however, which was turning out to be very different from what had gone before. He was doing the things he wanted to do. Financial independence can certainly be liberating, and Jerrold agreed that it was so; but he had just as certainly not waited for it in order to move on.

The Trouble with Education

Transition

In July 1955, Jerrold traveled to Washington, D.C., to receive the department of Defense Certificate of Appreciation, the Department's highest civilian award. The citation, presented by Defense Secretary Charles E. Wilson, took particular note of two projects, Hartwell and Lamp Light, the latter having been completed only a few months earlier. As he had been on Hartwell, Jerrold was the director of Lamp Light, a major study of fleet and air defense that had originated in the work of SAC-ODM. Jerrold had managed to gather nearly one hundred outstanding participants—the "prima donnas who could sing"—and to get them to work harmoniously and productively while keeping to an unusually tight deadline.

By now he had refined his technique for running such a study and had become even more skillful at it than he had been on Hartwell, but the principles remained basically unchanged. Jerrold still insisted from the start that the problems had to be understood as fully as possible, using the widest interpretation of them. He would then make certain not to get in the way of those who had been brought together to find solutions. The approach sometimes seemed peculiar and idiosyncratic, but it was effective because Jerrold made it work by constant attention. The style might be puzzling at the beginning (meetings often had no agenda, for example, in order not to affect the outcome prematurely), but by the end, the participants themselves were often surprised at how much they had accomplished. (One of them, a Bell Labs physicist named Winston Kock, was so impressed that he subsequently dedicated a book on wave phenomena to Jerrold.) At the conclusion of Lamp Light, the participants presented Jerrold with a large silver tray with all their ninety-eight signatures engraved on it; it was a singular tribute to his leadership. There were no established prizes for the summer studies, of course. The tray and the certificate came as close as anything could to that sort of recognition.

Nevertheless, when Jerrold traveled to Washington in that hot July, he had already decided to end his involvement in summer studies, and to step down as director of the Laboratory of Nuclear Science. He had just turned fifty and felt that it was time to pause, take stock, and perhaps make changes. Once he decided that changes were in order, he did not hesitate to make big ones. Clearly he needed a fresh challenge. He had developed a personal philosophy by this time that expressed this feeling: "If everybody's doing it, I seem to want to do something different. And it's not out of orneriness, but it's just out of the practical view that if everybody's doing it they don't need me."[1]

Jerrold had been director of LNS for ten years. It was now a mature, well-established, and important part of MIT, and the administration of it no longer offered much of a fresh or creative challenge. He had met whatever crises had arisen, large or small; year after year he had managed to secure the substantial budget that the laboratory required, and he had dealt on a daily basis with a host of smaller, but no less important problems as well.

It is the beginning of any enterprise that is generally its most exciting time, and for this laboratory the beginning was over. Whatever problems might arise for it in the future, solving them would not be as challenging and satisfying as getting the operation started in the first place. Leona remembered a minor incident from about that time, recalling a Saturday afternoon when she and Jerrold were walking in Harvard Square. They had met one of the senior group leaders from LNS, who took the occasion to pour out to Jerrold the tale of some minor administrative woe. Jerrold promised to take care of it on Monday. There was no more to it than that, but Leona remembered Jerrold's wistful comment at the time, reflecting the wish—it seemed at the time only a passing wish, she recalled, but it was almost certainly something more—for a day when he might have no caretaking problems to deal with.

He needed new interests, and the need extended beyond his professional life. In September, 1955, while he was still mulling over what to do next, he and Leona were invited by a friend to look over an unusual Cape Cod property that had come on the market following Hurricane Carol a year earlier—a twenty-room house that had been badly neglected, on eight acres of land, located on Tobey Island in Buzzard's Bay. They bought the house as a vacation home, and spent the next several years reconditioning it.

Jerrold took up sailing, one of the pleasures that came with living by the sea, and also began to discover for almost the first time the pleasures of recreational reading. He had not previously been a great reader, but he now developed the practice of steeping himself thoroughly for a period in some one particular subject or writer in which he was interested: at one time it was the Bible, at another it was Shakespeare, and when he was learning to sail, it was books on sailing. Until his interests moved on to something else, he would read only in that literature, immersing himself in it night after night. Even his figures of speech would often reflect what he was reading; his colleague David Frisch gleefully recalled an example of his Shakespeare period: "Jerrold's literary humor was absolutely non-pareil . . . they were discussing this guy, this student, in those days we had time to talk about them one by one, and Bruno [Rossi], who was always 300% loyal to anyone that ever went near him, told about the wonders of this guy's thesis. Jerrold got more and more irritated sitting next to him, and he whispered to me "I hate it that he brings down the rate of usance for us merchants here in Venice." It couldn't be any more apt, for Bruno was a Venetian Jew to begin with . . . it had everything."

Notwithstanding all his new interests, Jerrold was not at all ready to enter semiretirement and become a squire-by-the-sea, or anything remotely like it. Mostly, he thought about going back to his laboratory, and by November he was ready to take the step. He wrote to Julius Stratton, the MIT Provost, and to George Harrison, the dean of science, resigning his directorship of LNS. His letter was unusually introspective:

Dear George and Jay:

For the past several months I have been trying to make up my mind about how I want to conduct the remaining years of my academic life . . . I have, in the last few months, reduced to a low level the amount of energy and time that I spend on consulting and on government committees.

I find that the work of my molecular beam laboratory, instead of going stale and inappropriate to the time, is more than ever what I want to concentrate on. Since experiments of the sort that I am doing require that the experimenter and the apparatus be almost one, I want to spend as much time in the laboratory as is consistent with my duties as a teacher.[2]

Jerrold's laboratory had become an increasingly exciting place. He and Rainer Weiss were working on the big clock at that time and were still hopeful that they could be successful with it. If they were, Jerrold planned to take two clocks to Switzerland. With one fountain-clock on a tall mountain in the Alps and the other near sea level, he would be able, given the clocks' extreme accuracy and stability, to measure the minute effect on the clock rates of the very slight difference in gravity at the two sites. It would have been by far the best test of general relativity as of that time, and the first terrestrial test of the theory.

Robert Vessot, who later became a world expert on the precise measurement of time, recalled a colloquium talk given in 1955 by Jerrold at McGill University:

Zach was a spellbinder . . . he had this idea of using the fountain clock to test general relativity. It was fantastic. I wound up going with him and John King to the airport . . . several months later I was invited to MIT . . .

The four years there were tremendous. Zach had this knack, this talent, of inspiring others to outdo their own capabilities. . . . we worked, often till 2 AM or later. We loved it. . . . Zach had the gift of getting people to work like dogs and to enjoy it. And it really was a joy, there was no other word for it.

There were many other things going on in the lab besides the development of the cesium clocks, and the atmosphere was lively and stimulating. The excitement and happiness that had been characteristic of Rabi's lab in the thirties, and to which Jerrold had contributed so greatly, were in good measure recreated in Building 20 at MIT during the fifties. Enthusiasm carried Jerrold and his students along. They continued with the program of measuring magnetic moments and added a set of experiments to verify precisely the Breit–Rabi transitions in cesium as a test of whether quantum mechanics holds up in such cases. John King carried out impressive experiments to test the electrical neutrality of uncharged systems like the H^2 molecule, and he was able to set a new limit on the absence of charge in such systems. Vessot worked on the problems of using cesium to build an ultrastable low-temperature microwave oscillator; these experiments involved introducing helium as a buffer gas into a microwave cavity made of superconducting lead. This was all highly sophisticated physics, very much at the cutting edge of research.

In his letter to Stratton, Jerrold had stated as clearly as he could that he wished to be fully engaged with his laboratory and his teaching, and for a time he realized his wish. He was involved with everything that happened in the lab. Vessot, looking back more than thirty years later, recalled his sense that Jerrold had "the gift of a simply tremendous insight. . . . he had a marvellous insight for tinkering." Whenever there was a problem, he always seemed to know exactly where it was.

And yet all the time this was happening, behind all the excitement and enterprise of research, something else was brewing in the background. Quietly at first and then with rapidly increasing vigor, Jerrold became engaged in an activity entirely new for him but one that would eventually pull him completely and irrevocably away from the laboratory: education and its reform.

It is difficult to know precisely why educational reform so appealed to him. He liked teaching, of course, but all his experience had been at the college level, while nearly all of the projects into which he would soon throw himself passionately and completely involved reform at precollege levels. Surely there was more than one reason. Serving on SAC-ODM had provided one stimulus: "The American military would come in [to SAC-ODM meetings] and complain that the Russians were getting ahead of us, that we had to do something about education, about teaching—getting more engineers, more scientists—this, mind you, all before *Sputnik*. . . . I kept saying that the major trouble was the cultural level of the country. There was a committee set up to try to figure out what should be done—and never mind the names—they didn't recommend anything."[3]

Jerrold's opinions of those who could see a problem, examine it, but then not do anything about it were always less than charitable. His frustration with SAC-ODM on this issue alone might have been sufficient to get him started, but in addition, he was teaching undergraduates at the time, and it bothered him greatly that they had so little appreciation of physics as a primarily experimental science: "What was interesting to me was to try to persuade the MIT physics students that physics was not just theory, that physics *really*—underline *really*—was what you do either in a laboratory or by observation. And that theory is important, but only a piece of what goes on. . . . An experiment, people say that's what you do with your hands; the truth is, you do it with your head."[4]

This was a point of view about which he had always cared deeply and was always prepared to defend, even when it was not under attack. He was ready to enter into science education as a reformer, an apostle of hands-on experience for students and teachers alike. He felt he could see what was missing from science education and thought he knew how it had to be changed.

There was at least one more factor for him to consider, which reflects on the nature of a rapidly developing field like modern physics. He recalled later: "I was running the Laboratory of Nuclear Science at a time when it was just shifting from physics in the one- to eight-million volt region to very high energy physics: namely, the physics of three hundred- to one thousand-million volt machines. That would take a lot of thinking . . . " Physics was moving on into new areas. Experiments at unprecedented high energies would reveal new and complex phenomena; the main thrust of exploratory physics was moving further and further from his own area of expertise. For the first time he questioned whether he could keep up with the younger people in the field and still manage to do all he wanted to do. More accurately, he wondered whether he still wanted to invest the amount of personal energy it would take. On the whole, he told Leona, he thought not.

A Simple Memorandum

Do you realize that I had graduate students for only ten years? From 1946 to 1956. That's the sum total. And in 1956, I said, I've got a lot of experiments to do, and they're good. Will somebody else do them if I don't? And I said, Yes. They'll get done. In the eye of eternity, the stuff I want to do will get done, but education is a mess.

—J.R. Zacharias[5]

When Jerrold sent a memorandum to James Killian in March 1956, entitled "Movie Aids for Teaching Physics in High Schools," he made sure his proposal was on a sufficiently grand scale to warrant the attention not only of MIT's president but of anyone to whom he might show it. He wrote: "In an effort to improve the teaching of high school physics I want to propose an experiment involving the preparation of a large number of moving picture shorts. In order to present one subject, say physics, it is proposed that we make 90 films of 20-minutes duration, complete with text books, problem books, question cards and answer cards. Each of these points requires some discussion." (The memorandum is reproduced in appendix B.)

Jerrold had never himself made any sort of film before, and he knew little about the process, but he knew enough to know that quality would be essential: "Success or failure depends to a large extent on having the entire apparatus of the experiment really right. Like a high fidelity phonograph, one must have besides the machine a good piece by a good composer played by an artist. The room must be good, not too noisy, and the people have to want to listen, but that all depends on the piece."

Naturally, he knew how to begin: with a summer study. Only a few years later he described how he thought about them: "The basic concept of the [summer] study is that a fairly large group of able men, from differing disciplines, are likely to gain new insights into a problem if they themselves are made responsible for defining that problem, and are encouraged to seek solutions by means of discussion and debate. Thus, a summer study almost automatically breaks into two phases: the first during which the participants convince themselves and each other that they know what the problem is; the second, when they program steps to meet it."[6]

Jerrold said once that every planning activity he ever ran after 1950 was modeled after Hartwell. Beginning anything was straightforward, if you proceeded in Hartwell fashion: "You start in the upper left-hand corner of the blackboard . . . "

He was casual about the cost of what he was proposing to Killian: "The financial side of such an enterprise becomes obvious on considering the amount of film that will have to be thrown away before a first rate reel is accomplished. Copies of the film are easy. But the efficiency of Flaubert was not high."

Jerrold's attitude about the cost of the project was realistic. He would not yield on quality in this context any more than he would have done in building a piece of physics apparatus. He remarked often that quality costs money, and he believed it was only sensible to expect to have to pay for it. He thought it best to acknowledge that right from the beginning; otherwise, a tendency to cut corners and make compromises would inevitably develop. He was not willing to compromise when it came to quality.

On one of his trips to Washington that summer, Jerrold showed a copy of his memorandum to Alan Waterman, the director of the National Science Foundation; he and Waterman knew each other well from the latter's days as chief scientist of the ONR, when Jerrold had served on ONR's Nuclear Science Advisory Committee. The memorandum immediately interested Waterman intensely.

The National Science Foundation, then just five years old, had as part of its charter an obligation to deal with science education, primarily for the purpose of producing more and better-trained scientists and engineers. A portion of the Foundation's still meager budget had been designated for such purposes. A small series of summer institutes had been initiated by NSF two years earlier to train science teachers, but so far, promising initiatives dealing with new content or methodology had not arisen.[7] In the case of high school physics curricula, it had been fifty years or more since anything had changed. Waterman and his staff looked forward rather glumly, to spending scarce funds on what they suspected would turn out to be mostly unrewarding activities.

The concern about the state of education was widely shared. Bowen Dees, then head of the Fellowships section of NSF, remarked later; "Even when you were working on fellowship business, if you talked to a professor for more than thirty minutes or so, the topic of the miserable state of teaching in the high schools would open up."[8]

There was growing pressure on the foundation to act. In 1954 President Eisenhower had set up a Special Interdepartmental Committee on the Training of Scientists and Engineers, and the studies made by this group had revealed a serious shortage of trained scientists and engineers in the work force. The committee emphasized the shortcomings in American science education, especially in comparison to the Soviet Union; these were worrisome, since the cold war was becoming increasingly technological in character. The committee report led directly to the first White House

Conference on Education at the end of 1955, and this was followed in turn by a strong message on education to the Congress from the president. Congress clearly expected the foundation to get the message as well.

Jerrold's proposal probably seemed to the men at NSF like the answer to a prayer. It was unorthodox—without precedent, in fact—and presented some interesting new challenges for a government agency. Could the foundation support activities like film and textbook production, normally handled by the private commercial sector? Should the government be in the textbook business at all? Were there not serious ethical issues involved in that? One has only to replace the word *science* in the proposal with the words *political science* to see what sort of quandaries such issues represented.

Nevertheless, the proposal came from a recognized scientist of top-notch quality, known and admired by Waterman and his associates, and it bore the endorsement of the leaders of no less an institution than MIT. Jerrold had an impressive track record; when he undertook a project, it succeeded. Waterman and his colleagues decided on balance to take the risk, trusting that all policy problems could eventually be resolved.

The assistant director for scientific personnel and education at the NSF in 1956 was Harry C. Kelly, an experimental physicist who had worked at the Radiation Lab during the war; he had met Jerrold then but knew him only slightly. The summer training institutes for science teachers were his responsibility, as were questions relating to content and curriculum. At Waterman's urgent suggestion, Kelly immediately arranged a meeting with Jerrold at the Cosmos Club in Washington, and they met there on the evening of the very day that Waterman had first seen Jerrold's memorandum. In an interview some years later, Kelly remembered that in addition to himself, Jerrold and Waterman, Lee DuBridge and several others had also been present and that "they were all enthusiastic about this [education] thing." He himself was equally enthusiastic: "We started the summer institutes, and so on, and then we were looking for some way of attacking the course content thing, and we were kind of lost. And I was worried about one thing. If it ever got to be mediocre, it would be really a dangerous thing for the rest of the country . . . And it just happened that Zach was interested in the same thing. So we just jumped on it, like that. A physicist from MIT, respected guy, boy, you're our man." Jerrold recalled vividly the message Kelly brought him: "Harry Kelly said one very important thing; he said 'Zach, you've got to do this, you can't just turn it over to other people.' He said, 'You've got to use up all the money we've got, otherwise it'll be invaded by second- and third-raters.'" Jerrold continued: "[When] Kelly said, 'You've got to do it,' I said, 'If you want to persuade me, you must persuade Killian and Stratton first' . . . Kelly said that he and Waterman would come up to MIT and have a meeting with [them].

I remember Stratton's words: he said to me, 'Well, what are you doing this year, Jerrold?' I said, 'I've got a very simple arrangement with the physics department, I teach full-time when I teach and I do research full-time when I do research. And he said, 'What are you doing this semester?'—this was in August—and I said 'I'm teaching this first semester and research next semester.' At which Jay Stratton said, 'Well, you're teaching high school.' And I said, 'Okay, I'll give it a three months' try.' And I've been doing it ever since."[9]

The timing was right, certainly from the point of view of Waterman and the NSF, and the prospects for funding were good. Arousing interest in these ideas was proving to be relatively easy. Nevertheless, what Jerrold had in mind was still only a project, not yet a cause. It would become that only later.

Gathering the Forces

I have a funny combination of arrogance and humility, by which I mean . . . I seek to bring around me people who are a lot smarter than I am.

—J. R. Zacharias[10]

Those of you who know Dr. Zacharias know when he asks you to undertake some mission you will in the end agree, and delay is only a waste of time and words. Dr. Zacharias is a man who, if he were to discuss the weather, would finish not by just talking about it, but he would be doing something about it.

—I. I. Rabi[11]

Instead of sketches of circuit diagrams and bits of equipment, instead of formulas and tables of numbers, Jerrold's notebooks now began to be filled with lists of people. He was setting off to see what could be done about the state of education in this country, and he was starting in his own characteristic way: by gathering together the people he knew could help.

He was singularly well equipped to find the people he wanted: "I [had] an extraordinarily extensive acquaintanceship, from Los Alamos, and I worked at the Bell Labs, we had 4,000 people at the Radiation Lab . . . I just knew all kinds of people."[12] and was confident that he could recruit them. The question of their availability would arise only later, and often not at all, for he had developed an extraordinarily valuable knack of getting people to join with him in his ventures. Even those who were much in demand elsewhere, who were busy and otherwise involved, could be drawn into his projects; as he often said, it was the people who were "too busy" that he really hoped to get. He could convince a listener that exciting and important opportunities lay waiting; it was all the more remarkable because he had no special eloquence. He could at times fashion a sentence so involuted that, like Mark Twain's speaker of German, there seemed little hope of his emerging whole at the other end of it. Nevertheless, his message got through: he was offering the opportunity to be involved in extremely important problems and to help decide what should be done. The appeal was more than most people could resist. Later many of them could not be exactly sure how he had managed to be so persuasive, but no one ever claimed that the opportunities had been less than he had represented. Afterward, a surprising number said that he had changed their lives.

Not everyone could afford the degree of commitment that Jerrold's projects demanded. Some may have found themselves too deeply engaged in other projects that they found more compelling. Probably timing was a critical factor too. His appeal was particularly strong, according to colleagues' recollections, if it came at a moment when someone was trying to decide what to do next, what new thing to try. At such moments, Jerrold could offer a new, unexpected path and could hold out the promise of something worthwhile at the end of it. "What have you got better to do?" he might ask; but he would ask it in such a way and at such a moment that the question demanded an honest, soul-searching answer.

He began close to home, with his own physics department. With Francis Friedman, he had been co-teaching an undergraduate course, "The Experimental Foundations of Quantum Mechanics" (but invariably referred to by MIT undergraduates by the less romantic name of Physics 8.05). Although Friedman was fifteen years younger than Jerrold and although he was a theoretical physicist who had never carried out any experiment, a strong resonance developed between them; they communicated easily and well, finding in each other a ready mutual understanding. In the process of teaching together, Jerrold had come to admire Friedman greatly. Jerrold recalled: "Friedman and I had a very good lecture system. When one of us was lecturing the other one would sit somewhere in the room, not clear in the back, but somewhere in the room, and interrupt when he didn't like what the other one was saying . . . it takes a certain amount of gall to stand there with the chalk in your hand, talking away and have someone in the back of the room say 'But Jerrold, you're getting it all wrong!' Or, 'But Francis, my God, why don't you say it simple, hm?' And he says, 'Here's the chalk.'"[13]

Friedman quickly enlisted in Jerrold's new education project. From Jerrold's point of view, hardly anything could have been better. Friedman was considered by all of his colleagues to be an outstanding physicist, possessed of an acute, analytical, and independent mind. He came from a wealthy and cultivated New York family; his father had studied for the rabbinate but had become a banker and a collector of antiquities. Prior to Friedman's involvement in this new education program, he might well have been regarded as something of a dilettante. He was capable of hard work but according to those who knew him seemed to lack creative drive; his considerable abilities were more those of a critic than of an originator. He published little at a time when a high rate of publication was the key to many sorts of rewards; but his colleagues were inclined to attribute that to the very high standards that he imposed on himself. Once he had joined Jerrold's project, however, he threw himself into it with great zeal and energy. It was as if he had found his career line at last. Among all of the brilliant and highly creative men and women who responded to Jerrold's call, Friedman would be the person on whom Jerrold would come to rely most heavily for critical and independent judgment.

By late August 1956 a proposal to initiate educational reform had been submitted formally to NSF by Julius Stratton on behalf of MIT; it consisted almost entirely of Jerrold's memorandum, reproduced nearly verbatim. Waterman had promised rapid action and, in anticipation of a favorable response, Jerrold convened a small meeting at MIT on a Saturday early in September. Those who came included Ned Frank, Jim Killian and Rabi, in addition to Jerrold himself and Friedman. Rabi at the time was a visiting institute professor at MIT, so it was convenient for him to be involved in these early days, and, of course, Jerrold valued his advice highly. Killian brought along Edwin Land, the inventor of the Polaroid process, a man whom Killian admired greatly.[14]

They met from ten in the morning until four o'clock in the afternoon; it was Labor Day weekend and the halls of MIT were empty and quiet by the time they left. By that time, they had decided to make a pilot film on Newton's laws; to prepare a teacher's manual; to establish a curriculum committee chaired by Francis Friedman, to map out a two-year course in physical science; and to aim for a target date of February 1, 1957, to make recommendations for ways of implementing their ideas.

The members of the group quickly recognized that the problems of secondary science education could not be remedied solely by a set of films, however brilliant they might be, and spent considerable time in subsequent meetings identifying the broader needs and discussing how they might be met. The group was aware of a report, for example, that had been prepared in early June by the respected physicist and teacher Walter C. Michels, of

Bryn Mawr College, for the Joint Committee on High School Teaching Materials.[15] In it, Michels publicly expressed his dismay by refusing to rank the various existing textbooks in their order of merit, as he had been asked to do, because he was so completely dissatisfied with all of them.

Item by item, the list on Jerrold's blackboard of possible activities began to grow. The enormous scale of the problem began to emerge as well; the numbers describing it were formidable and daunting. By their second meeting, they had a tentative estimate of the audience: "the entire secondary school population [of the United States] that takes chemistry (600,000) and physics (300,000)." In an interview some years later, Bowen Dees (who, in addition to having been head of the Fellowships Section at the National Science Foundation, had been Harry Kelly's right-hand assistant there) spoke to these magnitudes in connection with the training institutes that the NSF had been sponsoring: "One of the things that held us up was the sheer size of the problem. While the high schools obviously needed help, there was a feeling within NSF that an attack on the problem when so little money was available would be almost trivial and worthless. The argument to do something finally won out, however."[16]

Jerrold was not at all intimidated by the numbers. Education, he noted, was about a $24 billion enterprise in the United States, but he was audacious enough to believe that it could be moved by good ideas, for only a tiny fraction of such a sum, less than a few tenths of a percent, in fact. The group of scientists he was gathering would produce those ideas. His little committee was beginning to think of education from a systems engineering point of view, he said, and that was something that had never been done before.

At their first meeting that September, they chose their name—the Physical Science Study Committee (PSSC)—and established themselves as its steering body. They considered in those first meetings others who might be added to the group, and determined to meet on a weekly basis until further notice. By the next meeting on September 14, Edward Purcell of Harvard had responded to Jerrold's call, as had Martin Deutsch of MIT. Both became members of the steering committee; Purcell even volunteered to make a movie on the nature of light. Jerrold was able to announce at that meeting that not only had the NSF earmarked $300,000 for the project, pending formal action, but that $1.5 million was being set aside for the year following "for a practical program, should one be recommended."[17] It was a substantial part of the NSF budget for education and a clear indication of the importance the foundation attached to the project.

It was also very clear that the project had to draw from a larger pool than "Greater Cambridge"—and not only to justify its federal funding. If it was to have a nationwide impact, it had to have national scope from the beginning. Jerrold agreed to recruit at Cornell, Cal Tech, Bell Labs, and the University of Illinois, where there were already some signs of interest. Further, he and Land would seek to draw in a group of chemists. In this they were unsuccessful; the chemists were the only ones who turned him down.

At Cornell, he first got in touch with Hans Bethe, who had been director of the Theoretical Division at Los Alamos during the war. Bethe expressed his own interest and suggested that Philip Morrison, also at Cornell, be approached. Morrison was one of a remarkable group of physicists who had obtained their doctorates under Robert Oppenheimer before the war; he had also been at Los Alamos and had, in fact, flown in one of the first reconnaissance flights over Japan only a day or so after the Nagasaki atomic bomb had been dropped.[18] Jerrold had met Morrison shortly before that, he recalled, on a flight from Albuquerque to Wendover Air Force Base in Utah, in the belly of a C-45 aircraft—a wartime encounter which was only mildly out of the ordinary in that extraordinary time but one that impressed itself deeply on Jerrold.

Morrison described how Jerrold had gone about recruiting: "I believe . . . he called about a dozen persons he knew, who were in key positions, influential positions, among physicists in a number of universities and industrial establishments . . . and he put to them the proposition that he had determined to try to do something about high school education in physics and would we help. And what he asked was that we find a group of people to have dinner. He would pay for the dinner and we'd come together and have dinner and try to recruit from this group a number of persons who would stick with it and do the next step in the form of sending him a document with ideas of what ought to be done."[19]

Morrison was both a prodigious worker and a gifted writer who had just published, with Bethe, a particularly lucid and valuable text on nuclear theory. He would quickly put these exceptional talents to work for PSSC and would become, alongside Friedman, a primary intellectual source for the project.

Those who joined the project were not all trained scientists, nor were they all people Jerrold had previously known. His appeal turned out to extend well beyond the domain of science, and his personality turned out to be as fully effective among nonscientists as it was among scientists.

One nonscientist who joined was Stephen White, a journalist who had covered the postwar development of atomic weapons for the *New York Herald Tribune*—as he put it, from "Hiroshima to Bikini"—and he had developed a wide acquaintanceship among the Los Alamos scientists, including

Oppenheimer and Rabi.[20] He had spent some time in advertising work and
a year writing scripts in Hollywood. Rabi saw White in New York from
time to time and thought that he might have something to offer to this new
movie-making enterprise. No one was more keenly aware than White of
how limited his film experience was, although it was no doubt more than
that of anyone else who came on the project in those early days. In the late
summer of 1956 White was unattached, having just returned from writing
a film script in Paris. Rabi told him about Jerrold's ideas. There was no
funding for them yet, Rabi said, nor were many people yet involved, but
he believed his friend Zacharias could use some help. Would Steve go up
and see him?

White came to visit Jerrold at Tobey Island that August, with the house
still under repair. He found a warm welcome amid the chaos, and he and
Jerrold hit it off. Jerrold told him about the PSSC project, saying, as White
recalled, that he was giving it "one of his full times." White decided to give
it one of his own "full times," without pay at first or even travel expenses.
He began to commute regularly between New York and Boston, to work
with Jerrold. White was among those who would later say, speaking of
Jerrold, "He changed my life."

White was a generalist, a manager, and an expediter, and he could write.
He was a fertile source of ideas and was very well qualified to serve as a bridge
between the world of the scientists and what to them were the curious and
unfamiliar worlds of filmmakers, publishers, and even advertisers. He made
PSSC his principal commitment from the first; in time, others would develop
commitments equally large, but in this, White was the first.

By the end of that September, Jerrold's recruitment efforts were bearing
fruit. The names of Vannevar Bush, Morris Meister, and Henry Chauncey
had been added to the steering committee. Bush at this time was the highly
influential chairman of the MIT Corporation, and his participation gave
additional evidence of the importance MIT attached to this work. Meister
was the principal of the Bronx High School of Science, a highly regarded
school devoted to the scientifically gifted children of New York, who would
be an important component of the audience for the program. Chauncey was
the president of the Educational Testing Service, representing yet another
important aspect of the educational enterprise. If the work of PSSC was ever
to be accepted, new ways of testing and evaluation would have to be
developed. It was essential to have representation from that quarter.

By the end of October the working group had expanded to twenty,
most of them on the steering committee. Groups at CalTech, Illinois,
Cornell, and Bell Labs were discussing the problems and preparing reports

for a conference to be held at MIT December 10–12 at which it was hoped they could come to definite decisions about what to do.

PSSC Gets Going

I said, Come to the meeting with your mind made up as to the way you want it, and we'll battle it out from there. So the walls bulged and the ceiling went up and down and . . . we ended up with a reasonable outline of the way we wanted it to go in about three or four days; I think it was only three, but it felt like four.

—J. R. Zacharias[21]

Every project that Jerrold had ever organized ended by going beyond its original terms of reference, and PSSC was no exception. Jerrold's projects did not, after all, start off looking for solutions to problems; they started off by identifying the problems and they always found more of them than had originally been anticipated. Accordingly, when the first PSSC conference took place in December 1956, the purpose was not simply to discuss which films to make or how to make them. Encouraged by Jerrold in the weeks before the conference, the study groups had already gone well beyond the framework of his original memorandum to Killian. In any case, that had only been a starting point. Jerrold said later that he had not even discussed the memorandum with any of his colleagues and that he had mainly intended to be provocative.

By December, many of the participants had done some preliminary investigation and were ready to go further. There was little argument about the depressing quality of what they were intending to replace. Walter Michels was there to repeat the views he had expressed in his report, and some of the study groups had taken the opportunity to look at some high school textbooks for themselves; the consensus was that they were poor. Philip Morrison put it this way: "The books seemed very bad and it seemed quite clear that there was something we could do . . . when I heard the question [posed by Zacharias] and looked at the books, I agreed that the public education in high school was very much behind the times and very much against the spirit that I thought was good for science . . . it seemed to me an important public duty to try to do something about that."

But the problem was not only textbooks. The entire approach could be seen to be in trouble; high school systems, undoubtedly struggling to do their best, were caught in a kind of trap. The report of the conference identified the nature of the difficulty: with not enough time to do everything well, the high schools did nothing well: "The amount of accumulated physical knowledge has grown rapidly, but the time available for teaching it in high school has remained the same . . . Since the course cannot illustrate the development of ideas for shortage of time, it is filled only with the results of physics and laws to be learned by rote . . . It becomes hard to understand, and of limited interest. To enliven it, technological applications are often added and thus the bulk of material to be learned is further increased."[22]

What was needed was something new. Rote learning would have to diminish sharply, and first-hand experience with simple apparatus would have to be introduced. Modern physics, as Jerrold had insisted in his original memorandum to Killian, was concerned with fundamentals. It had to do with particles and the forces between them, and with their motions, not with pulleys and levers. Modern physics dealt with atoms and molecules, and with stars and planets; the machines and engines that were central features of the existing physics courses were important but only as special applications of the science. On this there was the broadest agreement possible: if something had to be dropped for lack of time, it would be the applications. The fundamentals of the science must remain. Consequently, hardly anything from the old system was worth salvaging. Everything needed to be changed: the material, the methods of teaching, the attitudes, and the expectations. For three intense days they argued and debated how to go about it.

There were two opposing points of view, championed principally by Rabi and by Morrison. Rabi favored basing the new course on only a few themes, each developed in considerable detail, demonstrating the enormous power of what he liked to refer to as the "culture of science." Morrison preferred a broad and encompassing approach that would show the many different kinds of argument and the many different kinds of phenomena that are successfully unified in physics. Morrison recalled the outcome: "The decision was to vote for them both. So, in fact, the first part, which I wrote, is the enriching one. And the second part, which came very much from what Rabi talked about, is based on light and the ripple tank, and, from his point of view, it took off from the idea of the refractive index. That was the thing he emphasized."[23] The meeting was "noisy," Morrison said, but instructive and highly enjoyable. Jerrold described it as "that first big stormy conference." But they ended by agreeing on what to do and how to go about it.

What they agreed upon was a proposal for an entirely new course in physical science, to be based on four texts that would be drafted in the coming months. Altogether the course would be less inclusive than what it replaced; some of the traditional material would be omitted. There would also be an extensive set of monographs, perhaps fifty or sixty of them, for students would also need a rich supply of collateral materials to read, not only on technology but on biography and the history of science as well. There would be kits for students as well as for the classroom, containing materials and instructions for building and using simple devices: lenses, prisms, and whatever else might be necessary to enable a student to make a telescope or a camera or something of the sort. There would be manuals for the teacher, of course, and there would be the films, which would take up about one-fifth of the available classroom time but which would be available for students and teachers to view as often as they found useful.

Work started immediately; if people had other serious commitments elsewhere, they adjusted. Jerrold's policy of looking for people who were already busy paid off wonderfully well. Incredibly, the first draft of volume 1 of the text was completed by that summer of 1957. It was called *The Universe* and it was written by Philip Morrison. Jerrold was deeply impressed. "Morrison was of tremendous importance," he said. "He drafted some of the tough stuff, which broke the ice."

Volume 2, on *Optics and Waves*, more representative of Rabi's views, was well along by that summer, although Rabi did not participate in the writing, and so were the other two texts: one on mechanics and one on electricity and atomic structure. The course had shape and was beginning to have substance. PSSC was well under way.

Observation, Evidence, and the Basis for Belief

My first question always is "How do you know?" The American public's first question is "Who said so?"

—J. R. Zacharias[24]

It is possible to make a pretty good weighing device, in the form of a balance, using a couple of soda straws, a few pins, and a moderate amount of ingenuity. Jerrold liked to say that he could weigh a fly's wing with such a device, and he probably could have. A reasonably careful soda straw balance such as a high school student might make would probably not do quite that well; perhaps he might only be able to measure to the accuracy of, say, half a dozen flywings.[25] But that is still an impressive degree of accuracy, and any student who struggled to achieve it would almost certainly emerge with a new respect for the meaning of measurement.

That appreciation was the real point of PSSC: not accuracy for its own sake but rather for the understanding that it made possible. The new course was intended to present knowledge about the physical world in comprehensible and interesting ways; equally, it was designed to demonstrate, explain, and whenever possible to give first-hand experience in the ways that knowledge can be obtained. When successful, such experiences might provide a student with a sense of the playfulness that often characterizes good science and with the delight of discovery—perceptions almost totally missing from any public view of science. At bottom, the goal of PSSC was to make students familiar with the various modes of scientific reasoning. The facts of science would serve principally as the means by which the goal was to be reached. Certainly one of the most fundamental purposes of the course was to help the students know when they knew something and when they did not. This is what the members of the committee meant when they spoke of wanting to develop an appreciation of science: they meant developing in the student the ability to distinguish knowing from opinion; to understand the meaning of probability; to recognize that uncertainty necessarily accompanies all observation and measurement, and to reject the false certainty of dogma.

This is not to suggest that those who worked on PSSC physics sat around much of the time discussing epistomological questions. They tended to be practical people, and they were generally more concerned with practical matters. Nevertheless, the questions of whether and how well one knows what one thinks one knows are never far from a physicist's mind. The importance of such questions normally becomes evident early in a physicist's education, even if the questions themselves are not explicitly raised in the classroom.

The goal of PSSC physics was thus a cultural one, as Jerrold frequently pointed out, rather than a purely technological one. In the minds of the committee, what they were trying to do was analogous to developing an appreciation of music, art, or literature. It was commonly understood, they said, that one has to work at the liberal arts, to devote time and effort to them in order to get anything out of them. Educated people recognize the value and importance of the liberal arts, but only rarely do they appreciate science in the same way. People who would never countenance illiteracy in the liberal arts incline to a vast, unthinking tolerance of it when it comes to science. The committee had observed that parents, confronted with a child's inability to succeed at science in school, all too often react with an indulgent, "That's OK—I never understood science either."

Beyond fostering an appreciation of science, working on PSSC offered something else peculiar to the times. Jerrold reflected on this some years later:

Having lived through World War II, Hitler, Stalin, Joe McCarthy . . . it was perfectly clear that you had to get used to the notion that you have to understand why you believe what you believe . . . where is the evidence? Show me . . .

The reason I was willing to do it [PSSC] was not because I wanted more physics or more physicists or more science; it was because I believed then, and I believe now, that in order to get people to be decent in this world, they have to have some kind of intellectual training that involves knowing [about] Observation, Evidence, the Basis for Belief.[26]

Only a scant three years earlier, in 1953, such matters had struck close for Jerrold. MIT then was deeply troubled by the difficulties some of its faculty members had experienced when called before congressional committees to testify concerning their political beliefs and actions. Naturally, their difficulties were problems for MIT as well: the institute came under considerable pressure at times to abandon its principles. As one measure, President Killian had appointed a faculty committee to offer advice and counsel to those colleagues who might want it; the committee had been chaired by Jerrold, who thus became a close and personal witness to the damage McCarthyism caused. He was harsh about it; "I hated Joe McCarthy," he said. He believed that someone who did not act in accord with the spirit of science [he would offer several of the witnesses in the Oppenheimer case as prime examples] should simply not be called a scientist. The spirit of science was a demanding one: "We live in a world of necessarily partial proof, built on evidence which, though plentiful, is always limited in scope, amount and style. Nevertheless, uncompleted as our theories may be, they all enjoy, in a sense, the benefits of due process of law. Dogmatism cannot enter, and unsupported demagoguery has a tough time with us."[27]

PSSC physics had no political content. Jerrold was quite specific about that nearly twenty years later: "We didn't want to mix social thinking with a physics curriculum, and I wouldn't to this day . . . We didn't want to mix our political opinions, our Weltpolitik, with physics."

But science does represent a way of looking at the world and of thinking about it that seemed salutary in those difficult times. As the country began to emerge from that unhappy period of intolerance broadly symbolized by Joe McCarthy, it was little wonder that so many scientists were willing to join forces in developing PSSC physics. Even those who may have given small thought to the political realities of the time must have felt the prevalent anti-intellectualism and must have welcomed the prospect of fostering rational thought in schools. For the scientists, PSSC was like a breath of fresh air.

There were many other features of PSSC that were attractive to scientists. Improvising inexpensive and ingenious equipment, such as the soda straw microbalance, thinking through how to make atoms and molecules real and believable, filming complicated but interesting physical phenomena—all had enormous appeal. But these were challenges of a sort that physicists might easily have found elsewhere, in their laboratories, their own classrooms, and their workplaces. They did not altogether account for the optimism that infused the project from its beginning. Perhaps the physicists sensed that they were involved in something special and important and were glad to be engaged in it.

Two decades later, reflecting on the curious and abrupt turn his life had taken, Jerrold commented on that flash of insight that he had experienced while on the President's Science Advisory Committee. Listening to the military complain about how the Russians were getting ahead of us and fresh from thinking about how to lead a decent life in a troubled time, he had suddenly felt that these problems were rooted in the poor kind and quality of education that was available:

I believe that education is, democracy or dictatorship alike, of utmost importance. You say, Isn't it true that a large part of American anti-intellectualism comes from people who have been graduated from colleges? But I didn't say that colleges are educational institutions . . . Most of the people there don't understand Observation, Evidence and the Basis for Belief.

Now, I'm not saying that an education in physics, in experimental physics, is enough, nor is it necessary . . . one of the most liberal minds . . . was that of E. M. Forster, who was a non-scientist, he really wasn't interested in science. . . . [But] pick up the volume called *Two Cheers for Democracy* and you'll see an extraordinary mind.

Jerrold's reference to Forster seems a curious one in the context of PSSC physics. Forster was, after all, primarily a novelist and an essayist, who did not write about science at all. But Jerrold saw some relevance in the novelist's profession: "What you have to do is to get to the public in such a way that they understand that there is some evidence. Now novelists understand this; a novelist . . . is capable of developing, fictitious or non-fictitious, a stream of evidence about some important facet of life." Jerrold liked to point out that even if Forster was not a scientist, he knew what he believed and he knew why he believed it. He was irreverent about tradition and custom, which in itself appealed to Jerrold, for he was irreverent of them himself. Forster had continued in the same essay:

The people I admire most are those who are sensitive and want to create something or discover something, and do not see life in terms of power, and such people get more of a chance under a democracy than elsewhere . . . These people need to express themselves; they cannot do so unless society allows them the liberty to do so, and the society that allows them the most liberty is a democracy.

Democracy has another merit. It allows criticism. . . . I believe in the Private Member [of Parliament] who makes himself a nuisance. He gets snubbed and is told that he is cranky and ill-informed, but he does expose abuses which would otherwise never have been mentioned. . . . Occasionally, too, a well-meaning public official starts losing his head in the cause of efficiency and thinks himself God Almighty. . . . Well, there will be a question about them in Parliament sooner or later, and then they will have to mind their steps. . . .

So Two Cheers for Democracy: one because it admits variety and two because it admits criticism. Two cheers are quite enough. There is no occasion to give three.[28]

By the time Jerrold became interested in Forster, he was already a veteran of many struggles with well-meaning public officials opposed to his attempts to reform the established world of education. Perhaps he wryly identified with Forster's cranky Member of Parliament. But Forster had restated Jerrold's own beliefs. Jerrold believed that a scientist must not only accept criticism but must welcome it. "A scientist," he wrote, "lives his professional life with a spectrum of confidence in his beliefs, all the way from dead sure to very iffy, even to admitted full-bodied ignorance. In a sense, every scientist has to be his own best critic, his own best adversary, and he must submit every idea he works on and publishes to a fast, ruthless and sophisticated criticism."[29]

The PSSC group made no explicit effort to enunciate such ideas, although they were close to the heart of the program. It was better to teach by example than by precept; object lessons were better than sermons, and Jerrold probably had less patience for sermons than most. More likely, if anyone had asked him to state the PSSC philosophy in those hectic days at the beginning, Jerrold, who often said that he "lived by the aphorism," would have replied with his favorite motto, written by another of his favorite authors, James Thurber. It was the moral from one of his then recently published *Further Fables for Our Times*, and it came as close to being a PSSC philosophy as anyone needed: "It is better to know some of the questions than all of the answers."[30]

Into Orbit

One of the most captivating things you can do is to create something success-ful. If you want to really lose your life into something, be successful at it and it'll kill you. It's got to; there's no question.

—J. R. Zacharias[31]

Once the December meeting had ended and the general shape of the program had been hammered out, all resemblance between the PSSC and a summer study disappeared. In the case of a summer study, recommendations would be made to its sponsor, to be carried out or not, depending on policy decisions made elsewhere; PSSC, however, could only make recommendations to itself. The NSF might support them but could take no action itself. If anything were to be done afterwards, PSSC would have to do it.

The next step in the development of PSSC took place in the summer of 1957: a workshop at MIT, where the first coordinated work on the new course began. Many sorts of people joined the project around that time: scientists, filmmakers, high school teachers, and trainers of teachers. Some had unorthodox backgrounds, at least as far as a science project was concerned. Laura Fermi, recent widow of the Nobel physicist, accepted the job of producing what eventually became the impressive collection of small books called the Science Study Series. Jerrold predicted that no one Mrs. Fermi approached to write a monograph would be able to turn her down; certainly, very few can have done so, given the eventual output of monographs. Berenice Abbott was another example of an unorthodox talent with something important to contribute. She was a superb photographer, famous for her portraits made during the 1920s in Paris, as well as for her remarkable studies of New York City in the 1930s. For PSSC she began to make a series of stunning photographs of physical phenomena. These eventually became so well known as to constitute almost the symbol of the new course; many of them are still found in textbooks thirty years later.

Uri Haber-Schaim was one of the scientists who joined at this time. He was a young Israeli physicist, a former student of Enrico Fermi, with a junior faculty appointment at MIT. He signed on for the summer in order to make a little extra money and discovered that he had started a lifelong career in science education. Gilbert Finlay, assistant to the dean of the Education School at the University of Illinois, had expertise primarily in teacher training, and took charge of producing teachers' materials. Altogether, some sixty physicists, high school teachers, writers, editors, and filmmakers took part in the workshop. Some began a full-time commitment to PSSC at that time, while others were hired by Jerrold as part-time consultants for varying periods. The PSSC steering committee (see appendix C) had by then been enlarged to include, as Steve White put it, "the usual suspects." Growth of the project was very rapid, the work was hectic and concentrated, but there was remarkable agreement among the project members about what to do and a strong determination to do it. By the end of the summer of 1957, a workable, partly tested version of the course had begun to emerge.

PSSC was a large and unconventional undertaking. The simple structure of a summer workshop would not serve for the long term. The project was going to involve a lot of people and a lot of diverse activities, and it was quickly clear that it could not be operated out of any one person's hip pocket, any more than it could ever have succeeded as any single individual's creation. The more permanent structure Jerrold chose was reminiscent, not surprisingly, of a research laboratory. He established a number of more or less autonomous groups and saw to it that they remained in close contact with each other, working on the several texts, the equipment, and the films. The groups were to share a common administration, supplied by MIT, which would deal with budgets, space, support staff, and so on. Elbert Little, a science teacher at Phillips Exeter Academy, was engaged to be the first executive director of the project.

Jerrold was content to place Francis Friedman in charge of developing the texts. Friedman, who served as editor-in-chief and contributed substantially to much of the writing as well, was much more than an editor. Jerrold referred to him as "Lord-High-Everything-Else," and his influence was felt in every part of the project. The emphasis had shifted, and the texts and apparatus were to be the central part of PSSC; the architecture of the entire course depended on how the texts were organized under Friedman's direction. Jerrold trusted Friedman's judgment and was prepared to rely on him fully. "I knew the quality of his mind," he said.

Jerrold himself preferred to concentrate on the films. He had begun with the idea of filming the experiments of physics, and since that involved action, it appealed to him far more than the writing. He took charge of this part of the effort happily. In fact, he hated to write and was glad to leave it to others.

There was no question about who was in charge, however. If Friedman had authority—and he surely did—it was authority delegated to him by Jerrold. Jerrold's personality characterized the project, and his constant demand for quality and for thinking things through determined how it went. Philip Morrison referred to it as "Zach's project. Zach was responsible, because he raised the money, he saw the Government, he saw the foundations, and so on. . . . Zach was always full of ideas, especially schemes for supporting and making the project acceptable; he understood the social context of everything. Francis, I think, mainly took care of internal affairs—looking inward—of the integrity, the intellectual content."

As that first summer progressed, the accomplishments became substantial: the preliminary version of volume 1, completed by Morrison, was already undergoing revision and editing, and arrangements had been made for the introduction of the new materials on an experimental basis to about 300 students at eight high schools. Work was well under way to produce the second, third, and fourth parts of the text, which were to reach the experimental classes piecemeal; everyone hoped earnestly that the material would be ready by the time the classes were ready to use it.

Jerrold now further delegated responsibility. Uri Haber-Schaim, in addition to writing text material, would oversee the development of the laboratory apparatus. The apparatus, simple to make once invented but difficult to invent, was to be developed concurrently and in close collaboration with the writing. Haber-Schaim had been a theorist and was a good writer, who had turned into "kind of an experimenter," according to Jerrold: "He [Haber-Schaim] didn't invent the apparatus, but he insisted on trying it out. And if Uri could work it then it was all right for the kids . . . and gradually, he took it over. I was more than happy to have him do it."[32]

The film program was much slower getting started, a source of considerable frustration and disappointment to both to Jerrold and Steve White. Nevertheless, by August 1957, an agreement had been reached with Encyclopedia Britannica Films, Inc. for the production of a series of films. A very preliminary version of one film, *The Pressure of Light*, had been made, but Jerrold was dissatisfied with it, referring to it as "a tolerably good home movie." Beyond that, they had experimented in making a few short filmstrips and had drafted several scripts for longer productions. They clearly had a lot to learn about filmmaking, and the many hours Jerrold spent that summer looking at whatever films already existed convinced him that there was almost no one from whom they could learn it. Those who knew how to make films did not, as a rule, have the necessary comprehension of science and could not get straight what needed to be shown; conversely, the scientists, himself included, knew little about how to present science in films.

The work proceeded sporadically and rather ineffectively until November 1957, following the first summer workshop, when Jerrold convened a meeting of a large group of filmmakers and physicists for a general discussion of the whole difficult problem. By the end of that meeting, approximately twenty-five physicists had expressed willingness to take part in a film program, and they went to work in earnest. A vacant neighborhood movie theater was rented, and the work of converting it into a film studio began.

The project was proceeding at a very satisfactory rate even before the November meeting, and there was little doubt that the NSF would view their progress favorably when the next round of funding was considered. But on October 4, 1957, the Soviet Union launched *Sputnik*, and the nation suddenly seemed to shift gears. What before then would have been allowed to proceed at a moderate pace and at a moderate level of funding was suddenly expected to move faster.

The predominant American reaction to the Soviet achievement was alarm. The *New York Times*, reporting the story the next day, noted ominously that the satellite had already flown over the United States at least four times and would continue to do so regardless of American wishes. One of the front page articles about *Sputnik* said: "Leaders of the United States earth satellite program were astonished tonight to learn that the Soviet Union had launched a satellite eight times heavier than that contemplated in this country."[33] The American reaction was intense and even angry. Perhaps that was not so surprising. The nation had been harangued for years about the menace of imminent Soviet superiority; every right-wing politician seeking power and virtually every Pentagon general seeking increased military appropriations had underlined the threat. To be sure, the Soviet Union had not helped; only the previous August, it had warned that it had the ability to send a nuclear weapon over intercontinental distances. Here, then, was proof that the warning had not been empty bluster. Here, at last, was the event every hard-liner needed to beat the country about the ears.

President Eisenhower was principally concerned with reassuring a startled and worried nation. He did not yield to panic, nor would he acknowledge that the Soviet exploit demonstrated any degree of military superiority. He would not even concede that the Soviet accomplishment meant anything at all. He referred in his memoirs to the need "to find ways of affording perspective to our people and so relieve the current wave of near-hysteria."[34] Refusing to commit the country to a missile race, the president said: "Now, so far as the satellite is concerned, that does not raise my apprehensions, not one iota. I see nothing at this moment . . . that is significant in that development as far as security is concerned."[35]

Such reassurances were not enough to satisfy the public, however, and some additional action was required. The Science Advisory Committee was therefore at long last brought in from the cold, to help in these circumstances; formally, its name was changed to the President's Science Advisory Committee (PSAC). Rabi was then the chairman of the committee; he and Lee DuBridge, his predecessor, made the suggestion to Eisenhower to upgrade the committee to help deal with what they, at least, saw as a crisis. The president accepted the suggestion and on November 7 announced that he was establishing the position of special assistant to the president for science and technology; three weeks later, on November 29, he announced both the transfer of PSAC to the White House and the appointment of James Killian as his first science adviser. Rabi generously offered to step aside from his post as committee chairman, and Killian also took over that position. Science evidently was being given a new importance in government, and scientists were now to have direct access to the policymakers.

These developments made a significant difference to the PSSC program and to much that followed. If Eisenhower had decided to respond to *Sputnik* with increased military expenditures, support for education might have remained at its earlier, relatively low level, but he chose a different path. When he wrote about this later, he said: "This was a period of anxiety. *Sputnik* had revealed the psychological vulnerability of our people. . . . One beneficial effect to us was that the Soviet achievement jarred us out of what might have been a gradually solidifying complacency in technology. It caused us to give increased attention to scientific education in this country and ultimately to all phases of education."[36] Eisenhower clearly meant what he said. In his message to Congress in January 1958, he recommended a fivefold increase in the appropriation for NSF programs in the field of science education, which he described "as among the the most significant contributions currently being made to the improvement of science education in the United States."[37]

The direct result, then, as far as PSSC was concerned, was that the project never again was obliged to worry about adequate funding. More than that, the PSSC experiment attracted great public interest. Reports on its progress were carried in newspapers across the country. Jerrold became the resident expert on education on PSAC. Furthermore, since Rabi, Edward Purcell, Edwin Land, and Killian himself, in addition to Jerrold, were simultaneously members of both PSAC and the PSSC steering committee, it was clear that PSSC could continue to count on high-level support. This cross-linkage between PSAC and the PSSC steering committee was striking. It would have been difficult to find another private enterprise the directors of which formed so influential a board as did those of the Physical Science Study Committee.

Additional funding for the program arrived that fall. The Ford Foundation made a grant of $500,000, the Alfred P. Sloan Foundation made a grant of $250,000, and the NSF awarded $442,000 in addition to its original grant. Everyone realized fully how important the project had become on a national scale. The freedom from budgetary constraints meant that if there were a valid reason for doing something, it could be done. It was a heady feeling for the members of PSSC to know that the importance of what they were doing justified almost any expenditure. Perhaps not since the Radiation Lab during the war had any group been able to operate so freely.

In later years, the impression arose that PSSC originated in response to *Sputnik*, a notion that everyone involved in PSSC, and Jerrold in particular, was eager to correct. No foreign stimulus had been necessary to awaken Jerrold and his colleagues to the sorry state of affairs in the secondary schools. That awareness had sprung from their deep and prescient concerns about the educational welfare of the country. PSSC was launched into its own trajectory fully fourteen months before *Sputnik* went into orbit. By the time *Sputnik* intruded into American skies, the first draft of volume 1 of PSSC Physics had already been produced by Philip Morrison, and classrooms had begun experimenting with the course. Remarkably, Morrison's pre-*Sputnik* draft already included discussion of the possibility of an artificial earth satellite.

Although the text Morrison had drafted concerned itself with measurement of all kinds and had even mentioned satellites, it did not say how one might weigh an educational program against a satellite. Certainly, neither Jerrold nor any of the others who worked on PSSC ever had any doubt about which weighed more in the balance of national well-being.

How to Make a Movie

The "Pressure of Light" had a good story line.

—J. R. Zacharias[38]

In spite of the widespread coverage in newspapers and magazines, not everybody understood very clearly what PSSC was about. The editor of *Educational Screen and Audiovisual Guide*, a trade journal for those interested in filmed teaching aids, evidently felt that the usual product was somehow being threatened. In an editorial entitled "Misguided Scientists," he said, in part:

Between fifty and sixty "movies" on physics are to be produced in the next two years. Where are they to be produced? Hollywood. Who is guiding their production? A top Hollywood director-producer is serving as chief consultant of the film program. Why? Well, the executive director of the Science Committee is alleged to have said "we hope the films will be more interesting than the usual educational films." Maybe that's the reason the Committee was guided towards Hollywood.

We'd like to ask the Committee a question. DID YOU ASK MICKEY SPILLANE TO WRITE THE NEW PHYSICS TEXTBOOK FOR YOU? . . . Are your opinions of existing educational films based on scientific evidence? . . . Who has misguided you into the notions that the laws of physics will be better understood when mouthed by Bill Board's Bumpkins or Uncle Jim's Animal Cousins?

It was true that Jerrold, with Steve White, had discussed producing educational films with Walt Disney, and it was also true that Frank Capra, a highly successful producer-director of sophisticated Hollywood film comedies (who had also made training films for the army during the war), was on the PSSC steering committee. Hollywood was, after all, where most of the movie-makers were, and it was not surprising that the PSSC committee looked there first for advice. But these efforts had not led anywhere;

Hollywood had little to offer that they could use. Whatever curious visions the writer of the editorial may have had as he railed about "misguided scientists," he would not have recognized any Hollywood style in the Watertown studio of PSSC, nor would he have found much of it White's store-front office facing MIT. "It had been a used-tire shop," White said, "and as long as I occupied it, it continued to look like a tire shop."

Jerrold asked White to head the production end of the film division of PSSC. Naturally, he intended to select the subjects for the films himself, as well as to decide who would be in them, and he, together with Fran Friedman, would review the scripts; but it was to be White who would be responsible for production, including locating and hiring most of the people who knew how or could learn to make useful films.

The first contract they made was with Encyclopedia Britannica Films, (EBF), which attempted to make the first few films. Both Jerrold and White soon realized, however, that the arrangement was not working. John M. B. Churchill, an independent filmmaker and one of the first people to bring to the project any successful experience with science films, still vividly recalled the difficulty twenty years later: "EBF [Encyclopedia Britannica Films] was doing the production. And I think it was pretty clear between Steve White and me that it would not work . . . EBF could not make films that would please Zacharias. It was rather amusing to sit in Zach's office and listen to them even try to talk to one another . . . Zacharias wanted to have the physicist in the film and the EBF people knew, as surely as someone could know anything, that that wasn't going to work." EBF had its own tried and tested way of making films and was not about to alter it for PSSC, even if PSSC was the client paying the bills. The arrangement had poor prospects for longevity.

What were these films like, to cause so much controversy? Were they complicated or technically difficult to make? In fact, they were not but the films were to be made for new and different purposes. They were not simply to be didactic lectures, as most science films before then had been. They were to give the audience of students some idea of the excitement of doing science. If the films were hard to make to Jerrold's satisfaction, it was because they were meant to break with tradition, and neither EBF nor the editorial writer quoted above were prepared to accept that.

Breaking with tradition meant that there were no tracks to follow, and under such circumstances, among individuals who were talented, creative, and vocal, it was probably inevitable that personality conflicts would arise. The experience of Albert Baez, which he reported with admirable candor some thirty years later, is revealing.

Baez, trained just after the war at Stanford University as an x-ray physicist, had afterward devoted himself to undergraduate teaching at nearby Redlands University in California. In due course, Baez had attended the November, 1957 filmmakers' meeting and had been able to make one or two useful suggestions, largely because he was one of the very few at the meeting who had ever thought before about using film to teach physics.[40]

Within a week of Baez's return to California, Jerrold telephoned him to invite him to join the PSSC film project. Baez recalled Jerrold turning on what Leona always called his "snake oil":

Oh, [said Zach] that was a great idea of yours. I can just see it. I can just see that molecule zinging around in there. Come and work with us. Right away. We need you. Your country needs you.

To me [continued Baez] the invitation seemed very appealing. From a little place like Redlands, places like Stanford and MIT represented high adventure. Leonard Schiff [the Stanford physics chairman] said, Al, I would just like to let you know that I could never work with Zacharias. So that gave me some sort of warning, I suppose, but I was so enthused by the idea of going that I immediately started making plans.

Baez joined the PSSC project the following summer, and at first everything lived up to his expectations fully. But the excitement did not lead directly to a technique for making movies. Jerrold could recognize what he didn't want; he had not yet been able to articulate what he did want. Baez had had a little experience in TV studios, still in an almost embryonic state, where the approach to lighting, for example, was quick and casual; the professional filmmakers continued to bring in big lights and to take what Baez felt was an inordinate amount of time lighting a set. His insistence on a more casual approach was becoming an irritant.

More and more, he recalled, he found himself arguing against decisions already taken, philosophies already adopted. Finally, he said, "I was becoming a nuisance. They stopped inviting me to meetings. They called me the 'studio physicist,' which meant I would be in Watertown all the time, and not hanging around MIT and available for those meetings. And I began to have arguments with Steve White, and I worked up some resentment at Zach, because White was Zach's right-hand man, and Zach supported him, even on some questions of physics."

Clearly, the arrangement could not last. Whether Baez's ideas about filmmaking were good ones was not the issue. In the presence of extremely strong personalities such as Jerrold, Steve White, Francis Friedman, and all the others, there was no room for someone to develop a personal vision different from Jerrold's. PSSC had to be a corporate effort.

Jerrold evidently could see what was happening and did make an effort to salvage the situation by proposing to Baez that he begin to travel around for the project, giving demonstration talks about PSSC. But that did not work; Jerrold, at the height of his own self-confidence, underestimated what the lack of his full support would mean to someone who both valued and needed it. Baez believed "he was gracefully trying to find a way to pull me out of the film business, because he realized I was having trouble with Steve and the others. But I didn't take it that way. I felt put upon. I came to work on films and I was being pulled out, so little by little, I was getting angrier and angrier. And I became more and more eager to leave the project.

So—it was a bittersweet experience for me. But there's no question about it. Without Zach, the whole project would never have worked."

Nearly sixty films were eventually produced by PSSC. Each film had its own character, of course; nevertheless, the story of one film serves as a partial illustration of what the others were like. Jerrold had determined early on that they should make a film demonstrating and measuring the pressure of light. The film would serve as a good example of what he hoped to accomplish and might become a model for others that would follow. The experiment itself was a simple one to understand, yet the simplicity was deceptive, for it required sophisticated techniques and could not have been carried out in a classroom.

The idea was to measure the push resulting from shining a bright light on a thin piece of kitchen aluminum foil suspended in a vacuum. To minimize friction or other resistance, the foil would be suspended by a quartz fiber so fine as to be nearly invisible. The light from a 20-watt automobile lamp would be concentrated on the foil by a large lens; such illumination would be about thirty times more intense than direct sunlight. The total force of the light impinging on a square inch of foil would still only be about 10 micrograms; equal to about a fifth of a fly wing, it would be quite difficult to measure.

The film had a good "story line," as Jerrold pointed out, and it held to the same story line throughout at least three versions. The story begins with a discussion of a Crookes' radiometer, a device commonly available in gift shops and toy stores and often claimed, erroneously, to demonstrate the pressure of light. The radiometer consists of an evacuated glass globe a few inches in diameter, within which a four-vaned pinwheel is carefully balanced on a needle point. Placed near a light source, the pinwheel spins with satisfying but misleading vigor.

In all three versions, the film opens by demonstrating that the pinwheel spins the wrong way; it turns into the force rather than away from it. Light pressure clearly cannot be the agent. As the film proceeds, the glass globe is further evacuated with a vacuum pump, removing yet more of the residual air: the Crookes' radiometer then stops working. It is the interaction of the pinwheel with the unevenly heated residual air that caused the motion.

The film then sets about measuring the very small pressure of light. The scheme for doing it must have appealed strongly to Jerrold. The little piece of kitchen foil is now displayed, suspended in a high vacuum by its quartz fiber. The foil is free to rotate back and forth about a vertical axis, twisting the fiber as it does so. The fiber provides a restoring force, just as a coil spring might do, so that the foil, once set in motion, continues to rotate back and forth, oscillating first clockwise, then counterclockwise, then back again. Fiber and foil constitute what is called a torsional pendulum. Such a pendulum is used sometimes as the regulator of a certain type of decorative mantelpiece clock. Like any other pendulum, it has a natural frequency and can be set oscillating by a series of repeated small rhythmic pushes in resonance with that frequency, each adding a small amount to the effect of all the preceding pushes. Resonance is easily accomplished in this case simply by turning the automobile lamp on and off at the proper synchronous rate. After a time, the oscillation of the foil becomes large enough to measure. The pressure of light necessary to produce the observed oscillation can then be calculated. It was a small but elegant demonstration, which in the final version took just 23 minutes to show. It involved equipment that, although sophisticated and difficult to construct, was easy to understand, and the phenomenon itself was easily visible.

They initially tried to make the film with an actor, but he apparently understood neither the script nor the experiment. In the second version, EBF tried it with a professor of electrical engineering: "a wizard engineer," according to Jerrold; "a warm human, not a gnome or a quiz kid or a drip." Perhaps it should have succeeded. Unfortunately, said Jerrold, "his enthusiasm for this particular apparatus was not as great as mine, because I built it." The lack of enthusiasm evidently showed, and the matter was too important to Jerrold for him to settle for something that was not right. He was excited by the concept of this film and meant it to convey the power of using such films in the classroom. He wrote, "At the time of making this film there were only a few score of people in the world who had witnessed . . . the direct effect of a force exerted by a beam of light. But that statement is not quite correct. Every one of you has at some time looked up into the heavens at the stars. The big stars are held in balance by the pressure of light. . . . Thus, this seemingly trivial effect . . . turns out to be one of the major structural elements of the universe."[41]

The film did not come out right until the third version, a full year after they started, and by then EBF was out of the picture and Jerrold himself was in it as the experimenter in the film. Even then, Jerrold felt it should have been better. Why was it turning out to be so hard to make movies?

Filming in what Albert Baez refers to as the "classical mode"—using big lights, professionally built sets, and Hollywood cameras—can almost never be done in real time. Filming has to be interrupted frequently: to eliminate an unwanted stray reflection of light glinting off a glass vial, to correct an unexpected cough or stammer on the part of a speaker, to move the cameras and lights to a new place for the next few moments of filming. There may be ten times as much film shot as is kept. Movies are made by splicing together successful bits and pieces; if that is done well, the viewer will be unaware of the process, and the film will appear to be a seamless whole.

The process of filmmaking therefore requires that the experimenter become an actor, but this is not the same as requiring an actor to become an experimenter, although for some scientists it was almost as difficult. The experimenter in the film must work from a script and must memorize lines so that he can repeat the same actions and the same words as often as required. He must do this naturally and convincingly—even though the "take" is out of any logical sequence. It would be hard to find a better way to make normally pleasant and well-spoken people behave like sticks of wood. That kind of acting is not easy for an amateur, and all the scientists involved in making the films were rank amateurs. Even Jerrold, who surely had more clearly in his mind than anyone else what he wanted to achieve, had considerable difficulty.

These were no doubt some of the reasons, and perhaps the principal ones, that Britannica had wanted to leave the scientists out of the film altogether. They thought it would serve just as well to film the phenomena, adding a voice-over commentary later, or else to use an actor as what is called a "talking head." The scientist thus need not be seen in the film at all. According to EBF, the whole process would have been a lot easier that way.

But Jerrold felt it would also have been a lot duller. He believed that for a movie to be engaging, someone has to be present on screen behaving as though he belonged there. And he evidently had another purpose as well: to help dispel certain widely held but unfortunate stereotypical ideas about scientists. He hoped to demonstrate that scientists are ordinary people, not the wild-eyed, heavily accented caricatures of Middle Europeans one might see in grade B movies; that science does not require supermen operating on a level so abstract as to be out of reach. He recalled twenty years later what his policy had been: "We wanted a diversity of physicists. We wanted no physicist who spoke in broken Hungarian and carried a briefcase. If someone had forced us to use [someone like] Edward Teller [who had a strong accent] on a movie, we would have stopped the project rather than do it."

Some years later, after PSSC had made a number of successful films, he relaxed this constraint, and some of his European-born colleagues began to appear in the films. But plainly at that earlier time, he hoped that students watching these films would conclude that if a scientist was just like anyone else, then anyone could be a scientist—perhaps even oneself. At the very least, the film would serve to make scientific thinking seem less alien.

Jerrold, together with Friedman and White, never felt they had succeeded in getting any of these ideas across to those with whom they had contracted for making the films. During the course of the first year, they became increasingly aware that the arrangement with Encyclopedia Britannica Films would have to be altered or terminated. The company had failed to make any film that measured up to PSSC expectations or to accept any suggestions for making the films better. EBF's method of operation offered little flexibility; film was shot and then sent to Chicago for editing, and there was almost no communication between editor and cameraman, much less between editor and producer and director. EBF, in the final analysis, never seemed to comprehend what PSSC was trying to do.

The parting was not pleasant, White recalled. Jerrold had determined that PSSC would have to make its own films, but EBF was offered the opportunity to distribute them; that, after all, was where the profits lay, and it seemed to be a fair offer. But EBF would not settle for half a loaf, nor would either Jerrold or Steve White yield on the filmmaking. "That's where we come out, then," White remembered saying. "We'll arrange for the distribution ourselves." Two days later, said White, a truck from EBF pulled up to the studio and gutted it, leaving behind bare walls. PSSC had to start all over again, this time on its own.

It fell to White to get a new studio going right away—crew, equipment and all—and he set about it without delay. He knew that he was not capable of running a studio himself and had no wish to try. Instead, he contacted old friends in New York, among them several who had become network television producers of documentary films. They helped him find directors, cameramen, and editors, and they advised him on how to reequip the studio quickly. One of those willing to assist was Kevin Smith, a highly regarded cameraman on the executive staff of the CBS network. White hired him as a consultant for a few days, and then, as Smith recalled it: "[White and Zacharias] backed me into a corner and offered me a job. I sure as hell didn't intend to come up here [to Boston] . . . I came up on a one-year leave of absence and I didn't go back."

Kevin Smith was another good example of the sort of unconventional but talented person who was susceptible to the appeal of this increasingly unconventional project. On the record, he was an unlikely choice for anything to do with PSSC, but Jerrold soon came to appreciate his value. Smith had graduated from college as a political science major in 1930, and, like so many others in the early days of the depression, discovered unemployment. He drifted for a time and then wound up in the theater, where, to his own surprise, he spent more than a decade as a professional dancer, even at one time operating his own ballet school. Eventually he became assistant to the well-known choreographer Agnes de Mille, before moving on once again, this time to film.

When Steve White found him and introduced him to Jerrold, Smith was the CBS staff man for a major commercially sponsored television series on science, "Conquest." It was one of the few examples of science filming that Jerrold had seen that he thought showed promise, and since seeing it, he and White had wanted somebody from that CBS crew. When they talked to Smith, they found that he held views about filmmaking similar to their own, with the same insistence on going after the best. Smith remembered the advice he gave them: "You can't contract this work out, not in the innovative mode you're in. You need to own your own film production business. It has to be directly responsible to you . . . because you're the management. So there can't be middle people in this protecting the film people from what you want. . . . You need your own unit, your own crew."

That was just what Jerrold wanted to hear. Although Smith had never run an entire filmmaking operation before and had in fact never even worked in a film studio but only on location, both White and Jerrold sensed that he was the man for them. White appointed him head of the studio.

Smith remained head of the studio throughout the PSSC project. Under his guidance, the studio evolved into a kind of intellectual center for the project. Because it ran screenings and because there were interesting things happening there, the studio became a place to which people converged. After PSSC ended, Smith continued as head of the studio for another decade and a half, continuing to produce, on behalf of later projects, some of the most remarkable educational films ever made. He estimated that by the time he retired in 1974, the studio had produced about 800 films.

Barely a month after EBF had left a studio bare to its walls, White had obtained equipment, and a small and energetic staff was in place with someone competent in charge. Filming began again on *The Pressure of Light*. This time, said White, "we got it right."

One film, not made to be part of the PSSC series, deserves description because it conveys so well the light-hearted, creative spirit that came to exist in the studio. The film was called *A Million to One*, and Ed Purcell was the experimenter. The PSSC equipment group had improved upon an invention called a dry-ice puck: a simple disk that could float almost without friction on a thin layer of gaseous carbon dioxide, using the same principle as the British Hovercraft, which rides on a cushion of air. The carbon dioxide is obtained from an evaporating piece of dry ice carried in a container attached to the puck. The whole structure weighs perhaps 2 kilograms, but friction is reduced to such a low level that the puck can be set skimming across a smooth surface by a very small force.

It is best to use Purcell's own words to describe the film:

You see, Zach had always said—Zach was so impressed by these pucks that he kept saying, "I'd like to see a cockroach pulling one of these pucks," or something like that. Unknown to him, the film crew, after they finished our film [on inertial mass] . . . went down to New York, and there was a flea circus on Forty-Second Street, run by a Professor So-and-So. They made a deal with this guy and they made a little three or four minute film strip. This guy actually harnessed a flea to one of these five pound pucks, and they had this movie where the flea actually pulls this damn thing. It's really marvelous.

There's a clip from my movie in which I am very solemnly saying, "It takes very little force to move this, and I'll show you how little the force is." I pick up a soda straw and blow through it at the puck and get it moving. A very solemn thing. Then all of a sudden they cut to the marquee of this flea circus with a band playing. . . . Tremendous! . . . That's a wonderful thing. As I often say, it's the movie where I play straight man to a flea. They had made this as a surprise for Zach . . . so every time classes use the inertia film now they always want to get "A Million to One" to add on.

Clearly, getting it right involved more than simply having the right script or the right experimenter. When the attitude was right, the film had a good chance to be so as well. Within three years, they had made nearly sixty films, a splendid achievement.

A Corporation Like No Other

The problems of the PSSC project were not restricted to what went on in front of the camera or to the film division alone. They were not simply conflicts about who was in charge or whose creative judgment would prevail. Serious administrative difficulties emerged as well, as formidable as those on the creative side. MIT had originally accepted the administrative responsibility for PSSC, but once things got going, the commercial dimension of

PSSC—the production and distribution of films, textbooks, and apparatus—was not compatible with the institute's other activities. For example, the NSF as an agency of the federal government, was not permitted to set up a nonprofit business in competition with existing corporations; PSSC would therefore have to rely on commercial manufacturers, publishers, and distributors for the services it needed. But materials developed with taxpayers' dollars had to be in the public domain and could not be reserved for the use of a single manufacturer or publisher. What would be the incentive for such companies to participate? These were problems that required the most careful attention, but they were not the sorts of problems MIT normally had to deal with.

Another difficulty peculiar to the film division arose from Jerrold's insistence that only the highest quality would be acceptable. He was confident about the texts; he had gathered the best, most experienced people possible under the leadership of Fran Friedman, whose judgment he trusted implicitly. When it came to publishing books, he felt they knew how to do that. In any case, although he did not plan to participate directly in the writing of the texts, his close relationship with Friedman ensured his continued involvement with them, and his own judgment would continue to influence them. It was difficult sometimes to tell where Zacharias left off and Friedman began.

When it came to the films, however, they were all on unfamiliar ground, PSSC and MIT alike. Jerrold knew enough to know that quality required paying the talent. Kevin Smith insisted on a union shop for that reason; Steve White fully agreed. They were doing business, after all, with cameramen, script writers, lighting experts, and so on, and it was necessary to pay such people at competitive rates, even though they could not hope to match Hollywood and New York figures. The MIT salary scale was not sufficiently elastic; even with special dispensation from the union (the International Alliance of Theatrical and Stage Employees—IATSE), the salaries White and Smith proposed paying were off the MIT scale. By the time the first full year of operation had passed, White, Smith, and Jerrold had enough headaches to force Jerrold to seek a drastic solution.

He proposed to Killian and to Julius Stratton, who had just succeeded Killian as president of MIT, that the answer lay in a friendly parting of the ways. PSSC required its own administration, distinct from that of MIT. Killian and Stratton agreed, and in 1958, Educational Services, Inc. (ESI) was officially established as a private, nonprofit organization, the purpose of which was to carry out educational research and development. Since it was very much Jerrold's creation, it was he who assembled its board of trustees.

Killian agreed to serve as chairman of the board. Stratton became one of the incorporating members of the new company. A president was selected for the new corporation: James E. Webb, a former under secretary of state and director of the U.S. Bureau of the Budget.[43] On December 1, 1958, the entire PSSC project was moved over to ESI.

Even this did not spell the end of all administrative problems, for PSSC had developed a free-wheeling approach that remained difficult for administrators to deal with. ESI made several false starts; Steve White recounted an incident when one of the early financial officers of the new corporation had attempted to fire him. White, then in charge of the film division, had made a substantial cash commitment concerning a film without going through administrative channels, although whatever channels there were were hardly well established. There had been no other way to accomplish the purpose, White remembered but he failed to convince the financial officer. He recalled:

I was summarily fired . . . [which was] absurd. But I was very near the edge of my nervous system with the problem of the studio and its staff . . . so I simply said "All right," and went back to my desk and began cleaning it out. Whereupon I had a call from Jerrold. Could I come right over? There was something he wanted to talk about.
"I can't," I said. "I've just been fired."
"Who fired you?," Jerrold asked. I told him.
"Fire him back," Jerrold said.

Eventually the problems were brought under control with the recruitment by Jerrold of Gilbert Oakley, who had worked with Jerrold at Lincoln Laboratory. Oakley was a capable and tolerant administrator, and he was at home among scientists. Originally from Maine, he had never quite lost the Down East accent that gave away his origins. Like Kevin Smith and Steve White, Oakley's background was unusual: he had worked in a bank as a young man and then had worked for a yacht builder. After a time he had gone to work for a geophysics research group operating out of Wood's Hole, Massachusetts, eventually becoming skipper of their research vessel, the *Atlantis*. When, as he said, "the place got too highly organized," he moved on to MIT and its newly formed Lincoln Laboratory. The laboratory had given him a choice at that time, he said, of becoming the administrative aide for a man he described as "orderly, meticulous, [who] planned everything ahead," or for Zacharias: "Zach was pretty helter-skelter, all over the map, and I figured he'd need somebody to pick up the pieces . . . so I was with him as long as he was head of that division." Jerrold was undoubtedly capable of organizing his own life, but he always had too many irons in the fire. He needed someone else to see to the details. At LNS, he had had Mac Hubbard; now he would have Gil Oakley. Jerrold had the vision and the drive to create an ESI, but had he tried to administer it first hand, it would probably have been a mess, and he knew it. He always needed someone to "pick up the pieces" and had the good sense to choose Oakley.

After Lincoln Lab, Oakley had gone to work for Hycon, Inc., and had spent about a year and a half in Southeast Asia making a telecommunications survey for the company. When he returned, he found a letter from Jerrold inviting him to join ESI. He accepted and not long afterward found himself running the place as a kind of executive officer. Only someone as unusually experienced as Oakley could have dealt with Jerrold and his odd crew of scientists, filmmakers, writers, and technicians, but Oakley found arrangements that worked. He accepted Jerrold's style and worked in consonance with it. He recalled: "Zacharias would never draw an organization chart himself and he would tear up any one that he saw. And I fell in with that. You knew how things ran. . . . If a person on the college administrative side or on the professional side was talking to me about employment I would be very cold-blooded with him, and say 'Look, if you're expecting to know what you're going to be doing tomorrow, today or week-after-next, forget it, go work for the John Hancock Life Insurance Company or something, because this is not the place for you. It's always going to be in a state of flux.'"

The idea of using a group of the nation's top scientists to address the problems of secondary education was an entirely new one and, to make it work, ESI had to be unlike any previous educational corporation. However new and different any idea emerging from PSSC might be, ESI stood ready to try to make it work by whatever means were available, unconventional or not. The style of ESI was loose and permissive of everything except shoddy work. It appeared to be inefficient, although it was much less so than it seemed. There never was an organizational chart in those first years; people simply knew who was going to do what, or else they worked it out by arguing in the corridors. Most of the scientific staff was on leave from other places. Oakley recalled that this was sometimes a mixed blessing, but at least he didn't have to lose sleep worrying about what would happen to such people if the money dried up. They would have places to which to return. One of the nice things about it [he said] was that we didn't think we were building a permanent institution.[11]

Just as it had been mainly Jerrold, Fran Friedman, and Phil Morrison who had set the intellectual tone of PSSC, so it was primarily Jerrold, Steve White, Kevin Smith, and Gil Oakley who set the corporate style of ESI. Each organization had been created as a result of Jerrold's boldness and persistence.

As Jerrold pointed out, neither enterprise would have been as good as it was without people like Friedman, Morrison, Oakley, Smith, and White. There is little doubt, however, that without Jerrold, neither would have existed at all.

Proliferation

Upgrading educational programs is a never-ending process. People ask me, When are you going to finish this high school physics program? I say it will never finish. It will die if it finishes—and then this decade's new programs will be next decade's dead programs.

—J.R. Zacharias[44]

By the end of 1962, anyone examining the state of education in the United States would have had to be able to identify, in addition to PSSC, the acronyms BSCS, CBA, CHEMS, ESCP, SMSG, and UICSM, all of them well-established NSF-sponsored programs aimed at improving science and mathematics education in the high schools.[45] What these programs had in common, besides NSF funding, was the use of professional scientists and mathematicians, often drawn from the highest level, working alongside high school teachers and other educators to prepare new materials. Anybody familiar with PSSC would have sensed familiar patterns in them; the principles of modern curriculum development pioneered by PSSC and accepted by the National Science Foundation had turned out to be readily applicable elsewhere. It was hardly surprising, of course, for as Kevin Smith pointed out, "PSSC was an experimental program for the funders as well as for the people doing it. . . . We tended to take the leadership in management, and the National Science Foundation would then adopt this as policy, because they could see what we were doing as setting standards. . . . Several years later, policies that were applicable to PSSC became permanently set [in] concrete at NSF . . . you'd say, 'But I don't want to do it that way.' But that's the way you'd have to do it."[46]

The Physical Science Study Committee, shaped and constituted from its beginning by Jerrold's ardor and insistence, had succeeded in making work on pre-college education respectable for scientists. As much as anything else, this was a major contribution to the world of education. Without PSSC, the NSF would have been far less likely to have pioneered in educational reform, and without the success of PSSC, would hardly have continued with it. To be sure, in the days following *Sputnik*, there were other sources of funds, but they were of a different kind. In 1958, Congress had appropriated considerable sums of money for science education and it had passed the

National Defense Education Act (NDEA), in response to *Sputnik* and President Eisenhower's strong message on education. Under Title III of that act, substantial funds flowed out of the U.S. Office of Education, filtering down through the educational hierarchy to the local level, but those funds remained largely within the existing educational structures, which had rarely attempted innovation and were often resistant to it. Furthermore, Congress is an inherently political body; since it was the source of Title III monies, political considerations would always enter and influence decisions about what to do and where to do it. Where PSSC could draw inspiration freely from a world of ideas, and where it could determine content and insist on standards of quality, the suppliers of Title III funds could not.

It was the Physical Science Study Committee, therefore, that was the flagbearer: the committee carried the standard of change and showed what was possible. Nevertheless, by 1962, when many of the goals the committee had set for itself seemed to be within reach, the funds available for it began to diminish. More than $8 million had been spent, but much of that had been spent on the development efforts rather than in teacher training. The payroll alone, Jerrold recalled, had been "approximately a million dollars a year for four and one-half years." Not enough remained for the expansion of the PSSC efforts to educate teachers, although that was widely seen to be essential.[47] As Jerrold said, "the national effort... had begun to run out of steam in the early '60s."[48] The committee nevertheless had accomplished much. In 1962, the program was being taught by about 2,000 teachers to approximately 80,000 students, and these numbers were still increasing. Eventually the number of teachers approximately tripled, and at least 30 percent of all high school physics students in the United States were enrolled in the course.

PSSC had made as full use as it could of the NSF summer institutes to reach that many teachers; Jerrold afterward described the strategy:

The PSSC had a built-in multiplier. It was based on the principle of exponential growth, in which I have tremendous confidence: 3 gets you 9, 9 gets you 27, 27 gets you 81; and by doubling, tripling, quadrupling, and so on, you can build a good program from a solid base into a nation-wide effort. Bacteria know about this; so do rabbits... It is the exact opposite of dropping a billion dollars into an educational system running at 100 billion dollars and expecting anything more than minor ripples . . .

The first summer we started with 25 teachers. . . . The next summer, we assembled five summer institutes with 50 teachers-in-training at each. We had ten excellent teachers of teachers [from the initial group] to work with them. . . . During the third year we had 20 summer institutes [and] 1200 teachers teaching. But now, for the fourth summer, we had run out of one of the nutrients, namely money . . . so we kept running summer groups at the level of 1000 teachers per summer . . . until we had gotten to 6,000 teachers out of the 12,000 we had to reach.[49]

Was PSSC successful in what Jerrold would have called "the eye of eternity?" It might be argued that the initial goals were overly ambitious and that the program failed to meet them fully. For example, the number of high school students electing to study physics did not increase at all. The PSSC course was a difficult one, for despite all the efforts of the hundreds of people who worked on the new course, physics remained a difficult subject. It continues to this day to be selected mainly by an elite fraction of the high school population.[50] Furthermore, it is a difficult subject to convey without some special training for the teacher. All too often, it is taught by someone whose own background is either meager or nonexistent. The hope that the course might prove to be teacher-proof was too optimistic.

A more useful assessment follows, however, from considering the overall effect of PSSC, which was enormous and continues to the present. The flowering of educational reform efforts in the sister sciences of physics has already been mentioned; a study prepared for the National Education Association on the curriculum development projects active in 1962 listed no fewer than ten projects in science education and another twelve in mathematics and was nevertheless incomplete.[51] It did not, for example, include such a project as the Elementary Science Study just then getting started at ESI, nor did it recognize that a particular activity at Harvard University was about to become a major program.[52] The report also failed to take note of any foreign activities, for they were not within its scope, but by 1962 there were already several important ones.

PSSC also induced substantial changes at the college and university level. Jerrold pointed this out himself as one of the most important direct achievements of the program. The developers of the PSSC course must have seen that coming almost from the beginning, although it had not been mentioned explicitly as a goal. Jerrold wrote: "What we accomplished in the natural sciences was to speed up the learning process by about two years, for the better students who were reached by our programs. The clear proof of this statement lies in the reaction of the major colleges and universities when the better-trained students arrived at their doors. They had no choice but to . . . redesign their first year science courses to make them substantially more sophisticated."[53]

Thus, even if PSSC did not produce more physicists (Jerrold said later that that had not been one of his own personal goals), it led to better-educated ones. It placed great pressure on university faculties to improve what they offered, which led to new programs and institutions dedicated to that purpose. For example, a committee formed at the University of California succeeded in creating the widely used Berkeley Physics Course, which clearly drew much of its inspiration and its style from PSSC. And in 1960, reacting to Jerrold's proposals and stimulated by the success of PSSC, MIT established the Science Teaching Center, under the direction of Francis Friedman. A. P. French, later one of Jerrold's colleagues at the center, pointed out:

Francis L. Friedman, Philip Morrison and [Jerrold] Zacharias—were among the leaders in pressing for similar efforts at a higher level . . .
 With the massive backing of the National Science Foundation, both groups embarked on large-scale projects: the Berkeley group, to produce a set of texts and accompanying laboratory materials; the MIT group to devise a more diversified program. . . . Notable features of these projects were an increase in conceptual and mathematical sophistication [and] an injection of substantial amounts of 'modern physics' (relativity and quantum ideas) into the first year curricula and an organized use of many different kinds of learning aids.[54]

The proliferation of new approaches to the teaching of science extended down the educational ladder as well as up. It was reasonable for questions now to arise about what children were learning in junior high, and in elementary school. That was, after all, where children acquired the attitudes and habits of thinking that they would take with them to high school and beyond. Once PSSC had broken the ice, many who were interested in education began to see that problems existed at every level, including the elementary ones. By 1960, Jerrold and his colleagues had begun to think about new ventures at these other levels. Only a few years earlier, Jerrold's sole concern had been with the behavior of atoms and molecules; now, he found it necessary to think about the behavior of children.

Jerrold's life, or at least its outward aspects, changed sharply during those few years. He came to play a far more visible role than he had ever done before. Nothing in the past had ever brought him the degree of public attention that accompanied his ventures in education. Not even the atomic clock, which had been widely covered by the press, had resulted in so much sustained attention in newspapers and magazines all over the country.

He was emerging as a public figure, one of those people about whom others like to tell stories, who seem to be a bit larger than life. His epigrams and mottoes were repeated with relish within a widening circle; the many newspaper articles on PSSC generally featured him as the program's principal creator and printed his opinions. He was photographed for *Life* magazine by the noted Alfred Eisenstadt, and he was quoted in *Time* and *Newsweek* on the virtues of experiment. His reputation as an innovator in education spread.

He enjoyed the public attention, but it had little personal importance to him, for he had no great regard for the public's taste in heroes and put small value on becoming one. There were praises that did matter, of course. He took great pride in the award of the Oersted Medal in 1961, given by the American Association of Physics Teachers for outstanding contributions to teaching. The public attention was welcome enough and useful, however, for it meant that PSSC would find easier access to local communities than otherwise it might have found; in addition, Jerrold would have readier access to foundations and other sources of funding. This was important, of course, both for keeping PSSC going and for locating support for any new projects that might be launched. The publicity would help to open doors for him.

The fame of the PSSC physics course flourished abroad as well. Almost from the beginning, great interest in the course had been shown overseas, particularly in less-advantaged countries just beginning to face the problems and hopes of independent statehood. The potential of the course was quickly perceived, and its first translation into another language had occurred at the very first workshop in the summer of 1957. One of the participants in that workshop had been a young Thai, in this country on an exchange fellowship. He provided a carefully handwritten version of the text, which was then photoreproduced, since there was no print shop to be found with a Thai font. Other translations followed, under more official auspices, and the course eventually was used in at least a dozen different translated versions in perhaps twice as many countries. (There was even a dubbed version of *The Pressure of Light*, with Jerrold speaking flawless Italian. Those who saw it said it was rather startling.)

In October 1977, those who had been associated with PSSC were invited to a reception given by James Killian, then the honorary chairman of the MIT Corporation. The reception, held at the MIT Faculty Club, was to celebrate both the twentieth anniversary of the Physical Science Study Committee and the sale of the millionth volume of PSSC *Physics*. The book had by then gone through four editions.

In 1988, eleven years after the millionth volume, a sixth edition was in use and a seventh edition was being prepared. The first part of the text—the part that was originally written by Philip Morrison—had been made the basis of a junior high school text, under the guidance of Uri Haber-Schaim; that part was dropped from the high school version. Other parts of the text continue to be modified in accordance with feedback from classroom users. Teacher training courses are still available but supported only by tuitions; they are no longer subsidized by any government agency or by private foundations.

At the peak of its acceptance, the PSSC text was used by nearly half of the high school students studying physics; this fraction soon leveled off to about a third. After the Harvard Project Physics course was introduced, the PSSC share diminished as the Project Physics share increased. Today both courses together continue to account for about a third of the market.

Jerrold's simple memorandum had turned out to be like the small pebble that starts an avalanche, and some of the boulders it dislodged have never stopped bouncing. It took many people and many programs to alter the educational landscape, but PSSC had led the way. Was it a good program, then, or was it merely first? Perhaps Jerrold should be given the last word: "Let me say a word about 'good.' It is so easy to fall into the trap of the word 'better.' . . . you have to work a lot harder to make an educational system good than you do to make it better. The good has to be set up on an absolute scale."[55]

8

Three Projects

I must say that it is thrilling to witness the new young administration in the White House and the Executive Office Building—literate, clear-headed, fast-moving and courageous, who need and are eager for every help that the intellectual community can muster. It is a privilege and a pleasure to be able to witness a New Frontier in the decision-making processes of government.

—J. R. Zacharias (1961)[1]

The Eisenhower years ended with the election of 1960, and the United States prepared for a new era and what promised to be a new style of government. There had been many wholesome developments under Eisenhower, particularly in the fields of science and education, but overall the period had been difficult. There had been little international tranquility. The United States had rearmed, and the Soviet Union had reacted with yet more arms. The nations of the world had again and again prodded and tested each other and themselves: in Hungary and in North Africa, at the Suez Canal and in divided Berlin, in Southeast Asia and on the islands of the China Sea. The political order had shifted and changed; new nations had come into existence, and new coalitions had formed. Every nation had sought to define its role for the decades to come. The testing would continue; from time to time it would erupt into bitter hostilities as stability continued to elude those who sought it. By 1960, it seemed, the world had settled down to a durable and uncomfortable state of limited conflict, held to a standoff only by the threat of nuclear weapons backed by policies of deterrence through massive retaliation.

The time of the Eisenhower presidency had been one more of struggle than of hope, in spite of occasional optimistic interludes. At home in the United States the situation had not been easy either; the 1950s had witnessed cruel political harassment and repression. A close and prescient examiner might have found a few positive signs perhaps: the civil rights movement in the South, for example, began in that period. But the examiner would have had to look hard. For the most part, and for most people, the fifties was not a time for much optimism.

The election of John F. Kennedy in the fall of 1960 signaled the beginning of a sharp change. The beginning was a bit tentative: Kennedy had won the presidency by only a little more than 100,000 votes, and the election might easily have gone the other way. But that had not happened, and soon a new spirit seemed almost to enchant the country. Hope returned—if not for everyone then at least for the nation's youth, for even if the cold war could not be brought to an end, idealistic young men and women might still find something positive to do in the Peace Corps, and even if racial equality was still far away, there was still something that people could do in the growing civil rights movement to bring it closer. The idea of helping others was hardly new, but Kennedy made it a central part of his dialogue with the nation, and the idea took on a fresh vitality for the few years of his presidency. The change was not primarily a political matter; Republicans as well as Democrats felt the shifting mood.

Jerrold was in a good position to watch these changes take place and even to affect some of them. In 1958, in accordance with the rules he had helped establish, he had moved off PSAC, although he remained an active member of several committee study panels. After Kennedy's election, he was reappointed to membership, which gave him a kind of grandstand seat as the country adjusted to Kennedy's adventuresome style.

It was a style with which Jerrold was very much in tune, for people like himself were encouraged to become engaged in the public interest. Jerrold's own involvement with matters of public policy had begun five years earlier, but he felt much in sympathy with the changes of the Kennedy years. His projects in the field of education were a straightforward attempt at public service, although he would probably have rejected that description as pretentious and more self serving than public serving. In his own view, he had simply seen problems he felt he could do something about, and so he had taken action; he was not a contemplative man, inclined to find a broad philosophical context for his own actions. He simply understood that it was possible sometimes to do something worth doing. Perhaps if he had been much younger, the same spirit might have moved him to drop his raccoon coat and join the Peace Corps.

During the period 1960–1961, Jerrold and some of his colleagues, including Francis Friedman and Philip Morrison in central roles, started three important new projects in science education. Only one of them, the development of a new college physics course, would have been easy to predict as a natural consequence of PSSC physics. The other two were pioneering undertakings, bold adventures into both the minds of small children and into the unfamiliar cultures of Africa.

All three developments took place on a very large scale and ran for years. That they all began during the same short time span seems in retrospect remarkable, and serves as a comment on Jerrold's undiminished energy and enthusiasm in the sixth decade of his life.

Africa

We cannot wait for 200 years; we can only wait for a short time, and for that reason all the new techniques in education must be applied to the purposes of teaching and learning, so that the experience of a large number of years can be telescoped into a short space.

—Chief S. O. Awokoya, Federal Adviser for Education, Nigeria[2]

In August 1960, Jerrold was invited to address the International Conference on Science in the Advancement of New States, held at the Weizmann Institute in Rehovoth, Israel, and he and Leona traveled there with their daughter Johanna. Israel was one of the nations that had begun to experiment with the PSSC course early; consequently Jerrold had been invited to speak as a recognized authority on science education. His views were widely sought. Nevertheless, athough PSSC had spread to other countries, Jerrold's own personal experience was still limited to the United States. His remarks to the conference focused on that experience.

Earlier that summer there had been several small, informal meetings at MIT where he, Francis Friedman, and Philip Morrison, together with a few colleagues from the biological sciences, had begun discussing the problems and possibilities of starting a new education program comparable in scope, perhaps, to PSSC physics but at the elementary school level. They had become convinced that such a program was necessary and had formed some preliminary ideas about how to teach young children. These ideas were still germinating as he spoke in Israel to a group drawn for the most part from the developing nations. Although these ideas not yet sufficiently formed to be included in his address to the conference, they undoubtedly affected the way he listened to the remarks of others.

The voice that Jerrold heard most clearly at this conference was that of the Reverend Solomon B. Caulker, a native of the soon-to-be-independent country of Sierra Leone, in West Africa. He had been educated in the United States at the University of Chicago and at the Union Theological

Seminary in New York and at the time of the conference was vice-principal of Fourah Bay College in Freetown, Sierra Leone.

The first papers of the conference were concerned with nuclear reactors and nuclear power, as well as other aspects of high technology. Certainly these were subjects important to developing nations for which the lack of electric power production presented an almost insurmountable obstacle to economic development. But Caulker felt that other matters were far more pressing for the African states. In his own remarks to the conference, Caulker said:

I come from a country, ladies and gentlemen, in which, while it is of great interest to talk about nuclear physics and fusion and all these things, and we do read about them and follow what is happening in the world, it is of even greater interest to know how to save so many of our babies, for in Sierra Leone, eight out of every ten babies who are born die before they are one year old. . . .

The average life expectancy age, by and large, is in the low thirties as you know, compared to many of these advanced countries where the average age has gone to sixty, or near seventy. . . . And this Conference would be going a great way if it opened to us the doors to say: "You can have *life* as we've had it, in terms of good food and good health." So that, interesting as nuclear fission is (and I'm not saying here that we shouldn't listen to something about it), from the point of view of the pressing peril of the moment, this is the real thing.[3]

Jerrold had been deeply moved by these views when Caulker had expressed them to him privately. Now he impressed the entire conference and infused it with a new spirit. Caulker went on to speak more specifically of Africa's difficulties in a way that could not fail to engage Jerrold's professional interest as well. After first acknowledging that Africans themselves would have to solve many serious problems of hesitancy and self-doubt deriving from years of colonialism, he continued:

Professor Zacharias has mentioned [in his address] something I said to him about our problem of the teaching of science. It's very easy for you, very many of you, who have been raised in the Western outlook, in Western philosophy of history and so on, to take this for granted. But in our own thinking, first of all, one of the most difficult problems of the African people in these under-developed States is to understand that there is any relationship physically between cause and effect. This is a primary problem: whether typhoid is caused by drinking dirty water or whether it is caused by someone who has bewitched you; whether your babies are dying because you are not feeding them properly, or whether they die because someone who hates you has put sickness on them. . . .

Is there any hope? This question is of far more importance to me, and I am hoping that towards the end of the Conference, I can go home and say there is a possibility that these things do change.

For someone for whom Observation, Evidence, and the Basis of Belief formed the title of a personal credo, Caulker's appeal was powerful. In the days that followed, Caulker reemphasized his views to Jerrold, with whom he had many informal conversations during the remainder of the conference. What Caulker heard in return must have been encouraging to him, for he said, as the conference closed, "When I leave here on Sunday, it will be light not only physically but metaphorically . . . because of the dedication of the scientists who have come here to share with us their will, their insight, their research. I no longer go home feeling that we are isolated in our problems."

Caulker did not make it home to deliver his message of hope; he was killed in an airplane crash in Dakar, on August 30, 1960, on the way to Freetown from the Conference. Jerrold learned of his death from a fellow passenger's newspaper, while he, Leona and Johanna waited at London's Heathrow Airport for their own flight home; the news shook him badly enough so that he did not mention it until they had all returned to Belmont. For some time he took no action; he said later that he had felt helpless and saw no useful course to take. Caulker's remarks had had a lasting effect, however, and Jerrold finally saw that he would have to act.

He began as he often did, by seeking advice and help from James Killian; Killian, in turn, invited Detlev Bronk, the highly respected president of the Rockefeller Institute, to join the discussions. Steve White acted as the staff man, adding his own ideas as well as doing such writing as was called for. None of them knew very much at that point about the problems of Africa, or of any developing nation, except to know that the problems were many and severe. A summer study was what was needed, Jerrold said, and they began to work toward that goal. They formed a steering committee and applied to foundations for support: proposals to the Ford Foundation and to the International Cooperation Administration (ICA) were successful. (The members of the steering committee are listed in Appendix D.) The Ford Foundation made a grant of $110,500 in February 1961; several months later, the ICA, operating with unprecedented speed, made an additional grant of $122,500. The summer study—the first step to something, no one was sure what—could take place.

In the summer of 1961, a group of forty-eight scholars, teachers, and education administrators, including fifteen Africans from Ghana, Nigeria, Sierra Leone, and Uganda, met for six weeks at Endicott House, a beautiful suburban estate owned and operated as a conference center by MIT. The meeting came to be known as the Endicott House Conference and was a landmark event. Once again, as he had done so often, Jerrold brought the principles of systems engineering to bear on a gigantic, complex problem.

The selection of the African participants in the meeting was delegated by the steering committee to those of its members with extensive African experience—principally L. John Lewis of the University of London, assisted by W. O. Brown of Boston University and Francis X. Sutton of the Ford Foundation. At that time the Ford Foundation had its own direct interest in Kenya. It was financing a science center in Nairobi, Kenya, and Ford program officers were located as well in other English-speaking countries such as Nigeria. The men they recommended were leaders of the most influential educational establishments in their respective countries. They were an impressive group, thoughtful and eloquent, dignified and proud, usually appearing in imposing native dress. They were among the educated elite of their countries, an important part of the very small number of such Africans then to be found. They were all deeply committed to bringing tropical Africa out of its centuries-old poverty and superstition.

Jerrold ran the Endicott House meeting in his usual enigmatic summer study style. There was no agenda, of course, and because of the relative isolation of Endicott House, there was almost no escape. The African participants, trained in a British colonial environment and used to much more structured situations, felt keenly uncomfortable at first, as perhaps did some of their American co-participants as well. One of the Africans, T. T. Solaru, manager of the Oxford University Press branch in Ibadan, Nigeria, expressed the general view of Jerrold's approach:

We had attended other conferences on education, but it was a strange meeting here on June 19. We all thought, "Well, when we get there, we'll see a worked-out program, Monday, Tuesday, Wednesday, just like the conferences we knew before." But when we came here, we had a word of welcome from Zacharias and he more or less said to us, "I brought you here to study African education, but where we are going, I don't know."

Now isn't that a curious man who brought us together and didn't know where we were going? Isn't it usual to decide in advance where you want to go? I thought that was a dangerous assignment.

Nevertheless, Jerrold's technique worked, as it always seemed to do when he held the chalk. By the time the meeting ended, a structure had been sketched out for an ambitious African Education Program, including five new education development programs and an indigenous counterpart to ESI, located somewhere in Africa, to organize them. It was to be called Educational Resources Africa. The outcome moved Soleru to add to his remarks: "But as things went on a program did emerge. It was not a program that had been prepared and laid out before us and to which we had to conform. We ourselves laid out a program and then we found that program expanding and expanding. We had a job in six weeks to keep up with it. I don't think we have even finished keeping up with it."

It would be difficult to overstate the eagerness with which the Africans took up the challenge of planning educational reform once they understood and accepted that the Endicott House Conference was not simply another case of an advanced Western power imposing its own ideas through the mechanism of well-intentioned foreign aid, nor was it a case of competition between the West and the Soviet-led Eastern bloc. (Chief Soleru said at the conference, "I don't know why I call them 'East' and 'West.' They are in the North as far as we are concerned.") Those were situations they had all seen too many times before, and against which they could rarely afford to complain. They all knew at first hand of examples of aid programs either inappropriate or too rich for their countries to digest. Endicott House offered something different: the opportunity to think problems through for themselves and to determine both their own needs and their own order of priority. For some of the Africans, it was the first time that their views about their own countries had been taken seriously outside the continent, in what they privately referred to as one of the "overdeveloped states."

The Africans brought an extraordinary measure of hope to those early plans, as well as a measure of impatience sometimes bordering on desperation. They responded eagerly to the examples and the teaching demonstrations that they saw and heard. Chief Steven O. Awokoya, federal adviser for education in Nigeria, eloquently expressed the intensity of the need: "Time is not waiting for us if we are to survive in this struggle for existence. . . . We feel in Africa that we've got to study science and technology or die. There are quite a number of us who feel that the survival of the African nations depends on the extent to which they can telescope the study of science and technology." Awokoya was keenly conscious of the hurdles they must clear as they sought to leapfrog into the twentieth century. He told the conference about some of them, as Solomon Caulker had done before him, in Rehovoth:

Most African children believe that the rainbow is the excreta of the boa. There is a superstition that if you are able to find out where the rainbow touches the ground you will get the excreta of the boa, and if you can manipulate it properly you can become the richest person on God's earth. . . .

When I was a child, if I went into a forest and shouted "Whooooo!" and the sound was reflected from the leaves, I believed—and children like me believed also—that evil spirits of the forests and mountains were repeating the shout. It was a frightening thing to go alone into a big forest or to a big mountain and to shout, because the spirits might take one's voice away. . . .

All these [ideas] constitute the background of the average Nigerian child or African child that comes to school.

The hopefulness of the Africans was matched by a growing sense among the Americans that change was possible. The year 1961 was the first of the Kennedy years, the time during which the Peace Corps and the Alliance for Progress took shape; the national mood was more hospitable to foreign aid programs than it had been for a long time. Why not an African education program? If Africa wanted help, America would gladly provide it.

The Endicott House Conference decided to start with a mathematics program, which seemed to offer the quickest way in. Newly developed American materials might be adapted, and everyone was confident that little cultural bias would arise; everyone agreed that mathematics was the most universal of the disciplines and surely would transcend cultural differences.[4] A mathematics program would require no equipment more costly than a piece of string or a few wooden blocks, which was important when even the price of a piece of string might prove too much. Philip Morrison, who was an active participant in the conference, spoke about the role of mathematics and was quoted in the final report: "It is a plausible position that the integers in their infinite array contain some sort of model for all the kinds of order of which the mind has yet learned in the natural world. All that order, all that richness of relationship which enables the induction of regularities, the establishment of new relations and the rise of novelty, the quick test of truth, lies immediately at hand with nothing but chalk and the mind to carry into the classroom."

W. T. (Ted) Martin, chairman of the MIT mathematics department and a member of the steering committee, was asked to take charge of creating an African mathematics project. As soon as this was approved by the meeting in its final session, Martin and Steve White headed for the telephone. Half an hour later, White recalled, and before the meeting had actually ended, they were able to announce the formation of the African Mathematics Steering Committee and that one of its members, Patrick Suppes of Stanford University, would be en route to Africa within the week to begin conversations with local authorities. Another mathematician, David Page of the University of Illinois Arithmetic Project, who had impressed the entire conference with a brilliant demonstration of teaching techniques with a local group of children, would head for Kampala, Uganda, in September. It was very fast work, but the immediacy of the African appeal was not to be denied.

The Minds of Children: The Elementary Science Study

Now what I want is Facts. Teach these boys and girls nothing but Facts. Facts alone are wanted in life. Plant nothing else, and root out everything else. You can only form the minds of reasoning animals upon Facts: nothing else will ever be of service to them. This is the principle on which I bring up my own children, and this is the principle on which I bring up these children. Stick to Facts, sir!"

The speaker and the schoolmaster, and the third grown person present, all backed a little, and swept with their eyes the inclined plane of little vessels then and there arranged in order, ready to have imperial gallons of facts poured into them until they were filled to the brim.

—Charles Dickens, *Hard Times* (1854), bk. I, ch. I [5]

The members of the Endicott House Conference would almost certainly have been unwise had they tried to begin the African program with elementary school science, although they considered the possibility briefly. Such a move would been premature; new and exciting ideas were just then emerging about how to introduce genuine science to small children, but no one yet could claim much experience.

Elementary school science was a curious arena in which to find professional scientists. They were often regarded by school authorities as interfering amateurs, and they were hardly welcomed by theorists in the schools of education. Certainly neither Jerrold nor his colleagues had expected to be involved in this field. For example, Jerrold had been a member of (and Stephen White a consultant to) a panel of the President's Science Advisory Committee in 1959 and had contributed strongly to its thirty-five page report entitled *Education for the Age of Science*. While the report called on the nation to increase its efforts to improve science education, nowhere in it was there any mention of elementary-level education, whether devoted to science or not.[6] Most of the emphasis was on high schools and the improvement of teacher training, with a nod to college-level science education, and those were the issues on which Jerrold had expected to concentrate.

The idea that scientists could—or would—design a high school physics curriculum was hardly radical, even if it took professional educators by surprise. It was reasonable, after all, that scientists should take an active interest in physics at the high school or early college level, for that kind of physics was merely an early version of what physicists did professionally every day. But that scientists might bring the same kind of reforming zeal to the elementary schools was certainly much less foreseeable. Why should anyone imagine that these same scientists could invent a brand of classroom science suitable for children not yet 10 years old? What, after all, could scientists know about teaching little children, who had still to learn the multiplication tables and the names of the capitals of the states?

Yet in the spring and early summer of 1960, Phil Morrison, Fran Friedman, and Jerrold Zacharias had met to consider exactly that question:

how they might invent a suitable kind of elementary school science. Not professional science, of course, not molecular beams for third graders. But whatever new programs they might develop, they thought, they must manage to keep the essence of real science: the programs must be based on experiment and discovery, on learning by doing. The need for such a program had already become clear from the PSSC experience with high school students. The teachers and developers of PSSC had discovered that if young people had not learned by high school age to explore and experiment, they were unlikely ever to do so. One had to begin much earlier. In a reflective speech given in 1971, Philip Morrison spoke of what had been in their minds at the time: "It was natural that elementary science should appear to us as the place to go to try to prepare that instinctive concern, what Dewey called the 'sub-soil' of the mind for what we saw as science in the modern world."[7]

There were a few ideas to build on—John Dewey's among them—but very little experience. Nevertheless, they thought there were possibilities. For example, William Weston, professor emeritus of biology at Harvard (called Cap by everyone, students and colleagues alike), and Charles Walcott, a young associate of his in the Harvard biology department, had made some highly successful television films for use in elementary classrooms. Weston, at least, seemed to have part of the answer. Morrison went on in his speech to tell about it:

Cap Weston and Charlie Walcott, working together around television, had organized a highly successful early program in film in the classroom, called "The 21-inch Classroom," in which Cap Weston made an extraordinary personal impact—a man then 70, I think—on the schoolkids in the greater Boston area who had a 21-inch screen in the classroom, so that he became a figure who could not walk down the streets of any suburb without being attacked by dozens of little kids with their clam shells or their spider webs asking for special information from Cap, who appeared on television once or twice a week (while the children used the concrete material in the classroom).

They did not know how much of Weston's success came from his technique, which might be transferable to others, and how much from his own personality, which was extraordinarily warm and welcoming, to both children and adults. Weston seemed to be everybody's favorite uncle, and it remained to be seen whether others could succeed with his methods. His basic idea was one to which they were all completely receptive: that young children working with actual materials in the classroom could discover for themselves how things work instead of being filled with facts. If Weston's work could be extended—not on television but in the actual classroom—

children might be able to discover for themselves the basic canon of Observation, Evidence, and the Basis of Belief. No teacher would have to explain it to them; the teacher's role would be to allow it to happen, to encourage it, and not get in the way of it. It was the most natural idea in the world for anyone with experience of PSSC.

These concepts had indeed arisen at the Endicott House Conference in the summer of 1961. They were generally well received by the American participants but with a certain understandable wariness by the Africans, who were concerned about the risk of undermining the authoritative role of the teacher in African society. But the views Friedman, Morrison, Weston, and Zacharias expressed had seemed promising, and the promise outweighed whatever reservations the African participants may have had. Cap Weston, a participant at Endicott House, had put it this way: "We feel that in this teaching in Africa, or wherever it may be, it is no sense taking the receptive casing of the student and stuffing him with no matter how well mixed material, because then you have only something which when sliced can give back samples on the examination. . . . You cannot contribute to their development unless they can get some feeling of participating in these things, or taking part in working out these things by themselves. 'I found this out, I can go ahead and find out more'—that feeling of exploration and discovery."[8]

At the end of 1961, Educational Services, Inc. submitted a proposal to the National Science Foundation for the establishment of the Elementary Science Study (ESS). The ESI administrative staff drew up the budgets and so forth, and Jerrold himself shepherded the proposal through the foundation. Morrison recalled the beginnings: "[It was] Charles Walcott, a young biologist very much caught up in the problem, and Francis Friedman, an early mover, and also the primary science program in Africa which got us going. So [it was] Oakley, Zacharias, on the side with leadership; Friedman, I think, with Walcott and Weston, beginning to cope with the problem of what the subject matter would be; and a special deep assistance from the one person in the standard education schools in the country with whom we immediately formed the closest alliance, Gil Finlay of the University of Illinois." Morrison modestly neglected to cite his own vastly important contributions to the creation of an elementary science program, but they had been such that when a steering committee for such a program was formed, it was completely obvious and natural that Morrison would head it.

The basic idea of ESS was simple in concept and certainly not new. More than a hundred years earlier, Thomas Huxley, defender of Darwin's theory of evolution and a great scientist in his own right, had already expressed it: "In explaining to a child the general phenomena of Nature you must, as far as possible, give reality to your teaching by object lessons; in teaching him botany, he must handle the plants and dissect the flowers for himself; in teaching him physics . . . don't be satisfied with telling him that a magnet attracts iron. Let him see that it does; let him feel the pull of one upon the other for himself. And, especially, tell him that it his duty to doubt."[9] Huxley's comment could have been incorporated directly into the ESS proposal. Nevertheless, it was extraordinary that this concept should have continued to be so difficult for so many to accept.

At first, Jerrold had some reservations about how to implement the ESS philosophy, although he agreed with it wholeheartedly. But the program was the legitimate child, one might say, of PSSC, and he had no difficulty in endorsing it. Although most of the specific classroom ideas were not his (he was quick to point that out), he regarded them as expressing his most deeply held beliefs. He admired the ESS tremendously and in his capacity as academic director of ESI he became an ardent and tireless spokesman for the proposal. The planners had begun with support from the Sloan Foundation ($15,000) and the Victoria Foundation ($76,000), and in the fall of 1961 the National Science Foundation made an initial award of nearly $300,000. The signatories to the proposal had been Finlay, Friedman, Morrison, Oakley, Walcott, Weston, and Zacharias. It was a stellar lineup, well calculated to win support; but, of course, just as with PSSC, the continuation of any support would ultimately depend on the quality of the work.

MIT: The Science Teaching Center

Jerrold moved so easily among his many projects that it is easy to forget that his primary connections were with MIT. His concerns for MIT students had led him into education in the first place, and although he had become fascinated by new challenges, he still felt responsibility for MIT's undergraduates. By early 1959 he felt he could safely anticipate a successful outcome for PSSC, although a lot of hard work remained. He and Fran Friedman, therefore, even while still deeply occupied with the texts and films, began to discuss next steps. It seemed evident to them, and urgent, that colleges must be ready for freshman students who had taken the PSSC course in high school. Probably, they thought, a new college course would be needed, and they began to prepare a suitable plan.

On April 2, 1959, MIT submitted to the NSF a proposal by Zacharias and Friedman bearing the unwieldy title, "The Development of a Two-Year College Physics Course to Follow the One-Year High School Physics

Course Currently Being Developed by the Physical Science Study Committee." Jerrold was to be the project supervisor. The proposal represented a bold claim on a student's time as well as on his mind: it specified a two-year course in place of the conventional one-year version, dealing with topics more advanced than the usual freshman offerings. The best students would be attracted, they thought: students who had succeeded with PSSC in high school and were now ready to commit themselves seriously to college physics. Those were the students they wanted.

The proposal also stated MIT's intention to establish a new unit, to be called the Science Teaching Center.[10] "It is our intention to establish here at MIT something in the nature of an interdepartmental laboratory dealing specifically with the teaching of science at the undergraduate level. The general program of this organization, which may be tentatively be called the 'Science Teaching Center,' will be to foster a high level of excellence . . . to develop methods to help us nationally to deal with the very large number of students who will be seeking science education."[11]

The new curricular reform programs (PSSC in particular) had fitted well with the nation's growing awareness of the importance of science and technology, much of which had arisen from the well-publicized American efforts to gain the lead in the space race. Science had become popular once more, the career of choice, and Jerrold and Fran Friedman anticipated a flood of new students; they wanted MIT to be ready for them. Their prediction of an increasing science student population turned out to be correct, at least during the next decade, as the space race intensified and federal funds for science and technology—including ample fellowships and scholarships—continued to be plentiful. They hoped the Science Teaching Center would have a major role to play in bringing about a new public awareness of science education. The development of a new freshman-sophomore physics course could hardly accomplish that by itself, but Jerrold meant the development of the new course to be only the first program of the center; he expected that other programs would follow in the other sciences. Jerrold's dreams for the center were characteristically ambitious.

The MIT proposal went to Harry C. Kelly at the NSF, but this time the results were neither so immediate nor so favorable as they had been when PSSC had been proposed. MIT was no longer alone in the field; there were now other competing proposals for the NSF to consider. The American Association of Physics Teachers, for example, had submitted its own proposal, and others had expressed interest. Furthermore, the NSF had not yet decided what it wanted to do about college-level science education. Not least, the foundation was also concerned about the political implications of making MIT almost the sole developer of new educational programs. The NSF could not afford any suggestion of favoritism, and MIT had already received a large share of the foundation's education funds, starting with PSSC.

Kelly therefore wrote to the officials of MIT what he had already communicated privately by telephone to Jerrold: that the NSF would have to proceed diplomatically. "It might be proper at this point," Kelly wrote, "for a conference of physics professors from various colleges and universities to consider what should be done with collegiate physics, and to think about the scope and content of college physics courses. The result of such a conference might help the Foundation decide what future steps should be taken."[12]

Jerrold understood from Kelly that funding for such a conference was likely, although the NSF was not yet ready to fund a program. He immediately submitted a new proposal calling for a series of three meetings that he hoped would "crystallize the thinking . . . and result in the submission of specific plans of action." He offered long lists of possible participants in such meetings; they included physicists and educators from thirty-seven academic institutions, large and small, from several national organizations, and from industry and government. Jerrold's experiences with PSSC had brought many new people into his own network, and Friedman was able to supply still more names. Once again, Jerrold's ability to find people all of whom were busy but who nevertheless were prepared to participate in his ventures, was impressive.

At Jerrold's strong urging, MIT decided to go ahead with the Science Teaching Center, using its own resources. The center was formally established in September 1960 (as soon as Jerrold had returned from the conference in Israel), with Francis Friedman as its director. The meetings Jerrold proposed were indeed funded by the NSF and took place in late 1959 and early 1960; eventually, in March 1961, the NSF made a direct grant to the center for the development of the two-year physics course.

The center itself, as the original proposal had said, was to be an interdepartmental laboratory. Jerrold was eager for the project to succeed. He saw education as a vast territory badly in need of exploration and development, and he wanted MIT to be the pioneer. The center would provide the means.

Jerrold's concept of the center resembled more the Laboratory of Nuclear Science than it did a small laboratory devoted to research or teaching. It might start small (the initial proposal to the NSF had been for approximately $120,000 but its potential for growth was large. Jerrold thought he had reason enough to feel optimistic about it, for he continued to find strong support in the upper levels of MIT governance. The associate dean of science at that time was Francis Bitter, a close friend of Jerrold since their Columbia days. Bitter favored the project and backed it firmly. Jerrold felt that Killian, now chairman of the MIT Corporation, would also support it as he had the original PSSC proposal, and although Killian no longer involved himself with the details of MIT governance, his views would carry weight. Julius Stratton was then MIT president and generally sympathetic to Jerrold's ideas about education. Jerome Wiesner, perhaps Jerrold's closest friend, had just left MIT for Washington to become President Kennedy's science adviser; his influence was strong, both at MIT and in Washington, and Jerrold felt he could look to him for support as well.

But above all, his hopes and expectations for success lay with Francis Friedman as director of the center. As had been the case with PSSC, Fran Friedman would perfectly complement Jerrold; Fran would be the "inside man," providing the follow-through on their ideas, many of which he would have inspired himself. Jerrold would, as Al Hill liked to say, "just be Jerrold": raise funds, generate ideas, stimulate, persuade, poke, and prod. He would attract and recruit talent through his network and he would see to the general welfare of the center, both at MIT and in Washington.

It didn't happen that way. In time, the Science Teaching Center did absorb much of Jerrold's energy and occupied a central part of his life. But when that time came, Fran Friedman was no longer there to share it with him. In the language of the PSSC filmmakers, a different story line had developed. The final scenes were not played until much later.

A Visit to Lagos

MIT celebrated its hundredth birthday in the spring of 1961 with a grand centennial convocation. That spring, Jerrold was preparing for the Endicott House Conference on African education and did not involve himself with the centennial committee; as one committee member recalled it, Jerrold simply "kibitzed from the periphery." But as it happened, Jerrold had taken notice of the young man who was in charge of making the arrangements.

He noticed him again later, in the fall of that year, at the first U.S. Pugwash conference in Stowe, Vermont. Jerrold was a member of the U.S. delegation, and Leona had accompanied him to the meeting. The conference had been organized by the American Academy of Arts and Sciences, and it was the same young man, on loan from MIT, who had chosen the site and had made all the arrangements. His name was Arthur Singer, and in addition to being young, he seemed to be smart and energetic as well.[13] Jerrold, whose recruiting instincts were never dormant, made an on-the-spot judgment. Singer recalled:

Zach said, "We're having a meeting out there, somewhere in East Africa—we'll probably use the university at Kampala and would you go out and set things up the way you did for the centennial and for Pugwash? Would you be the advance man?" It turned out that the plans for that summer meeting were in the most elementary shape, nobody knew what the meeting would be like, except maybe Zach, and very little thought had gone into it. Ted Martin was closely involved, and Zach said, "You get briefed by Ted Martin and Steve White and then get on the road, get out there."

So I went to see Ted and Steve—that's when I first met them—and they gave me a list of names of African educators and English expatriate educators who were in Africa, and they bought me a ticket and sent me on my way.

That was typical of Zach. He took a liking to me and the way I worked, and decided he was going to bet on me to put this together.

Kampala's Makerere University, Singer found, could not possibly host a mathematics conference in the summer; it would be in session and its halls and dormitories in full use. He turned instead to the Lake Victoria Hotel, beautifully located in Entebbe, Uganda, 30 miles from Kampala and 50 from the spot where the Nile begins its long northward journey. The hotel could accommodate a large group, and with a magnificent view of the largest lake in Africa, would be an extremely pleasant place to work. Singer signed it up for the summer at once. The hotel served as headquarters for the program in the summer of 1962 and for several summers thereafter.

Singer also managed to make sure that a considerable number of the educators on his list would be available, at least insofar as he could get commitments from them without himself yet having funds to commit. He then traveled across the continent to Lagos, Nigeria, to meet Jerrold, who had traveled there with Leona, Ted Martin and his wife, Lucy, and L. John Lewis. It was already the spring of 1962, and Singer was anything but sanguine about the future: "First of all, I was feeling kind of inadequate to this assignment and I was also feeling as though this was an awfully big plunge to take, to get all these people gathered—we had no money at that time, it only came through later—we were planning something without the money and without the Africans understanding what we were about or even the man who was representing the undertaking—me—understanding very much either. I thought we were really flying by the seat of our pants."

Nigeria, the most populous country in Africa and partly because of that the most influential, was necessary to their plans, and Jerrold had come to Lagos, the seat of its government, to try to ensure Nigerian participation.

At that time, all African nations had centralized educational systems; texts, syllabi, and requirements were all selected by their respective ministries or departments of education. Unless the government was willing to commit itself to it, therefore, no curriculum development program could succeed. Consequently, in order to foster government participation, Singer had arranged a meeting between Jerrold and Babatunde M. Somade, the chief inspector of education, Government of Western Nigeria.

Singer was present at the meeting, together with a representative of USAID (U.S. Agency for International Development). Singer recalled his impressions: "The conversations were between Zacharias and Somade, it was as though no one else was in the room. Somade was an impressive man, tall and dark. He had great presence and exuded strength, and he was smart. He was very articulate. . . . He said to Zach, 'Tell me about your undertaking,' and Zach told him about it . . . and Zacharias told him how this new math needed to come to Africa, they needn't go through the same trial and error that had characterized American education. . . . Africa could leapfrog to get ahead fast, they could jump ahead."

Somade was a man with a reputation among Nigerians as a thorough-going bureaucrat, much concerned with his own power and influence. Since he had not been at Endicott House, he had not been exposed to the contagious enthusiasm that had developed there over the course of weeks. He decided that Jerrold, as an American, must know little about the African education system or of the British system on which it was based. According to Singer, Somade "could not be persuaded that Zach was a man who knew what he was doing." Singer recalled further:

When Zach was finished Somade gave a long speech. . . . He said, in effect, "Look, that's not the way to start. You're trying to get all of English-speaking Africa together in one major project. . . . These peoples are very different from one another, there are so many tribes and languages and dialects and tribal rivalries. You're covering the continent from Accra in the West to Nairobi and Dar-es-Salaam in the East, millions of people, many different traditions.

We need help, but the way you are starting is so ambitious. . . . You're going to fail. Try a small experiment in one location, working on a trial basis with a small group of African educators. Plant a little seedling, see how it grows

This was not at all what Jerrold had traveled so far to hear. Singer re-called the end of the episode:

When we left the meeting and got in the car, I said to Zach, "You know, that guy makes an awful lot of sense. . . . I've done a lot of advance work for you but I've done it holding my breath. I'm not at all sure it's going to work."

Well, he gave me the most withering look, I'll never forget it, as though I had slipped badly in his estimation. . . . Zach's theory of change didn't at all coincide with Somade's, and Zach said to me, "Never change course without a good reason—and we haven't heard any good reason." And he was not the least bit deflected.

Jerrold was not a diplomat and could not be one as long as it meant saying one thing and meaning another. He did not believe gradualism would work and he had not responded well to Somade's evident desire to keep developments small and under his own control. Jerrold apparently had allowed Somade to see how he felt about such advice. It was honest but not very politic.

Jerrold had been emphatic at Endicott House that no program should be attempted unless the Africans wanted it, and as far as he was concerned, that issue had been decided by people he had come to respect and trust. Somade was a bureaucrat of the kind one might find in any agency in the world; Jerrold did not intend to let Somade turn him aside. He was highly impatient to get on with a program. If the goals were big, so was Africa, and there was little time to sit around and talk about them. Jerrold's enthusiasm and impatience had even at times caused difficulties back in the United States; it was guaranteed to cause difficulties in Africa, where no one was familiar with his direct style. Africans, newly independent, were understandably sensitive to anything they interpreted as reminiscent of colonialism. The British had been in Africa for generations, and Africans felt they had had enough of being patronized by them; nevertheless, British methods had won a grudging respect and acceptance on the continent. Ted Martin spoke of a view of Americans frequently held by Africans and British expatriates alike: "We were seen as the worst kind of upstarts: we had money and we acted as though we knew just what had to be done." As matters developed, Somade eventually became a supporter of the program, traveling to visit ESI in the fall of 1962. Ted Martin reflected on the general difficulties of Americans in Africa at that time, and on Jerrold's experience in particular: "It was awfully easy to offend somebody in those days... and we didn't really know very much. I remember once in Zach's office somebody mentioned "Dar-es-Salaam", and I went home and said, "Dar-es-Salaam, Dar-es-Salaam, I don't even know where it is." And I had to look it up. . . .

I didn't have one-tenth of the ideas Zach had, so I could devote more of my mind to diplomacy and to compromise. If you'd had only Martin out there, you wouldn't have had any problem, but you wouldn't have had any solution, either."

Jerrold was eager to get home. The Science Teaching Center at MIT needed him. It was running under Francis Friedman's direction, but Francis was not well and needed help. In any case, Jerrold had never intended to involve himself substantially with the African mathematics program once it got going, and after his excursion to Africa (which included a stop in Nairobi) he left it to Martin and the mathematics steering committee. Later, he came to regret that he had not involved himself more in the substance of the program, but he had apparently decided that the role he preferred was that of initiator, and he was now ready to turn his attention to yet other projects, including an African science program. It was possible to find capable people to run things afterward, and Jerrold felt that often they did it better than he would have. As Martin, White, Oakley, and many others before and after had recognized, Jerrold liked to have someone to oversee details and implement programs; but no one was ever better than he at overcoming barriers and breaking past conventional ways of thinking. Africa was only one example. Not for the first time and certainly not for the last, he had become impatient with those who, as he liked to say, "had all of the brakes and none of the motors." As Ted Martin had suggested, however, if his impatience sometimes irritated his associates, it could sometimes push them toward solutions.

Francis L. Friedman, 1920–1962

Nothing of him that doth fade,
But doth suffer a sea change
Into something rich and strange.

—W. Shakespeare, *The Tempest*, I, ii, 375 (These lines are inscribed on Shelley's gravestone.)

Francis Friedman died of cancer on August 24, 1962. He had been ill for less than a year; the disease had advanced slowly enough to be cruel and yet rapidly enough to keep anyone from becoming reconciled to it. Over the months of physical decline, Friedman had been obliged to withdraw more and more from the projects that were so important to both him and Jerrold, and to which his contributions had been so central.

His death was a heavy blow to Jerrold. There were not more than a half-dozen people to whom he had felt so close: Leona and his daughters; his old nurse, Anna; Rabi, his earliest mentor and intimate friend. There were others with whom he had formed strong bonds of friendship, but his relationship with Friedman went beyond ordinary description. Edwin Taylor, a younger colleague who came on the scene shortly before Friedman had become ill, confessed he had not understood what he saw; in a brief memoir that he wrote later, he said he had been "terrified at the friendly ferocity of the intellectual engagements between Friedman and Zacharias."[14] Jerrold and Friedman had complemented each other remarkably well, each supplying something that the other needed, each forcing the other to higher levels of achievement. They had shouted at each other and bullied each other and together had achieved more than either could have done separately.

Jerrold had not previously had to deal with a loss of such magnitude. It was all the more painful for being unanticipated, all the more unforgivable for being of someone still young, someone still filled with bright promise, someone so close and valued. He had no ready relief from his sorrow; he had always been a private man whose serious emotions remained covered. He could laugh easily with others, but he could not weep with them. Friends sought to help, of course, but whatever release comes from expressing grief was denied to Jerrold; he was obliged to try to come to terms with it at some deeper, less articulated level.

There was a memorial service at MIT that autumn. James Killian spoke in eulogy, describing Friedman's high standards, his deep understanding of science and his integrity. Julius Stratton spoke of the irreparable loss to all the hopes and plans of the Institute, to which Friedman had so much still to contribute.

By contrast, Jerrold, speaking last, was almost terse. He referred briefly to Francis's brilliance, to his qualities, to "the keen edge of his exquisite mind." And then he addressed himself directly to Friedman's children, sitting quietly in the audience: "So, when you find yourself saying, as I so often have these past few days and will many times again, that I wish Francis were here to set me straight, try to recall just how *he* would bear down on any idea until he was satisfied that he understood and that others would, too."[15]

Later, when Friedman was posthumously awarded the Oersted Medal for outstanding contibutions to physics teaching (the same award Jerrold received in 1961), Philip Morrison also eulogized him, speaking of his "insight and taste."[16] But it was a grievous loss, not to be assuaged with words. The closest Jerrold could ever come to expressing his own feelings was the remark that became familiar to his friends and associates, and that he still repeated twenty years afterwards: sitting beneath the photograph of Friedman that had hung in his office since Friedman's death, he spoke quietly to a visitor. "I miss him every day," he said.

Moving On

Nothing could be the same after Francis Friedman died. MIT had to make adjustments, of course; Julius Stratton asked Jerrold to take over as director of the Science Teaching Center, and he agreed to do so, serving until a new permanent director was selected two years later. The directorship was not a position he had wanted, any more than he had ever wanted, say, to be president of ESI, where he had the title of academic director; he had had all he ever wanted of administration. But he felt that all the projects must go forward.

The Elementary Science Study also had to go forward, and Philip Morrison now recruited David Hawkins to be its first director. Hawkins was a professor of philosophy at the University of Colorado who had met Morrison at Los Alamos, where he had spent the war years as the official historian of the Manhattan Project. He and his wife, Frances, an innovative elementary school teacher highly respected for both her writings and her success in the classroom, had been invited by Morrison to a summer workshop-conference in 1962 to help get the ESS program started.

Hawkins's recollections of his first encounters with Jerrold are revealing of Jerrold's style:

Taking on that job proved to be a major turning point in my own and Frances' lives. I left ESS after two years—I was on leave from the Philosophy Department here [at Colorado]—but I've been involved in early education one way or another ever since. . . . In one way or another, Jerrold's often audacious promotional ventures made waves that have not damped out, even after all these years.

I didn't get to know him at first. He didn't intervene in our early work at all. But when I had to write a new NSF proposal, my draft of it went to him of course, and he sent back a veto. It was the first time I'd ever written one, and only a few months after I'd become involved. With help from Phil [Morrison] and others I redid it, and it passed his censorship. I think he was still quite skeptical about what we were up to, but also didn't trust his own judgment, not yet.

As a scientist, Jerrold remained skeptical. It was not that he did not believe in the ESS approach, for he surely did; but he was not yet convinced that they knew how to carry it out. Hawkins felt that as far as Jerrold was concerned, he still had something to prove, although he thought his wife, Frances, probably did not: "As soon as he [Jerrold] had met her he had immediately accepted Frances as a trustworthy and creative colleague, she passed his tests; while I was still on his questionable list, and stayed so for some time. . . . Later we really got together on a number of issues, and could agree (usually) or disagree (sometimes) with personal trust."

Jerrold's transformation into an all-out enthusiast for the ESS product turned out to be relatively quick and simple. "One of our early 'units,'" wrote Hawkins, "was invented by Dave Webster, *MEALWORMS*. When Jerrold saw the draft of that he was decidedly negative, however, and we had the making of a small storm. Someone . . . urged him to go and observe Dave teaching it before he made any judgments. Well, he did, and I think from then on, he was a convert."[17]

Hawkins, ably abetted by Morrison, ultimately was able to infuse ESS with an imaginative and almost poetic spirit, and with this infusion, the project prospered under the tolerant and encouraging management of ESI. Like PSSC, the Elementary Science Study needed to develop entirely new and unconventional methods of working; only a management philosophy as flexible as that of ESI under Gil Oakley could have allowed it to flower. But flower it did, first under Hawkins's direction and then under that of his successors. Hawkins had a natural eloquence, and the goals and purposes of elementary science education have rarely been better expressed than in his essays; but he was a practical man, too, with a realistic view of what ought to happen in the classroom.[18] The combination was effective.

The Elementary Science Study ran for nearly eleven years. When its history was written, a list was drawn up of those who had been associated with and contributed to the work of the project; it numbered nearly 360 persons.[19] The impact of the project on American education was large and continues to the present. A recently published sourcebook records dozens of projects and hundreds of publications, all offering versions of the ESS approach.[20] As David Hawkins said, "the waves from Jerrold's audacious ventures have not yet damped out."

Strange Streets

At the beginning of September 1964, Jerrold and Leona traveled to Rome. He had two reasons for visiting Italy just then. First, the Italian physicists, with the support of the Italian government, were celebrating Galileo's four hundredth birthday with special scientific conferences all over the country, and there were several that he wanted to attend. Second, and in many ways the more important reason, it was time to get an African science education program started, and Rome, equally convenient for both Africans and Americans, was a good place to hold a small planning meeting.

On September 3, 1964, according to a remark in his notebook, Jerrold set out for a walk along the streets of Rome. Starting from the Bernini Bristol Hotel at the foot of the Via Veneto he strolled more or less at random through the neighborhoods of the city and walked again the next morning. By noon, he was back at the hotel, ready for lunch and a scheduled meeting on African science.

James Aldrich, then a senior administrator at Educational Services, Inc., also set out that morning on a stroll through the unfamiliar but fascinating

Rome streets. Unlike Jerrold, he managed to lose his bearings and wound up taking a taxi back to the hotel—a distance, he recalled with embarrassment, of a block and a half. Like Singer in the mathematics program three years earlier, Aldrich was to be responsible for finding suitable locations for meetings and for making the necessary arrangements. He recalled that he had had a moment's panic imagining Jerrold's reaction had he arrived late at the very first foreign meeting he had organized.

The meeting took place successfully, in spite of Aldrich's nervousness. It was an informal session, to which Jerrold had invited some of the Africans who had been at the Endicott House Conference in 1961: A. Babs Fafunwa of Nsukka University, Nigeria; Bede Okigbo of the University of Ibadan; Akindele Osiyale, University of Lagos; John Gitau and Gilbert Oluoch of the Kenya Ministry of Education; and one or two others whose judgment he thought valuable. Several American members of the African science steering committee were present.[21]

Athough Jerrold was ready to start an African science education program, there remained questions in his mind that had not been answered by the experience with mathematics. Would the Africans be prepared to accept the open-ended, freely inquiring approach that had come to characterize American innovation? The African systems were closely patterned on the British model and remained strongly influenced by agencies such as the British Council and the English Nuffield Foundation. Was there any room at all in that mix for an American-sponsored program? Perhaps Jerrold remembered too keenly his encounter with Somade. How would free scientific inquiry strike the bureaucratic mind?

Whatever step came next, Jerrold would have to be the one to take it; he was prepared to seek the seed money that he knew might be available from the Ford Foundation but only if he could offer a plan with some promise of success. That would depend on the answers to his questions; there was no use, he said, of importing programs to Africa if the African education establishments did not want them and would not use them.

The Africans who came to Rome urged strongly that they proceed. The optimistic and enterprising mood that had been created at the Endicott House Conference continued to move them, and they were still confident that the problems, however large, were all capable of solution. They pressed for a new version of Endicott House, a major conference focused on science to be held in Africa itself. They wanted to bring together the educational decision makers of Africa—the ministry officials, the heads of the schools inspectorates, and the deans of the university education schools, with American and English educational innovators. What would emerge, they were certain, was consensus that ESI-style science innovation was the right path for Africa.

Did Jerrold persuade them, or did they persuade him? It was not clear, but a pan-African conference was worth a try, and Jerrold agreed to seek support for that winter. Short notice, thought Aldrich, wincing; the logistics would certainly not be simple. But he was learning, as Singer and White and Oakley had learned before him, that when Jerrold shifted into top gear, everybody else had better shift with him or be left behind.

Less than six months later, approximately sixty scientists, educators, and ministry officials gathered at the Central Hotel in Kano, northern Nigeria, for a ten-day planning conference on African science education. It was a remarkable location for such an undertaking; Kano was a still-feudal town on the southern edge of the Sahara Desert, surrounded by the villages of the Hausa and Fulani peoples, cattle tenders whose practices were little changed in a thousand years. It was a center of trade and arid-land agriculture; huge pyramidal stacks of groundnuts lay massed at the entrance to the city, awaiting shipment, sometimes by camel caravan into the Sahara. The city seemed to connect far more closely to the ways of Africa's ancient past than to any hopes for the future. Robed and turbaned men went by on horses or camels; the city's structures were of sun-baked mud and wattle, including even the elaborate palaces of the emirs. But for those who came there to discuss science teaching, Kano turned out to be exciting and filled with potential. It contained, for example, an enormous market, many acres in size, capable of equipping a hundred school laboratories, if only imagination and creativity could be mobilized. Fruit and grains, basketry and rope-making and weaving, antimony for cosmetics and pigments for color, all were there: there would be no shortage of substance for the teaching of science. To those who were strangers to the continent, Africa itself seemed like a single, vast and wonderful laboratory.

About half the participants in the Kano meeting were Americans, including Cap Weston and Phil Morrison; half came from English-speaking tropical Africa. There were a few people from the United Kingdom (notably from the Leicestershire school system, where a successful experiment was being carried out inspired by the same philosophy that motivated the equally successful Elementary Science Study). Observers from the Ford Foundation, USAID, and the British Council completed the roster. Jerrold ran matters expertly, convening the sessions with a large and handsome brass bell of local hand manufacture that he had acquired somewhere in the town.

Jerrold described the meeting some years later as "simply thrilling."[22] Fafunwa of Nigeria and Gitau of Kenya described small but ambitious starts that had already been made in Nsukka (Eastern Nigeria) and Nairobi. Philip Morrison spoke about the basic educational philosophy that had inspired ESS, and E. R. Wastnedge and L. Sealey, of the Leicestershire project, reported on that experiment. Classroom demonstrations were striking: M. B. R. Savage, a brilliant schoolteacher of Nigerian-Ghanaian-Scottish descent, had come to Kano a few weeks ahead of the conference and had worked for a total of only six hours with eight African primary school teachers.[23] Those teachers, in turn, having learned to teach several ESS units (in particular, the Batteries and Bulbs unit and the Pendulums unit), were able to give demonstration classes with local children whom they had not taught before. Their performances were stunning and absolutely convincing, as impressive to those who were familiar with ESS as to those who were not. There could be no doubt that science education for Africa's young children was full of extraordinary promise.

The Kano meeting fulfilled every hope that Jerrold had had for it; with the promise of funding from USAID, the African Primary Science Program was initiated. Barely four months later, the first large-scale science workshop meeting in Africa was held for six weeks in Entebbe, Uganda, with some sixty people in attendance; the African Mathematics Program had by then found other quarters that they found more congenial (at Nyali Beach, Mombasa) and under the pressure of time, ESI found it simplest that summer to stay with the Lake Victoria Hotel, lake flies and all.

The African Primary Science Program (APSP), born at Kano, turned out to be a great success, overall; it was a model foreign aid program. The Entebbe curriculum development workshop in the summer of 1965 was followed by an even larger one in Dar-es-Salaam, Tanzania, in 1966 and then by one in Akosombo, Ghana, in the summer of 1967, this last being held virtually in the shadow of the recently completed Volta Dam. In all of these workshops (which were directed by the author), the approximately fifty-fifty mix of Africans and Westerners that had characterized Kano continued to be used successfully.[24] Far more than in any other foreign-aid program, APSP was operated as an equal partnership between Africans and those offering assistance. Teacher training workshops followed, and new science teaching materials were introduced in a dozen Anglophone African countries. In 1971, control of the program passed to African hands. In an extraordinary instance of international agreement, a number of African states established the Science Education Programme for Africa (SEPA), fully pan-African in concept, with headquarters in Accra, Ghana. Under the directorship of Hubert Dyasi, SEPA took over the management of the program and the responsibility for raising the necessary funds, both from private sources and from African governments.[25]

Jerrold was vastly pleased by this program, especially by its flexible and imaginative adaptation of the ESS philosophy. It proved once again that the principles of science—the principles of observation, evidence, and the basis of belief—transcended boundaries of nation, culture, color, and creed. He wrote

From many standpoints the results of the [Kano] meeting were extraordinary . . . the conference recommended (and has resulted in) an international co-operative spirit, not only among English speaking African nations, but with other English-speaking nations in the world.
     . . . It is interesting to note that neither the processes of learning nor the topics the children undertake to learn were in any of the syllabuses envisioned in 1965, which we thought might constrain us. Approximately fifty topic units have been planned, tested, refined, retested, and are generally available for classroom use. For lower primary grades there exist one called *Dry Sand*, another called *Wet Sand*, and still another called *Wheels*. For middle and upper primary we have *Seeds, Small Animals, Ask the Ant Lion, Mosquitoes, Chicks in the Classroom, Torch Batteries and Bulbs, Making a Magnifier, Estimating Numbers, A Scientific Look at Soil, Measuring Time, Balancing and Weighing*, and *Pendulums*. However, no amount of talking about this kind of work is a substitute for doing it or watching it in the classroom. I believe that of all the curriculum projects I know about, The African Primary Science [Program] is by far the best.[26]

He could feel now that the obligation he had voluntarily undertaken after the death of Solomon Caulker had been met as well as might be possible. The task had taken him down strange streets in strange places, but had been remarkably rewarding. No one who had the good luck to be caught up with him in those days could possibly have disagreed.

Education and the Great Society

Difficult Years: 1963–1968

The "thousand days" of John Kennedy's presidency—January 1961 to November 1963—were days of change, a time of finding new directions. Bold steps were taken and perhaps some foolish ones as well. Historians will probably continue to have a difficult time assessing the Kennedy presidency, but it is clear that in that brief administration, the United States made some hopeful and imaginative new starts. Kennedy's legacy included the idea that individuals make a difference; people, in small groups as well as large, even on occasion singly, sometimes succeed in making the world a better place. Such beliefs, when they become part of how people think, have consequences far beyond the limits of government itself. Some years later, Rabi expressed this view when he spoke to Jerrold's students at an MIT seminar: "The machinery [of government] consists of human beings. . . . The national goals as laid down by the president and his people, which look to you corny, have enormous consequences when translated into thousands of separate actions."[1]

Inevitably the attempt to make a better world led to increased expectations among those who saw themselves with less than their share of the national well-being. There emerged into the public consciousness—gradually at first and then with great force—the fact that inequity and poverty severely afflicted large sections of the population. The problems of the inner cities could no longer be ignored; the voices of disadvantaged minorities began to be heard increasingly loudly. In many places a mood of restlessness and dissatisfaction began to be evident.

Lyndon Johnson, succeeding Kennedy in the White House, found himself with his hands full. The civil rights movement was rapidly gaining strength in the South, and in his first presidential summer, racial tensions exploded into violence in black ghettos in New York's Harlem, Newark, Chicago, and Philadelphia. Johnson, like Kennedy, attempted to take bold steps. In March 1964 he submitted to Congress an ambitious war on poverty program, calling for nearly $1 billion in new funds to break the cycle of poverty for the nation's young people, and in January 1965, following his decisive electoral victory over Senator Barry Goldwater, he presented to Congress his goals for what he called the Great Society.

These proposals called for additional federal support for education, health care, and the arts. Johnson called for congressional action and support for projects to improve the quality of life in the cities, to cope with regional or local pockets of poverty, to eliminate obstacles to voting rights, and to establish care for the elderly and disabled. Congress responded, in one exemplary instance, by establishing Medicare. Johnson proposed grants for public schools, and Congress responded by establishing regional education laboratories. Funds continued to be available to such agencies as the National Science Foundation, especially for allocation to programs that promised to improve the national welfare. There seemed to be no limit to what the country could afford to do if it decided to try to solve its problems.

Jerrold was in a particularly advantageous position to understand and take part in this movement. He continued to serve as a member of the President's Science Advisory Committee under Kennedy and then under Johnson. He had remained chairman of that Committee's Panel on Educational Research and Development since the days of Eisenhower. His service on PSAC and the view it provided of the nation's domestic problems profoundly affected his activities all during this period and gave to them a new dimension of strong social concern.

President Johnson's list of proposals in January 1965 did not deal solely with issues of poverty and injustice; it also contained proposals signifying other serious concerns. The country was still deeply troubled by the conflict with the Soviet Union, and there was still a widespread fear of the encroachment of communism. There continued to be the belief in many quarters that the arms race could be won. Together with his proposals for Medicare and for new initiatives in education, Johnson proposed with unconscious irony appropriations for the *Poseidon*, a new submarine strategic missile, as well as other considerable increases in the military budget. In 1965 the United States was evidently not yet through with the cold war and had not yet discovered that it could not afford to do everything.

In 1965, American involvement in Vietnam grew inordinately, from 23,000 military advisers at the beginning of the year to 184,000 soldiers and marines—no longer described as advisers, they were now full military combatants—by the end of it. The draft of young men was doubled in July, from 17,000 to 35,000 per month. Nothing, however, was done to eliminate inequities in the draft process that led to a highly disproportionate number of blacks being called to fight in Asia while college students, a large fraction of whom opposed the war and were becoming increasingly vocal about it, remained exempt from such service. It would have been difficult to invent a more divisive practice.

The problems turned out to be more severe and more complicated than anyone had imagined. Eventually the noble scheme for the Great Society was destroyed as the war in Southeast Asia continued without success or resolution while unmet domestic needs became more pressing and critical. Again and again the restlessness that characterized those years erupted into violence in American cities. Civil unrest spread from the ghettos to the campuses; the civil campaign against poverty and discrimination in which the government was prepared to join merged with the popular campaign against the war, in which the government would not join. The results were tragic for the nation.

Lyndon Johnson did not seek reelection in 1968, and Richard Nixon succeeded him in the White House. But that, as Jerrold would have said, is another story.

Social Concerns

The country's growing restlessness in the 1960s was mirrored in Jerrold's activities, for during this period he initiated or was involved in project after project, barely seeming to pause as he went from one to the next. But any impression of dilettantism would have been illusory; everything he did, every project he set in motion, was aimed at the single target he thought most important: education. Each project was motivated by the belief that much that was wrong with American education could be made right if only the right people could be gotten to work on it.

By now Jerrold had evolved a useful and effective formula for initiating educational reform projects. It was a method well calculated to get things going; it was simple and direct, consisting only of five steps:

first, decide where a suitable problem might have to be confronted,

then if necessary, look to private foundations (or, on occasion, to MIT or to ESI) for small seed money grants to get started,

and then bring together a few influential and knowledgeable people to plan a summer study,

and then   let the summer study define the problem and design a course of action,
and finally take that proposal to the appropriate federal agency or private foundation—or both—for the major grants that would get things going.

It was such a simple scheme that it might have seemed to some as if
anybody could do it—except that anybody could not. It required Jerrold,
with his ability to bring together first-class minds, with his reputation for
successful innovation, and with his access to the decision makers in Washington and elsewhere, to make it work. Above all, perhaps, it required
Jerrold's unfailing instinct for identifying the essential problems, the ones he
liked to describe using Robert Oppenheimer's phrase: "important, difficult
and urgent."

The technique was the same one he had used to initiate first PSSC and
then ESS. He had used it again with only minor variations to get the Science
Teaching Center started, and he had used it with great success in 1961 to
get the African education program off the ground. He used it yet again the
very next year to begin a social studies program.

The Endicott House Africa Conference had sought to move forward
on social studies, language, and humanities, as well as mathematics and
science. But there had been nothing on which to base programs in these
areas—no idea or project comparable to PSSC, or to the several existing
mathematics programs, or even to the not-yet-launched ESS, which was
already exciting but still in its germinating stage. Moreover, it was easy to
predict that programs in social or humanistic areas, being deeply rooted in
local culture, would inevitably be difficult and controversial, but in Jerrold's
view, that was no reason not to try.

Step one in the process could thus be taken; a problem area had been
identified. The next steps followed quickly. In the winter following the
Endicott House Africa Conference Jerrold convened a series of planning
meetings at MIT, setting up a small social studies steering committee with
himself and Frederick Burkhardt, president of the American Council of
Learned Societies, as co-chairmen. Other members included Elting Morison,
professor of history at MIT, and Francis Keppel, dean of the Harvard School
of Education.[2] (Francis Friedman, already desperately ill, had a deep interest
and came to the first meetings but could not continue for long.)

There followed another Endicott House summer study: two weeks in
the summer of 1962. It was as stormy and contentious a meeting as any Jerrold
had ever run, but out of it came a recipe for action. By the fall of 1963, funds
had been obtained from a variety of sources,[3] and the Social Studies
Curriculum Program was under way at ESI. In time, it became one of the
largest and most influential projects ever launched by that organization. In

the ESI quarterly report for the winter-spring of 1964, it was possible to report a number of important additions to the ever-growing ranks of Zacharias recruits; the new program was placed under the overall direction of Elting E. Morrison; Franklin Patterson, director of the Lincoln Filene Center for Citizenship and Public Affairs at Tufts University, became director of the junior high school component of the project. An elementary school initiative began under the direction of Douglas Oliver, professor of anthropology at Harvard; this component was later led by the psychologist Jerome Bruner, then also at Harvard. Several major film programs had been started by the beginning of 1964, under the general supervision of Kevin Smith's ESI film studio. Included among them were pioneering studies of the social behavior of baboon troops in East Africa and an extended anthropological study of the Netsilik Eskimos of Pelly Bay, Canada, covering more than a year of their lives. These films eventually were judged to be among the most successful of their kind ever made; because of their candor, they would also be judged to be among the most controversial ever introduced for school use. At least partly because of this controversy, NSF support for precollege education would later be sharply diminished.

The social studies program, which became known as MACOS (Man, A Course of Study), broke important new ground. Students at all levels were encouraged to come to terms with a new awareness of the social divisions in the country; for the first time, they could take a close look at how their nation had reached its current state. Through facsimiles of historical documents, contemporary newspaper articles, photographs, drawings, circulars, and announcements, they were provided with the raw materials of history and sociology and invited to experiment: to investigate on their own initiatives how history had developed and whether, for example, other outcomes might have been possible or desirable; to discuss with each other their own histories, considered as living, human stories rather than as the inevitable product of an unchangeable, somehow nonhuman process. It was no wonder the program provoked controversy. But in the turmoil of violence and potential violence that followed the murder of Martin Luther King, Jr., in 1968, the developers of this program had the skill and wisdom to produce within a few days a substantial study program on racial relations in the United States and to get it distributed. It was something tangible, and it relieved stress; students could take hold of the issues and feel less helpless. No other program had anything to offer the schools of the inner cities at that moment of strain.

Of course, the specific ideas that found implementation in the Social Studies Curriculum Project were not Jerrold's; clearly, he was beginning to range far outside his own area of specialization or competence, but his style had always been to gather together as many people as he could who did have ideas and then to get out of the way. What was attributable to Jerrold was the initiative to get things started and the establishment of an intellectual climate favorable to the creative ideas that would follow. It is not hard to see why those ideas found favor with him and why he continued to foster and encourage them, for here were his old friends, Observation, Evidence, and the Basis for Belief, brought to life in the study of history and sociology and literature. More than ever before, he was convinced that the patterns of thought of a scientist—not science itself, necessarily, but the way a scientist thinks—were the best antidotes to what was wrong with society.

Jerrold took every opportunity to transmit these ideas to others. He had become well-known in the world of education, to a degree that virtually guaranteed that anyone wishing to begin a new initiative in education—in any area whatever—would feel it almost mandatory to involve him in it, at least to the extent of having his views and advice and, in all likelihood, his assistance as well. Physics had taken the lead, under Jerrold's leadership, and it now seemed clear—to him, at least, for it was still early days—that education in social studies could be changed with similar techniques. Even elementary school education, formerly out of bounds for all but educational psychologists, would yield. The arts and the humanities also now felt the need for educational change, for the times seemed to call for such change. Perhaps there was something here that the humanists, too, could learn. Jerrold was called on for advice even concerning the humanities.

Thus, in October 1964, when an Art Education Seminar was held at New York University (supported by the U.S. Office of Education), it was not surprising that Jerrold was invited as a participant. He was the only natural scientist present. Indeed, there was something that he could contribute from his experience: it was to talk about the value of hands-on experimentation, which he felt to be as essential in art as in science. He was sensitive to the fact that he had no experience in art education, and he reminded his listeners that he could only suggest that certain approaches to curriculum reform in the sciences be examined for possible usefulness. But he felt strongly that those who would lead in curriculum reform should "get out into the classrooms to find out what needs doing, how to do it, and to know first-hand the gorgeous potential of children who are as yet unspoiled and are anxious to learn." It seemed to him imperative, he added, that teachers and students of every special subject—science, mathematics, art, or anything else—have laboratories or studios or workshops to use in teaching, for teachers must convey "the particular appearances, aromas, and textures, *the entire environment and atmosphere* . . . if special subjects are to really 'come alive' for students."[4]

He said nothing different to this audience than he would have said—and had said, many times—to audiences of scientists or anyone else interested in education. What had been discovered by the Physical Science Study Committee within the narrow domain of physics was of far more general application. It could be made to work throughout an entire society, provided only that that society place sufficient value on education.

In the summer of 1964 Jerrold was appointed by the president to a special Task Force on Education, chaired by John W. Gardner, president of the Carnegie Corporation. Lyndon Johnson had made education one of the central issues in his electoral campaign, and the report of the task force was intended to help define a national program. The report was ambitious; it attempted to come to grips with real problems—of exclusion and segregation of the underprivileged, of access and opportunity for everyone—in honest and realistic terms. "This is a fateful moment in the history of American education," it began.

For more than a decade, we have been engaged in a lively, argumentative reappraisal of our schools and colleges, and a search for new paths. The years of appraisal and innovation are just beginning to pay off in a clearer understanding of where we have failed and a surer notion of what our goals must be. . . .

. . . Education will be at the heart of the Great Society. And it will be education designed to serve the high purposes of that society—education that will enable every child to develop his talents, that will liberate and enhance every man and woman, that will create a sound moral and political community.

In the past decade, every informed American has become aware of special contemporary reasons why education is of overriding importance to us—the crucial role of talent in modern society, the rise of automation, increased leisure, the unemployability of untrained men, and the knowledge explosion. Economists are now telling us that the spending of money on education is one of the most productive investments a society can make.

These considerations need not diminish our awareness of an educational aim much more deeply rooted in our national consciousness—the requirement that each of us be sufficiently educated to discharge his duties as a citizen and to exercise the responsibilities that go with freedom.[5]

The task force had addressed itself to six issues, all of which either had been principal concerns of Jerrold or would soon become so. First, it attempted to deal with the urgent question of access to education: the opportunity to learn. "Even today," the report pointed out, "children of disadvantaged background are deprived of normal access to educational opportunity."

Second, the report spoke of developing new and better ways to learn;

it proposed "the creation of a nationwide network of large-scale National Education Laboratories," to be affiliated with universities, to conduct basic research on learning, and to "speed the dissemination of improved methods and practices."

Third, the report devoted a special section to the issues associated with higher education, especially those involved in eliminating the cost barrier to equal access. Not many years later, Jerrold would head a commission proposing an Educational Opportunities Bank, making specific and imaginative proposals directly on this subject. Also, the idea of small or lesser-known colleges' establishing ties with larger, more successful institutions was explicitly encouraged in this section of the report, and assistance for collaborative relationships was proposed.

In a fourth section, the task force had addressed itself to financial resources; the report recommended that "funds allotted to international educational activities be doubled as the first step in a new drive to promote international understanding." A fifth section recommended significant organizational innovations within the government itself—the creation of an independent Office of Education at the presidential level, for example—and the final section addressed itself to the financing of education, with a specific acknowledgment of the vital role of privately supported institutions.

It is not hard to recognize Jerrold's concerns everywhere in the report of the task force: "Education will be at the heart of the Great Society." It may not have been Jerrold who contributed that line to the report, but it was pure Zacharias, nevertheless.

The idea presented in the President's Task Force report of collaborative relationships between smaller and larger institutions had already been tried out successfully several years earlier—by Jerrold, on behalf of MIT. In late 1959, officials of Oklahoma City University had approached MIT for assistance in an attempt at self-improvement. The university had never achieved distinction in scientific or engineering areas. It was mired in an early 1900s approach to education, which had become highly inappropriate to its oil-rich, technologically oriented environment. It needed an overhaul of its curriculum, its plans for academic growth and modernization, and its buildings. The university was prepared to put substantial resources to the task, but it needed help. The initiative to approach MIT had come largely from one university trustee, Dean McGee, president of the Kerr-McGee Corporation.

Julius Stratton, MIT's president, had promptly referred the request to Jerrold, who agreed to lead a small MIT task force, including Jerome Wiesner and Ted Martin, to help Oklahoma City University design and implement what its community came to call the "Great Plan" during the following two years. In May, 1964 just before he joined the President's Task Force on Education, Oklahoma City University had conferred on Jerrold the honorary degree of doctor of science, in recognition of his contribution to that institution.

It was not the only academic recognition he received that year; the next month, in June 1964, he received the honorary degree of doctor of humane letters from Tufts University, in Massachusetts. He was asked to give the commencement address, which he entitled "In Defense of Committees." It was a reasoned argument about why summer studies (and similar enterprises) work, although he did not refer to them by that name; he simply called them committees. It was also an appeal to the graduates to become "committed," to take part in the great issues of the day. He took the occasion to quote Robert Kennedy quoting his brother, the late president: "The hottest places in hell are reserved for those who in time of moral crisis preserved their neutrality."[6] This was not far off the mark in expressing Jerrold's own point of view about engagement in the public interest.

Jerrold shared the day's honors with the noted pianist Arthur Rubinstein, among others; given his early relationship to music, that gave him a very special pleasure. The award of the degree was itself special to him because it acknowledged the breadth of all of his contributions to education. The citation read, in the curiously terse but ornate style of such documents, "Teacher, scientist, administrator, humanitarian, you have turned your amazing talents to the benefit of education at all levels, in all disciplines, and for the benefit of the disadvantaged as well as the advantaged. For this broad sweep . . . the Honorary Degree of Doctor of Humane Letters."[7]

The Deprived and Segregated

Students are brighter than you think.

—J. R. Zacharias[8]

Concern with the severe problems of the nation's minorities, its poor and its disadvantaged, was something Jerrold brought with him to the president's special task force. He had become aware of these difficult issues much earlier, and he had made them an essential part of the work of the PSAC Panel on Educational Research and Development (ER&D), of which he had been chairman since 1961. Perhaps his recent exposure to black Africans and their efforts to help themselves had led him to take a fresh look at his own society; he was always in favor of taking a fresh look at things, as part of what he called "getting my head straight." But more than likely, his increased awareness was simply a natural stage in his growth as a human being; for after more than six decades, he was still learning and still growing.

In September 1963, well before the special task force had been established, the Bank Street College of Education had mounted a seminal two-week conference, "Education of the Deprived and Segregated." Its co-chairmen were John H. Niemeyer, president of the Bank Street College, and Jerrold Zacharias; the conference was held at MIT's Endicott House. It was supported by several agencies within the U.S. Department of Health, Education and Welfare, but it had been conceived at the request of the ER&D Panel.[9] Jerrold had persuaded MIT to provide facilities and accommodations for the meeting.

A curious and racially diverse mixture of people, sixty in all, gathered at Endicott House in that early New England autumn. Among them were several physicists, a handful of sociologists and a church rector, a psychiatrist, several mathematicians and economists, a number of educators and welfare administrators, a novelist and a pediatrician, representatives of federal agencies, state agencies, and private foundations. What they had in common was a sense of the urgency and the complexity of the needs of "the difficult thirty percent"—those children and youth who, as the conference report stated, "because of deprivation and segregation, are not getting the kind of education that will prepare them to become effective adults in our changing world."[10]

The participants did not come as experts but rather to learn. Ralph Ellison, the author of the prize-winning book *Invisible Man*, spoke movingly to the point that the "difficult thirty percent" do indeed have a culture: a rich and resourceful one that helps them to survive in a strange and hostile environment.[11] That culture, he said, must be seen and understood as part of the pluralism of America.

We have been speaking as though it [America] were not made up of diversified cultures but were in fact one monolithic culture. And one which is perfect, the best of all possible cultures, with the best of all people affirming its perfection.

Well, if this were true, there would be no point in our being here . . . [and] the problem seems to me one of really scrutinizing the goals of American education.

It does me no good to be told that I'm down on the bottom of the pile and that I have nothing with which to get out. I know better. . . . Let's not play these kids cheap; let's find out what they have. . . . I don't mean that these kids possess broad dictionary knowledge, but within the bounds of their familiar environment and within the bounds of their rich oral culture, they possess a great virtuosity with the music, the poetry, of words. The question is how can you get this skill into the mainstream of the language, because it is, without doubt, there.

. . . We must recognize that the children in question are not so much "culturally deprived" as products of a different cultural complex. . . . One of the problems is to get the so-called "culturally deprived" to realize that if they take what we would give them, they don't have to give up all of that which gives them their own sense of identity. Indeed, the nation needs some of the very traits they bring with them[12]

"Let's not play these kids cheap." It was a powerful, simple and eloquent message, a call to recognize and reinforce the dignity and sense of worth of people whose circumstances of race and station were different than one's own. It was a message that Jerrold had already heard clearly at the first Endicott House Conference, when the message had concerned the African child. But it was a universal call, certainly transferable to the United States of the 1960s.

Jerrold reacted to Ellison's ideas on a personal as well as a professional level. In later years, he enjoyed giving people printed copies of lists of things about education that he considered important, such as the perennial observation, evidence, and the basis for belief. On one of these lists, labeled "Areas of Concern," he had included dignity along with far more pedestrian concerns like measurement and the decimal number system. He explained its importance to an interviewer: "Dignity . . . it's there just to remind you to be damn sure that everbody has a share in the enterprise. Okay?"[13]

Three major programs emerged as an immediate result of the Bank Street conference. Within a year, action had been taken to set up a Model School System in Washington, D.C., the Negro College Program, as it was then called, had been started, and an Educational Resources Center had been established by Bank Street College in New York City's Harlem. But the conference had been even more effective, for many of its ideas became incorporated into an influential report subsequently issued by the PSAC Panel on Educational Research and Development. That report was entitled *Innovation and Experiment in Education*, and a number of its recommendations found their way into Johnson's plan for the Great Society.[14] It was interesting, said John Niemeyer, to note how closely the new NDEA (National Defense Education Act) provision, Title IX, resembled the conference statement. The participants in the Bank Street conference, he added, felt themselves "in the vanguard of a great national movement."[15]

The Negro College Program was of great interest to Jerrold. There existed a chain of such institutions—Howard University, Fisk University, Lincoln University, and others—spanning the southern states; they were generally underfunded, neglected, and out of the mainstream of educational reform. They were frequently understaffed as well. Earlier Jerrold had become aware of some of their problems through his friendship with Herman Branson, a black physicist who then was chairman of his department at Howard University and later president of Lincoln University. That long and enduring friendship dated from the later days of PSSC, when Branson had made a film for the project. At Jerrold's invitation, Branson had been a participant at the Bank Street conference. Now, armed with the Bank Street report, he and Jerrold managed to persuade the American Council on Education to form a committee to organize a long-range program of assistance to these colleges. It would begin with a meeting "to discuss . . . five summer institutes for college teachers, which are intended to aid them in reorganizing and redefining their own courses so that the students at their individual colleges—approximately 50% of whom plan to become teachers—will be better prepared to teach the new curricula in the sciences, mathematics and humanities."[16]

The meeting was held on April 18 and 19, 1964, at MIT, with support from the Carnegie Corporation; the Negro College Program was formally established under the administrative coordination of ESI. Later that summer, at Pine Manor Junior College in Wellesley, Massachusetts, there was held a writing conference, involving some forty-seven participants, to develop new materials in English and mathematics primarily for Negro high school seniors in the South planning to go on to higher education.

The Negro College Program ran for more than ten years, characterized by many summer teaching training institutes and by faculty exchanges with predominantly white colleges. To some, these activities seemed to represent gradualism at a time when there was increasing demand for something immediate and dramatic. The Negro College Program was not much noticed, its undoubted success buried in the noise of ghetto uprisings and civil rights demonstrations. But it had been useful and, in a quiet way, effective.

Sketch for a Portrait

But such a day tomorrow as today,
And to be boy eternal.

—W. Shakespeare, *The Winter's Tale*, I, ii, 63

When Jerrold turned sixty, he and Leona celebrated by giving a party at the MIT Faculty Club. It was an important birthday, they felt, and they made it a suitably festive and elegant occasion: black tie (and silver slippers, as one guest put it), champagne, and dancing, with a midnight supper afterward. There were nearly a hundred and fifty guests, many from Boston and Cambridge and many from far away. They had touched many people in the course of their lives. There were recent friends as well as those of long standing; virtually all the periods of their lives were represented in one way or another.

Their granddaughters, Sarah and Katie Hafner, aged nine and seven, were there, in specially made long dresses. Katie remembers the occasion still: "The most vivid reminder . . . is the memory of his 60th birthday ball, a grand event that we tiny granddaughters were allowed to attend. I remember searching through a sea of knees for him at the party and using all my strength to wrap my arms around what I thought was his leg. I looked up into the face of a perfect stranger, and recoiled in horror and embarrassment. When I did find him, it was with special gratitude and warmth that I hugged him."

Jerrold at sixty was near the top of his form. He was trim and fit, more slender than he had been two or three decades earlier. He was a vigorous and active man in spite of periodic incapacitating attacks of gout—a long-standing ailment—and a chronic bad back. His own view of himself at this time was characteristically wry. When asked why he had not been willing to appear in TV programs as he had done in his movies, he replied: "I'm too square. I'm likely to wear a gray suit and a gray necktie and a gray shirt and gray hair and gray socks and gray shoes. The jacket designed in 1926. . . . I'm not fit for TV."[17] The image was highly inaccurate, for it failed to suggest the most important thing about him: the liveliness of his presence. He was anything but drab. The suit might indeed be gray flannel, but it was custom made, and the necktie, as often as not, would be red. Katie Hafner, from her child's perspective, caught a better likeness:

It was usually on sailboats that my grandfather's playful side came out. He had a comical way of curling the far reaches of his lips up and down without moving the rest of his mouth. After we'd been on the boat for a while, and my sister and I would begin to fuss and fidget, he would perform this trick for us. . . .

The most striking feature of my grandfather were his eyes. Sad and sloped at a steep angle, those eyes were like none I've known. I sometimes wondered how he had gotten such remarkable eyes and wished that mine, too, would point to the ground. To this day, when I see sloping eyes, I think it must be a sign of great intelligence.

Others, too, had more lively impressions of Jerrold than his self-description would suggest. A sixtieth-birthday souvenir book was prepared for presentation to him, and many of his friends wrote for it. Their words present a kind of sketch for a portrait; this is how they saw him.

A. P. French, now professor of physics at MIT and then Jerrold's colleague at the Science Teaching Center, wrote a letter for the birthday book recalling their first encounter in 1959, when Jerrold had offered him a job: "Here was a man who knew what and whom he wanted. Knew a little too well, perhaps. Would one find oneself in a constant battle to avoid being overpowered? Would the choice (though it could not be a choice) lie between submission and an exhausting struggle to assert one's own views, even one's separate identity?"[18] French did not then accept the job; he did not join Jerrold's group until three years later. His impressions, after working with Jerrold for three more years, were not of a man whose principal coloring was gray:

All who know Zach are familiar with the magical ingredients [of his personality.] First and foremost his volcanic energy, always seeking new outlets. Next (and a logical continuation of this metaphor) his instinct for applying pressure at places where substantial progress will quickly become apparent. His special blend (often but not always efficacious) of truculence and silver-tongued persuasiveness. His whole-heartedness in approval or damnation of ideas and people—where lesser people would not find it in themselves to be so categorical, would be less often wrong, but also less often right and far less often successful. His impatience with mediocrity or equivocation. . . .

There, I had thought, is a man I would rather keep at a discreet distance. But circumstances threw us into close proximity... where I had expected to find a man whose chief weapon was the bludgeon, I found one possessed of extraordinary delicacy.

James Killian also wrote a birthday letter to Jerrold, reflecting on their many years of association: "I have found delight [and] a sense of confidence in your plans and goals, and pride in your innovative contributions. Sometimes I have had to defend these plans and goals before people who were skeptical or critical because they didn't know you well enough to understand your ebullient advocacy . . . or were not aware of the good taste and high aims which were inherent in your proposals and work."

Jerome Wiesner, closest of friends, wrote recalling the many ventures into which he had been led by Jerrold: "You are still going strong and we expect that you will stir up many more exciting things in yet unpredictable areas of the human scene. . . . As I have thought about writing this, I have remembered the Radiation Lab, Los Alamos, Project Hartwell, Project Troy, National, Hycon Eastern. PSAC. the missile committee, disarmament, Hermes, PSSC, ESI, the Panel on Educational Research and Development, and most of all, MIT."

Fortunately, Jerrold had a sense of humor that kept him from letting such praise—of which there was so much—affect him unduly. He had his own ways of judging things that left him fairly free both of overdependence on praise and of hypersensitivity to criticism. He had normal human inconsistencies, of course, but his sense of the absurd allowed his friends to tease him about them. Helene Deutsch, the eminent psychoanalyst, was one of Leona's and Jerrold's close friends. She wrote to Jerrold on his birthday, reminding him of their first meeting twenty years earlier, when he had visited to bring greetings from her son Martin, then still at Los Alamos. She had served a special dessert for the occasion, she remembered:

With each additional serving of the delicious chocolate soufflé (with whipped cream) he let me know his positive response. . . . I recognized that before me was a man revealing his emotional sensitivity and his great intuition!

But . . . I was to learn another side of my guest's personality: he was a man of an iron and unbending will! It came with the coffee: he categorically refused sugar and from his vest pocket came a little box with saccharin. . . . [He said] "I am on a strict diet and I cannot and will not take sugar." It was evident that no power in the world, be it a woman's appeal or a man's force, would bend this will of iron.

Now I knew: here is a man of great richness of emotions and of unbending will.

The letter came accompanied by a beautiful small gold pill box—filled with saccharin tablets.

Franklin Patterson, then of Tufts University, offered as a birthday remembrance a number of vignettes from his encounters with Jerrold:[19]

• Near the beginning of PSSC, a somewhat nervous prophet spotlighted on the stage of Boston's largest theater, reaching out with little rhetoric and no pretentiousness to several thousand faceless teachers . . . in the symbolic darkness beyond the proscenium. . . .
• Collar a little too large, gray shocked head shifting with sudden quizzical movement, a physicist lion in a den of Daniels, chairing the 1962 quasi-riot charitably known as the Endicott House Conference on Social Science and Humanities Education. . . .
• On the Endicott House terrace, armed with one of the first Land color cameras, his face in mixed aspect of glee and some sort of wonder. . . .
• A man, having heard an eighth-grade teacher who really knows how to teach, suddenly possessed and on fire, bouncing up and down in his chair, saying "That's it! That's it!". . . .
• A non-listener to lectures, head on elbows [sic], eyes closed, asleep or disgusted? Or both?
• A joyous and Joycean speaker ad libitum infecting as staid an audience of Brahmins as Boston can produce with his high spirits and wit. . . .
• A dancer waltzing to the best music [those] good men and true could make, manifestly delighted with his partner, a lovely, solemn and very small granddaughter. . . . And a birthday celebrant, supposedly sixty summers gone, when asked how old he really was, saying, "You know, I've thought about that. My real age is nine."

Naturally, people who contribute to a birthday album can be expected to fill it with praise. It is worth noting, nevertheless, the language that his friends used to praise him. He is said to be *ebullient*, to have *volcanic energy*, to be both *truculent and silver-tongued*; what he does is *exciting, unpredictable*; he is *joyous, high-spirited, a physicist lion*. The portrait is that of an active and powerful man, strong-minded, still youthful in spirit, optimistic about the possibility of making the world a better place. "For Jerrold," said Martin Deutsch, "every day was a new day."

Fresh Ideas: CCCP and All That

The College

The initials CCCP stand for the Committee on Curriculum Content Planning, one of MIT's more ambitious efforts to redefine itself, to come to terms with its mission and to look at how it was doing its job. The fact that the initials, read as cyrillic characters, also stand for the Union of Soviet Socialist Republics, may be taken as one of the small wry jokes that delighted Jerrold, for the committee had been his idea, and he named it as well as ran it.

He had gone to Julius Stratton in late 1961 with a most curious request; he had asked that such a committee be formed and that he be assigned to chair it. Most members of faculties shy away from committee service, which they are, of course, constantly being asked to provide. The idea of volunteering such service, especially for a committee likely to be both time-consuming and controversial, does not arise naturally in the academic community.

But Jerrold had become worried by the rigidity of the MIT curriculum, so full of requisites and requirements, so uniform in its demands on students of diverse backgrounds, interests and abilities, so inclined to be dogmatic. "Requisites and prerequisites," he wrote, "of one sort or another are invented to destroy. It probably is just as well that the schools use *Silas Marner* as 'required reading,' simply because it is better to kill *Silas Marner* than something better. This may just reflect my attitude late in life toward the *Silas Marner* that I was required to study. . . . [T]he whole educational system, as it now goes . . . has become a system which tends to create mice instead of men. It tends to create people who are prepared to listen to authority and who end up asking for authority."[20]

He did not spare his own colleagues, those who were trained in the sciences: "It is so easy for a professional scientist to say, 'Here, I know the answers. Scientists have worked on these problems for centuries. Here is the way a planet moves around the sun. Learn the law that I learned from the man who learned it from someone who taught it to him.' Never mind that a planet is a lovely object in the sky and that one can observe a planet and wonder why the planets move around in the heavens."[21]

So he went to Stratton and said that he wanted to be chairman of a committee:

Dr. Stratton looked me square in the eye and said, "Jerrold, you are out of your mind." I said, "Wait until you hear what I want. I want the full teaching time (that means half-time at MIT) of eight faculty members hand-picked by me. I want to be able to pick the kind of people who are too busy. What I propose that we do is spend a year trying to understand the course requirements at MIT." Remember that MIT is a school of science and technology even for the undergraduates, although we take into account social science, history, and the humanities. So I was allowed to pick a biologist, a mathematician, a physicist (other than myself), a chemist, a chemical engineer, an aeronautical engineer, an economic historian, and a humanist. I did not want to interfere with the productive research time or creative time of these people, but only to use what they would ordinarily spend on teaching classes. And as I said, at MIT this means half time.[22]

The way he arranged for the CCCP to operate reflected Jerrold's own somewhat iconoclastic views on the best way to learn. He believed that one learns most efficiently and effectively about something—French language, building atomic clocks, sailing, quantum theory—by immersing oneself completely in it until it is mastered, at least as far as necessary to satisfy oneself. That was not how college students were expected to learn, however; he pointed out that MIT students, like their counterparts everywhere, were expected to take at least four or five subjects simultaneously, with equal seriousness and devotion. He believed that that expectation was unwise and unrealistic, and he decided that it would not do for the committee. He described how they went about it:

We first had the notion that we would try to learn something about each other's fields of competence and something about the specialties that were required of the students. We wanted to do this because we said we would try to write down, early in our deliberations, what we thought was imperative for an MIT graduate to know. And if we could get some common core of requirement in terms of specific knowledge, we would then be able to know what we should require of our students. So we went to work in a way which is characteristic of a professional intellectual, whether he be a scientist or a humanist or whatever. We went at it unremittingly and in depth. For instance, we decided first to study biology, so we read for a solid week, and then our biology member gave us a one-week lecture on the subject of modern microbiology. He could do it moderately easily because it is a subject that he knows in and out. . . . And when I say that we met for a week, we met at nine o'clock on Monday morning and let up sometime late Saturday afternoon. We did go home nights, and we did interrupt the lecturer. In fact, it is hard to remember now whether the lecturer spoke much at all. But it is this . . . in-depth, sharp method of learning which is so characteristic of people who try to do something. We did not say that we would study four subjects simultaneously, nor carry other burdens. During the two weeks of the month when we were neither studying a subject nor participating in our own classes, we were equally [dedicated] in going at our own researches or other duties.

Let me give you an example of how this might work. We wanted to study economics to get the flavor of how economic theory works. So we were assigned first a famous text written by one of the MIT professors, Paul Samuelson. Then three members of the economics department spent two days with us. And when they hit my office where we met, they came unprepared, not knowing what we would ask of them. It was simple. I said, "O.K., teach us the tax cut." And so we spent two days on tax cuts—one fluid model, two fluid models, three fluid models. This kind of theory is not very difficult for anyone who has any good feeling for servomechanisms, for feedback theory. And naturally no professor [at MIT], I suppose not even our expert on Faulkner, had any trouble with economics, simply because he understood feedback theory. I guess that feedback theory is in the air at MIT, not because of Norbert Wiener, but simply because feedback theory dominates so many things at MIT.[23]

Not everyone was sanguine that the committee would get anywhere. The biology member, Cyrus Levinthal, professor of biology, had known Jerrold only slightly beforehand and remembered that he had serious misgivings, not so much about the method of operation as about the style of their leader: "My conclusion [after my first] conversation was that it would be impossible for me to have a civilized conversation with this man, and since I had recently agreed to serve for a year on a committee he was running, the prospects of civilization did not look good."[24] To Levinthal's surprise and delight, it did not work out that way at all, for service on this particular committee turned out to be a remarkably civilizing experience. He went on to say why he thought it had been so successful:

The remarkable fact about so many of [our] sessions is that they were intellectually exciting. We could, in fact, discuss curriculum without talking about rules and regulations. We could talk about what was exciting to mathematicians without having to be side-tracked by teaching methods. We could talk about Faulkner novels without having to worry about how many credit hours of humanities subjects could be required. . . .

Jerrold likes to think that his committees work because they put in enough time to become experts on the subject. I have the feeling that it is somewhat closer to the truth to say that his committees spend so much time with each other discussing things of mutual interest, that they get to be experts on each other. Once this has happened, it is fairly easy for them to sit down and make sensible pronouncements about curriculum changes or anything else at hand.

The CCCP issued its report in May 1964; its recommendations for reform fell into three categories: "to improve the quality and relevance of the scientific and mathematical materials the student would find available; to find ways to allow the students freedom 'to make their own mistakes,' to do their own work and to find their own ways; to find improved mechanisms for professorial guidance: not simply one professor guiding with a firm hand, but several, acting as stimuli, goads and guides."

Much of the report was particular to MIT and its task of training bright, technologically oriented students to be creative scientists and engineers. Basically, the committee adopted Jerrold's idea that the institute be less prescriptive in the early stages of a student's career and encourage responsibility and choice on the part of the student. The institute would still be more demanding than most liberal arts colleges, but it could organize its demands with more care and understanding. The committee pointed to the extraordinary rate of growth of knowledge and the need to take that into account. As Jerrold remarked on another occasion, quantum mechanics, which had been beyond the grasp of most of the Columbia faculty when he had been an undergraduate there, was now taught to students in their sophomore year.

The first task of any university faculty committee is to make sure that its report gets read by the rest of the faculty, yet few documents are more unreadable than the typical committee report. The CCCP cleverly avoided this pitfall; the substance of the report was presented in the form of a dialogue among hypothetical characters: a dean, two engineering professors, a science professor, a "generic" professor, a guest (presumably representing the employers of MIT graduates), and a member of the CCCP itself. It was a brilliant device, for it became possible to air the misunderstandings and prejudices that existed, as well as the proposed remedies without their attribution to real persons. The report was widely read, and all of its recommendations were adopted by the MIT faculty. It was a remarkable exercise in university politics, successful largely because, as Levinthal finally put it, "No one can be with [Jerrold] without being utterly convinced of his intensity, his enthusiasm and his devotion to the tasks he has undertaken even when one is completely convinced that he is wrong."

The Medical School

Only a few months passed before Jerrold took up the next challenge: the question of medical education. Oliver Cope, then professor of surgery at the Massachusetts General Hospital, recalled how it had come about: "The idea of a Summer Study to reconsider medical education developed during a cruise along the coast of Maine in July, 1964. A biochemist, a psychiatrist, a lawyer, a physicist and a surgeon made up the ship's company. . . . Jerrold

Zacharias, the physicist, told of the progress in education in physics, mathematics and biology. . . . The absence of plans in medical schools to alter their programs . . . was obvious."[25]

What appealed to Jerrold was the possibility that medical education might be made less bookish, less formally structured, and more practice oriented; some of the motivation to learn might arise naturally from a student's early involvement with patients in various real but manageable situations, helped by experienced physicians. It might make possible a shortening of the years of medical training, with a consequent increase in the rate of production of new physicians. If the promise of the Great Society program of increased health care for an increasing population was to be met, the United States would have to deal quickly with its physician shortage.

Cope's interest was evidently engaged, and as a result, at a small meeting of concerned physicians (at the October 1964 meeting of the American Surgical Association), he arranged that Jerrold be invited to organize an appropriate summer study. It took the form of a conference held in July 1965 at Endicott House. It was supported by the Carnegie Corporation, the U.S. Public Health Service, and the Richard King Mellon Foundation, and it was hosted by MIT.

Ten working days were surely not enough for this particular summer study, but it was all that a busy group of thirty-five eminent physicians could be persuaded to give. They represented many specialties[26] and they worked hard to identify the problems faced by their profession and to chart possible ways in which medical education might be altered to meet the changing demands of the times. The report of the study pointed out that "more than 85% of all new physicians today choose to become specialists. Even the 15% who intend initially to enter general practice are quickly reduced as the general practitioner, for one reason or another, and by one means or another, finds himself impelled toward practice limited to fewer and fewer areas. Medical research has less and less direct association with medical practice. . . . The differentiation of function, with all its centrifugal tendencies, has made it necessary to call into being restoring forces if the central objective of medical practice—the care of the patient—is not to be forgotten."

The conference made a number of specific suggestions rather than recommendations; the participants understood that they had only had time to scratch the surface. But they had found among themselves enough common ground to urge that a new look be taken at the possibility of a better coordination—perhaps the integration—of premedical and medical curricula. In urging that means be found to shorten the time necessary to produce a physician, the conference emphasized strongly its concern for providing adequate and sensitive patient care. The report laid stress on the need for physicians well trained in the principles and practices of psychology. In short, if the medical profession needed talented researchers, it needed equally those whose talents and interests lay primarily in the doctor-patient relationship. Medical education could follow more than one path, therefore, and not everybody need be a specialist.

In the end, the report did not have the impact the physicians had hoped to achieve. One of the participants, Dr. Edward Kass, reflected on the experience:[27]

The study never really decided what its objectives were. . . . It was a time when organized medicine was moving increasingly toward specialization (and now toward sub-specialization) within the United States, while within Great Britain the system had moved in the opposite direction toward training a large number of practitioners and a small number of specialists. . . .

. . . It was out of step with the American psychology. Science was marching onward, knowledge was increasing logarithmically, and the capacity of the general practitioner . . . to keep up with it all in effective practice was obviously being compromised. . . . It fell apart because an increasingly enlightened citizenry keeps seeking "the best."

The growth of the National Institutes of Health with its increasing number of subdivisions symbolized this as well as anything. A few farsighted people recognized that the existing means for funding medical care was inherently unfair, in that the poor, the elderly, the incapacitated, were not receiving a reasonable amount of attention. These [considerations] were largely swept aside in the euphoria of the times that said that all of these problems would be solved by increasing prosperity and by increasing scientific advance which would lead to more and more prevention and less need for the doctor.

The public, in other words, might tolerate a broader education for its natural scientists or its engineers, but it wanted its physicians to be specialists. And as a result, what the Great Society promised in health care to the elderly and disadvantaged, neither the medical profession nor anyone else knew how to deliver; nor does anyone today, a quarter of a century later.

Things Change

Vietnam

The Cambridge Discussion Group consisted of a small number of individuals who met from time to time in the winter and spring of 1966 to discuss the steadily worsening problem of Vietnam. The membership in this informal group was neither fixed nor constant. It included John Kenneth Galbraith and Carl Kaysen; George Kistiakowsky and Jerome Wiesner, each of whom had been special science adviser, the former to President Eisenhower, the latter to President Kennedy; Jerrold Zacharias, Franklin Long, and Richard Neustadt. Others came occasionally or only once: Frank Lindsay, Eugene Skolnikoff, Henry Kissinger, Charles Wyzanski, Robert Benjamin.[1] Nearly all had had important government experience of one sort or another. Kaysen and Long, for example, had participated in the negotiation of the limited nuclear test ban treaty in 1962 and 1963; Kaysen had been an aide to the national security adviser at that time. Long and Zacharias had extensive service on the President's Science Advisory Committee, of which they were both still members. It was a group with considerable experience in the ways of government and a readiness to participate when it was appropriate.

The group had evolved partly out of an earlier association , Scientists and Engineers for Johnson, which had been active during the 1964 electoral campaign. Now, however, it was nonpartisan, for in 1966 it was time to be concerned not with electoral politics but with the enormous problem of Vietnam. What was to be done about that? None of these men had any official government standing. They met simply as private citizens concerned about what was happening to the country and ready to help however they could. Kistiakowsky and Wiesner drew up a short prospectus and arranged an initial, exploratory meeting; they wrote in the prospectus: "The meetings which we propose have as their purpose to clarify, through discussions, our understanding of the problem of the peaceful settlement of the South

Vietnam war and *possibly* to reach a consensus leading to some agreed-upon group action."[2]

It was unusual for a group of private citizens to meet in the hope of affecting affairs of state, but this was no ordinary group, and the times were extraordinary as well. The military circumstances in Vietnam gave little hope of an early resolution of the war, which, during 1965, had become substantially Americanized. American ground forces were in combat in South Vietnam, and the American bombing of North Vietnam—code-named Rolling Thunder—had begun. It was not clear where the limits to American involvement lay. The government of South Vietnam was highly unstable, and America might be drawn in still further. The Korean conflict, with its image of a million Chinese troops pouring across the Yalu River, was still vivid in everyone's mind, and the possibility of provoking either Chinese or Soviet intervention was deeply troublesome.

On April 19, 1966, Kenneth Galbraith wrote to President Johnson on behalf of the Cambridge Discussion Group, expressing their discouragement at the prospects they saw:

It appears that continued disorganization or the appearance of a neutralist government are sufficiently probable so that we venture to suggest contingent action in the event of either development.

Should South Vietnam cease to have a government capable of or willing to carry on the war, we trust that no steps will be taken to replace this with either a direct American or fully American-dependent regime. This would be to launch a long, painful and extremely costly venture which, given the past history of non-Asians in this part of the world, we could not expect to be successful. While it is not an ideal solution, we strongly feel that the only course of action would be a well-planned withdrawal.[3]

Galbraith commented on the B-52 bombing raids on North Vietnam, having already pointed out that the essential problem was the political one in the South: "Those members of our group who are closely familiar with air operations doubt strongly the general military value of the B-52 attacks in this kind of situation. They even question the wisdom of exposing the practical weakness of this weapon. You will see, I think, why we feel that any tendency to cover weakness in the South with air action in the North should be strongly resisted."

A few weeks earlier, three members of the discussion group—Kistiakowsky, Kaysen, and Wiesner—had met informally with Secretary of Defense Robert McNamara, in an attempt to convince him that if the infiltration from the North could be halted by other means, the bombing could be stopped.[4] They urged a summer study to make independent evaluations and to examine alternate technological means to accomplish the goal; Jerrold was proposed as director. On April 16, 1966, McNamara wrote directly to Jerrold, accepting the suggestion: "I understand that you are willing to try to assemble a group of people who have for many years been familiar with military technology, and who would work a substantial part of the summer on technical possibilities in relation to our military operations in Vietnam. I looked at the preliminary list of names and find it impressive, as befits the difficulty of our present situation."[5]

Arrangements were made for the study to be supported by the Department of Defense, through the Jason Division of the Institute for Defense Analyses (IDA); the project came to be referred to as the Jason summer study.[6]

It is clear from the documents that have been made public that what interested McNamara most was the possibility of an electronic barrier across Vietnam, dividing the North from the South and effectively stopping the infiltration of men and materials across the demilitarized zone.[7] The idea is generally attributed to Roger Fisher, a Harvard Law School professor, who had suggested it in a note written in January 1966 to a former Law School colleague, John McNaughton, then serving as assistant secretary of defense. In any case, it was an option considered seriously by the summer study.

Jerrold convened his summer study group on June 13, 1966, at Dana Hall, a small preparatory school in the suburban town of Wellesley, Massachusetts. They began with ten days of intense high level briefings on all facets of the war. The forty-seven scientists—"the cream of the scholarly community"—soon to be supplemented by another twenty from IDA, divided into four groups working on different technical aspects of the problems. In August, four reports were prepared:

The Effects of US Bombing in North Vietnam
Vietcong/North Vietnam Logistics and Manpower
An Air-Supported Anti-Infiltration Barrier
Summary of Results, Conclusions and Recommendations

The narrative of the *Pentagon Papers* suggests how important these reports were:

Several factors combined to give these conclusions and recommendations a powerful and perhaps decisive influence in McNamara's mind at the beginning of September 1966. First, they were recommendations from a group of America's most distinguished scientists . . . men who were not identified with the vocal criticism of the Administration's Vietnam policy. Secondly, the reports arrived at a time when McNamara, having witnessed the failure of the POL [petroleum, oil, and lubricants] attacks to produce decisive results, was harboring doubts of his own about the effectiveness of the bombing. . . . Third, the Study Group did not mince words or fudge its conclusions, but stated them bluntly and forcefully. . . . Moreover . . . [the reports] apparently had a dramatic impact on the Secretary of Defense and provided much of the direction of future policy.[8]

What the reports said was that the bombing of North Vietnam was ineffective. They were severely critical of the failure of military strategists to recognize that the bombing of the North had produced no useful result in the South; that the bombing was considered successful if it inflicted damage in Hanoi, while its real purpose, the prevention of supplies and men reaching the South, remained unachieved.[9]

The scientists studied the electronic barrier, in which McNamara was already interested, as an alternative to the bombing. The technical details of a multisystem barrier were provided: location, dimensions, measures to be taken against personnel infiltration on the one hand, against vehicles on the other; time scale and cost, use and maintenance. The barrier was not to be a static concept; there would be a constant need to upgrade it as the North Vietnamese discovered ways to defeat it: "We envisage a dynamic "battle of the barrier," in which the barrier is repeatedly improved and strengthened by the introduction of new components, and which will hopefully permit us to keep the North Vietnamese off balance by continually posing new problems for them. . . . Counter-countermeasures must be an integral part of the system development."

McNamara was pleased. He received the reports on August 30, and a week or so later a group of high-level representatives of the Defense Department, including Assistant Secretary McNaughton, came to Tobey Island for further consultations with a number of the summer study scientists.

Predictably, however, the military were not pleased. They argued against the substantial cost (nearly a billion dollars, according to the Jason estimate) and the diversion of combat troops and other resources from the main goal. Admiral U. S. G. Sharp, commander in chief, Pacific area, wrote, "Military operations against North Vietnam and operations in South Vietnam are of transcendent importance. Operations elsewhere are complementary supporting undertakings." He and General William Westmoreland had just completed a joint strategy paper, which included, in their advice on how to pursue the war, the following recommendation: "In the North—Take the war to the enemy by unremitting but selective application of United States air and naval power. . . . The movement of men and material through Laos and over all land and water lines of communication into South Vietnam will be disrupted. Hanoi's capability to support military operations in South Vietnam . . . will be progressively reduced."

When Secretary McNamara submitted the recommendation for a barrier, as a possible replacement for the bombing, to the Joint Chiefs of Staff for their comment, he encountered strong opposition: "The Joint Chiefs of Staff do not concur. . . . They believe our air campaign against North Vietnam to be an integral and indispensable part of our over all war strategy."

It was a serious rebuff to McNamara, who was persuaded that the bombing was not working, a position supported now by the report of the Jason summer study. The dispute intensified at high levels of government about the primacy of air strikes, and one may readily discern in the argument echoes of the same bitter struggle that had arisen nearly fifteen years earlier on Projects Charles and Lincoln—the same belief in the effectiveness of offensive air power, regardless of evidence to the contrary; the same alarm at the possible diversion of resources; the same unwillingness to look at the military problem in its entirety, which, in the view of the Jason summer study, meant measuring the effectiveness of the bombing of the North by its results in the South.

McNamara made a trip to Vietnam in October and on his return recommended to the president that the electronic barrier across Vietnam and Laos be built as an alternative to further escalation of the bombing. He argued that the barrier would persuade Hanoi that the United States was in for the long haul and that it would therefore be wise to enter into negotiations; it would help to convince the North Vietnamese that the United States could not be outlasted.

President Johnson did not authorize the barrier, but neither did he give the Joint Chiefs all they wanted. He authorized new targets for Rolling Thunder but refused to allow unlimited raids on the North. The internal policy debate grew still more heated. By the summer of 1967 the Senate Preparedness Subcommittee, chaired by John C. Stennis of Mississippi, was ready to take a hand. A full year after the Jason summer study, the Stennis committee urged the president to step up the bombing: "that the air campaign has not achieved its objectives to a greater extent cannot be attributed to inability or impotence of air power. It attests, rather, to the fragmentation of our air might by overly restrictive controls, limitations and the doctrine of 'gradualism' placed on our aviation forces which prevented them from waging the air campaign in the manner and according to the timetable which was best calculated to achieve maximum results."[10]

McNamara lost the argument, and the bombing continued. McNamara himself was soon replaced as secretary of defense. Rather than ending in 1967, the bombing went on well into the presidency of Richard Nixon.

Whether the electronic barrier would have been as effective as its proponents hoped cannot be fully known. From the end of 1969 to the end of 1972, a partial version of the Dana Hall recommendations (code-named Igloo White) was implemented in Laos along the Ho Chi Min Trail.[11] Most of the passive elements of the barrier were not included; the installed version included only that part which operated with prompt air support. Igloo White was far more complex overall than had been proposed by the Dana Hall study, and its highly sophisticated acoustical detectors were used almost entirely to identify bombing targets with—it was claimed—increased efficiency and reliability, thus justifying, according to the air force, widening the bombing campaign. This, of course, was exactly opposite to the original intent of the barrier.[12] Furthermore, the cost was prodigious, estimated at about $1 billion per year over its three-year run.

Many of the specific devices recommended by the Jason study were subsequently developed by the military for other applications and have led to what is frequently called the electronic battlefield. Dickson has pointed out that the evolution of "smart" laser-guided missiles, for example, which were so effective in the Gulf War of 1991, grew out of such development activities.

But if it cannot be known how effective a full-scale Jason-type barrier would have been, given the still early state of the art, neither can it be known with certainty what would have been the result if unrestricted bombing of the North had been permitted in 1965, 1966 or 1967. Even if the bombing had been a technical success, which many doubt was possible, it is likely that the level of international disapproval would have been so high that such a success would have been hollow. In either case, neither barrier nor the bombing could have solved the fundamental problem of the war, for that was above all a political matter, not to be fixed by bombing or by application of a more subtle electronic technology.

Operation Rolling Thunder failed to accomplish its stated purpose of stopping infiltration across the demilitarized zone. As the *Pentagon Papers* narrative put it, "1967 would be the year in which we relearned the negative lessons of previous wars on the ineffectiveness of strategic bombing."

In February 1973, a reporter for *Science* magazine included a comment on the Jason summer study in an article on the Jason group itself:

McNamara enthusiastically adopted the barrier notion after a series of lunches with some of the scientists and a 6 September trip to Zacharias's summer home on Cape Cod. . . . McNamara arrived in his own Air Force plane and was offered a drink; they spread out large maps of Southeast Asia on the coffee table and floor, while the scientists did the briefing.

Imagine them, marking out on these top-secret maps exactly where this thing would go, while dogs and children were running through the house from the beach. It must have been incredible."[13]

The article reported a number of opinions, not all favorable to the work of Jason, which were unattributed. Jerrold took exception to the reporting. He drafted a letter to the editor of *Science*:

My hospitality to Secretary of Defense McNamara on 6 September 1966 was less niggardly than reported by [your reporter] in Science. In addition to the drinks we served, two substantial meals were provided at short notice for about thirty visitors.

And [your reporter's] journalism is equally careless when she writes about the dogs that came running in from the beach while top secret maps were spread on the floor. The dogs in my household are fully cleared for participation in classified discussions (T[op] S[ecret], Q and COSMIC).[14]

Brave New World?

By 1968, when Lyndon Johnson announced that he would not seek reelection, the United States had nearly half a million troops in Vietnam, and the number was still growing. The war would not end for another four and a half years; the tragic waste of people and resources continued that long. The effect on the programs of the Great Society was calamitous.

Johnson's programs did not succeed in eliminating poverty in America or provide an ultimate resolution of the great national problems of health, education, and welfare. But in contrast to his Vietnam policy, his Great Society programs could not be judged to have failed completely. Although aching problems remained in the areas of civil rights, education, employment opportunity, and the decay of the cities, real and constructive steps had been taken and real beginnings had been made; a new path had been indicated, if only the nation had been able to follow it.

Between 1963 and 1968, federal funds for health increased by a factor of about three; for programs aimed at children and youth, by a factor of two and a half; for the aged, one and a half. Funds for poverty programs increased two and a half times during this five year span, and funds for education programs nearly quadrupled. Much important new legislation appeared. Some examples drawn from the newspapers of 1968 give the flavor of what the Great Society legislation had intended to achieve and did achieve:[15]

Civil rights An act of 1964 outlawed segregation in hospitals, restaurants, hotels, and employment and gave the federal government enforcement power through the authorization to withhold federal aid. An act of 1965 protected voting rights; an act of 1968 initiated fair housing requirements.
Health An act of 1965 established Medicare and Medicaid; other acts provided assistance for the training of nurses and physicians, for mental health programs and for prenatal and postnatal care.

Job opportunity   The Economic Opportunity Act of 1964 established the Job Corps; the Manpower Development and Training Act in the same year provided programs to qualify persons for new and better jobs.

Education   The Elementary and Secondary Act of 1965, strengthened in 1966, provided increased aid for quality education in the schools; $9.8 billion was authorized for the next two years alone. An act of 1965 provided liberal loans, scholarships, and construction funds for higher education, and in the same year an act established the Teacher Corps to train teachers.

In these and in other areas, such steps were admirable. Why then did the Great Society fall short? Social theorists are unlikely to agree on whether too little money was spent or whether too much was wasted; whether inefficiency was at fault, or bad social design; whether too much was expected or not enough. It is fair to point out, however, that by 1968, public attention was badly distracted, and the national will had become confused. Rioting in the inner cities combined with the costly, intensely unpopular war in Asia led to a sense of urgency that often obscured the need for a long-term commitment to social planning and reform. Such a commitment would have to last over decades or be permanent, but the American public wanted and needed something more immediate.

Education, in spite of substantial increases in funding, did not fare well, for along with the increased funding came greatly increased demands and expectations. The difficulties of desegregation were perhaps nowhere more intense than in the nation's schools and nowhere more bitterly contested. In Mississippi, Alabama, Newark, and South Boston, tempers ran high. Mobs erupted and neighborhoods burned. Somehow, education was expected to address all the vast and intractable social problems for which there were no visible solutions: the problems of unemployment, of men and women who were not qualified to join the work force in any but the most menial jobs; the problem of the decay of the American family as an institution; the problem of illiteracy, perhaps growing, certainly not lessening. And always in the background was that terrible war that nobody wanted, dividing the country and draining its substance. It was estimated that in a single year, 1968, the war in Vietnam had cost more than $30 billion—more than a $1,000 each second throughout that year and the next as well.

Partly perhaps because of its failure to solve these enormous problems, education began to lose favor in the nation's collective mind. Higher education, in particular, became an arena for real battle. Starting in 1964 with the Free Speech Movement at the University of California, Berkeley, student unrest spread to the campuses of the nation. Building takeovers by militant students and other forms of lawless protest alarmed the nation; the fact that similar student protests overseas were closely allied to a kind of highly organized, aggressive neo-Marxism was hardly reassuring. It brought little comfort to watch on television as thousands of students marched through the streets of Paris in the disturbances of May 1968, almost toppling the government.

Education was getting a bad name. The idea of the school or college as a place to gain knowledge and to learn how to think was rapidly being subverted. Student unrest was widely interpreted as a failure of educational authorities to exercise leadership, while among dissident students—and others, it should be noted—schools and colleges were being regarded as points of leverage, where social change and strong action should be forced, although there was practically never any agreement about what action would be appropriate. Times had certainly changed. After 1968, the demands on education to solve specific immediate problems implied a sharp diminution of interest in and support for other things. It became difficult to obtain funds for curriculum development, for example, unless a proposal could be shown to be relevant to these pressing problems. The nation's leaders had become less concerned about producing scientists and citizens for a bright and shining future than about putting out fires in the present.

By 1968 Jerrold and most other scientists had moved off the Washington stage. Richard Nixon dismantled the President's Science Advisory Committee and put nothing in its place. Science advising never again achieved as significant a role in the upper levels of government. The most scientifically sophisticated, technologically dependent nation on earth has not since had a useful mechanism to provide to its leaders day-to-day policy advice on science and technology.

For someone who believed deeply, as Jerrold did, "that the country will falter if public education fails," the prospects were dark. Years later, testifying before Congress, he commented on the failure to follow up on the educational innovations of the period 1958–1968: "Any farmer knows not to pull up the seedlings to look at the roots, nor to desiccate and impoverish the healthy plants as they are beginning to grow. . . . The jobs that remain to be done are ten to one hundred times larger than what had gone on in the first ten years."[16]

Dissent and Protest

We cannot emphasize too strongly that dissent and orderly protest on campus are permissible and desirable. American students are American citizens.

—Report of the President's Commission on Campus Unrest (1970)[17]

In his 1969 book, *The Crisis of Confidence*, the historian Arthur M. Schlesinger, Jr., commented on the campus unrest that had by then erupted almost everywhere in the United States. He entitled that chapter of his book, "Joe

College, R.I.P.": "The quality of the educational experience has begun to change [he wrote]. . . . For years adults saw college life in a panorama of familiar and reassuring images . . . big men on campus, fraternities and sororities, junior proms, goldfish swallowing, panty raids. . . . College represented the 'best years of life,' a time of innocent frivolity and high jinks regarded by the old with easy indulgence. But the stereotypes don't work anymore. Students today have a new set of rituals: demonstrations, strikes, sit-ins, interrupting classes, locking up deans, howling down eminent visitors, seizing college buildings, fighting cops."[18]

He might well have added raccoon coats, tea dances, and Packard touring cars to his list of Joe College symbols, for in his own undergraduate days, at least before he met Rabi and settled down to serious work, Jerrold had come close to being that carefree stereotype. But more than forty years separated that time from the present: forty years, a depression, two wars, a third bitter conflict under way, and no respite in sight.

The 1960s in America were characterized by violence. The reaction of the police in the South to civil rights activists was violent; the murders of John Kennedy, Martin Luther King, Jr., and Robert Kennedy were violent; the rioting in the inner cities was terribly violent and destructive; even the 1968 convention of the Democratic party in Chicago, an event that should have represented the best of democracy and the emergence of peaceful political consensus, was instead the scene of violent confrontation between demonstrators and police.

In 1969, Richard Nixon was elected president with only a bare plurality. By that time, students on many campuses were ready to join in vigorous protest: against the war, against racial injustice, and against what they saw as the abuses of the university. According to the report of the President's Commission on Campus Unrest, issued in 1970, this meant rejection of "what it [the new youth culture] sees to be the operational ideals of American society: materialism, competition, rationalism, technology, consumerism and militarism. This emerging culture is the deeper cause of student protest."[19] Then, on April 30, 1970, President Nixon announced the widening of the war in Southeast Asia to include Cambodia. Student reaction was immediate and widespread. Six days later, protest again exploded into violence, this time at Kent State University; four students died. By the tenth of May, nearly 450 campuses were either crippled or closed down by student strikes. Four days later, there was more violence: two students at Jackson State College in Mississippi were shot to death. By the end of May, nearly one-third of the nation's colleges and universities had experienced some kind of strike or protest.

On Thursday morning, May 7, 1970, in New York City, I. I. Rabi attempted to enter Pupin Laboratories at Columbia University and found his way barred by striking students with arms linked across the door. The newspapers offered an eyewitness account:

Dr. Rabi tried to argue with them unsuccessfully. Exasperated, tears in his eyes, he tapped his cane on the granite step and said: "What you are doing is wrong, and you are crazy. You are blocking my way. Do you want to fight with me? Would you fight with me?"

He took hold of a long-haired boy's arm and pulled him towards the sidewalk, where a group of 20 other students, who looked amused, were watching.

The youth declined to fight, and a moment later, he stepped aside to let a workman enter the building. Dr. Rabi sneaked in too, although those at the door tried to push him back out for a moment.[20]

There may perhaps be no better symbol of the frustration and the intensity of that time than the image of a slight, seventy-two-year-old man offering to fight a long-haired youth over access to a place so totally dedicated to the freedom of inquiry—a place, in fact, that in the past had often given hospitality to refugees from intimidation elsewhere.

Protest, marches, and demonstrations came to MIT as well as to Columbia—and indeed to almost every other major university in the country. At MIT, the principal issue was the institute's involvement with military research, which was then greater than that of any other university. In 1969, for example, $108 million of MIT's $218 million operating budget was received from the Department of Defense; only $17 million of it went for nonmilitary research, and the bulk of it went to its defense-oriented laboratories. The defense portion, in fact, was nearly four times as large as that of its nearest competitor. MIT seemed vulnerable to challenge.

The two laboratories receiving the majority of Defense Department funding were the Lincoln Laboratory and the Instrumentation Laboratory.[21] Both had been established by MIT at the request of the government: the Lincoln Laboratory to deal with problems of early warning, the Instrumentation Laboratory to develop guidance systems for the *Thor*, *Polaris*, and *Poseidon* missiles. More recently, the Instrumentation Laboratory had also taken on the task of developing guidance and navigation systems for the *Apollo* spacecraft. (The first Apollo manned moon landing took place on July 20, 1969, just as the connection of the Instrumentation Laboratory to MIT was undergoing its most serious scrutiny.)

The student dissidents, joined by many faculty, raised the fundamental issue of whether military research ought to have any place on a university campus, especially as the very nature of the research, as well as the results, had to be circumscribed by strict rules of secrecy. The student protest brought into question the entire relationship that had developed between scientists and the military since World War II, and they did so in a manner such that the question could not be ignored or set aside. That uneasy relationship had originally involved only small ad hoc efforts—the summer studies—but from these, somewhat unexpectedly, had sprung gigantic laboratories devoted to military research and development. Had it all gone too far? The original motivation for men like Zacharias, Hill, and Wiesner had clearly been to provide assistance to the nation in a time of cold war stress and confrontation. But in 1969 the country was involved in a war that many considered unethical and immoral, led by an administration in Washington that found no favor among highly cynical young men and women. To them, such a motivation seemed as unjustified and unworthy as it had once seemed deserving of praise. Often with great emotion, they charged the universities with selling out. To the more radical of the students, MIT seemed like little more than a branch of the Pentagon; since to many of them the Pentagon represented all that was immoral, then MIT must be immoral as well. The slogan of the day was "Make love, not war." MIT was cast among the war-makers.

The MIT faculty as well as its administration were certainly not averse to reexamining the relationship of the institute to the government. Whatever anyone's feelings may have been about the morality of Defense Department research, enough questions had been raised and enough uncertainty evoked that rethinking the issues seemed desirable. In April 1969, the faculty voted to suspend classes for a day to allow students and faculty members to discuss these issues, and in the same week the institute announced that it would refuse temporarily to accept any new classified research contracts while an institute committee studied the relationship of MIT to the two laboratories it ran for the government. That committee was to report later in the year, in October.

The questions that needed to be discussed by the entire MIT community were these: whether the institute had grown over the years into a state of excessive dependence on military funding; whether the necessarily secret nature of military research (apart from the question of its morality) did violence, either in principle or practice, to those values that a university must protect; what in fact was an appropriate role for a university in national security matters; and whether the education that MIT provided for its students had somehow suffered because faculty members were too involved with matters that might or might not belong at a university but in any case had little to do with education.

MIT professors were inclined to be thoughtful about these issues. Bernard Feld, who had once, long ago, been the "haus-theoretiker" in

Jerrold's beams laboratory and afterwards had carried out highly regarded work in meson physics, had spent much of his life concerned with questions of disarmament and peaceful coexistence. He objected, according to a newspaper report, to secret research as long as the two laboratories were part of the educational process at MIT: "There is no room in an academic setup for classified research. . . . I would draw the line on classified research involving substantial numbers of students and faculty." At the same time, he said, there was an advantage to having MIT people involved in defense work "because they would be capable of criticizing it if necessary."[22]

Feld's comment emphasized the ambivalence that many people felt about the role of MIT professors. On the one hand, there was a growing dislike for military work; on the other, there was the nagging question: if MIT people (and their counterparts elsewhere, of course) did not get involved, then who would watch over the enterprise? J. P. Ruina, then vice-president for special laboratories at MIT, agreed that these questions ought to be looked into but warned that if universities severed all ties with the military "the country would be left in the hands of the professional military and industrial group." Walter Rosenblith, a biophysicist and electrical engineer and at that time chairman of the MIT faculty, agreed: "For MIT simply to cast these things out on the waters is not in any sense responsible."

Jerrold's views were conservative, reflecting in some degree his own long experience with the summer studies, the value of which he did not doubt. He was quoted as saying, "I believe secret research, with the world as it is, still must be conducted somewhere. Some members of the academic community should still be involved in it. The question is, do the administrative arrangements for carrying on secret research require that the work be done at the universities. In 1969 it is hard for me to see that the universities are needed. The people, yes, but the organization, no."

Albert Hill was less compromising but was willing to take an honest look at what was going on. He did not believe that universities could turn a blind eye to the nation's difficulties, for they had an obligation imposed on them by the society they served. He summed up his views tersely: "I do not think universities American style can say they are above all involvement with the nasty world . . . [but] very clearly a look at the activities of these labs is overdue."

When it reported in October, the commission appointed to study MIT's relationship to its two defense-oriented laboratories recommended that the institute not sever its connection with either of them but that the emphasis of each laboratory be shifted to socially relevant problems: housing, employment, and so forth. The recommendation seemed to satisfy nobody.

The MIT administration doubted publicly that enough funds were available to make possible such a shift. The militant students doubted that the institute leadership was sincere. In November about a thousand students marched through the campus denouncing the military research laboratories as well as activities at the Center for International Studies, which they felt to be Pentagon inspired. There were sit-ins and other demonstrations, and threats of worse; Howard W. Johnson, the institute's president since Julius Stratton's retirement in 1966, felt obliged to obtain a court order banning the members of the November Action Coalition "from violence or the threat of violence during planned demonstrations . . . against defense research at the Institute."[23]

It is remarkable that President Nixon's Commission on Campus Unrest, reporting to the entire country in 1970, found little to say about the role of universities in military research, although the problem had become troublesome almost everywhere. Like many of the other issues that had torn and divided the country in that difficult time, it never was resolved; gradually, after the war in Vietnam ended and tempers began slowly to cool, this problem receded into the background, although it has never completely disappeared. Attention shifted to other matters; only a few years later, the country was trying to deal with the devastating revelations of Watergate.

There was always something.

Reason and Discussion

We hope of course that a new tune, inaudible to ourselves, is now being played to the young, and that one day it will re-echo and give them the strength which certainly keeps us going now. But on that point we can get no evidence, and never shall get any.

—E. M. Forster (1939)[24]

Those of us who lived through the Vietnam war in the largest college city in the world (Greater Cambridge, with 200,000 undergraduates) realize that it was not a matter of just who is in control, but a steady deadening of the spirit of the young people by what they called the establishment.

—J. R. Zacharias (1976)

Somehow, amid the drama of confrontation and the excitement of strikes, sit-ins, and marches, the business of education continued. It was not business as usual; change was in the air, and many students and faculty members on campuses everywhere sought to break free from the constraints of tradition and habit. There were many things, of course, that had driven students to protest against the universities, and not all of their reasons were national in scope. Students became increasingly eager to take charge of their own lives, perhaps because they felt so helpless in the larger political arena. Many, students and faculty alike, felt that the universities had become too impersonal as they had grown to enormous size; it had even been necessary to invent a new word, *multiversity*, to describe them. Students felt they had become of secondary interest and that for want of greater interest, their educations and their very lives had become overprescribed and inflexible.

Jerrold agreed with the proposition that an MIT education was too narrow and prescriptive. He campaigned for increased possibilities of choice and for a wider array of learning options: small seminars as well as large lectures, independent as well as structured laboratories, learning at variable, self-chosen rates as well as in lock-step with others. He argued for the development of one-on-one teacher-apprentice relationships, not restricted to matters of curriculum. He felt that if conflict was to be at least mitigated, it was essential to engage students in discussion of the pressing and contro-versial events of the day, and of all the history that had led up to them.

Jerrold had been appointed Institute Professor in 1966, a title awarded only to a few in recognition of their achievements. It meant that he no longer had departmental obligations and was free to do as he wished. He took advantage of this freedom. In 1969, he and Leona together organized a seminar for the purpose of discussing with students whatever important events might be selected. It began by meeting on Tuesday evenings, from seven to ten o'clock; soon that changed to meeting from seven until midnight; then they began to meet several times a week, and finally, every weekday night. By all accounts it was a huge success for those who participated. Of course, the seminar could not engage those who had moved beyond discussion, nor could it engage more than a tiny fraction of MIT students, but at a time when rational discourse was scarce, it was a valuable and positive contribution.

Many topics engaged the interest of the seminar. They discussed at length the case of Robert Oppenheimer, for example; when the thought-provoking Broadway play *In the Matter of J. Robert Oppenheimer* came to Boston, Jerrold and Leona took the seminar to see it, arranging to meet the cast backstage.[26] Jerrold subsequently invited the cast of the play to MIT to meet in a larger forum with other students and faculty; he arranged as well for his colleagues Martin Deutsch, David Frisch, Albert Hill, Philip Morrison, and Victor Weisskopf to attend. All of them had known Oppenheimer and were closely familiar with the case. More than 700 people packed the Sala de Puerto Rico in the MIT Student Center; the discussion was enthralling. This was in November 1969. Only days after it had been necessary to seek court protection for the institute from the demonstrations of the November Action Coalition, it had been possible at MIT to discuss rationally and intelligently, and above all peacefully, a controversial matter regarding the role of the military in America.

The seminar also debated in depth the meaning of a long paper by Andrei D. Sakharov that had been published in the West only the previous summer.[27] Sakharov was a Russian nuclear physicist who had been one of the principal theoretical architects of the Soviet thermonuclear bomb, and his essay, "Thoughts on Progress, Peaceful Co-Existence and Intellectual Freedom," had been circulating hand to hand in the Soviet Union: a samizdat paper, so-called, unauthorized by any official and probably illegal.[28] It offered a wide-ranging plan for cooperation and eventual rapprochement between the Soviet Union and the United States, and it looked forward to the eventual convergence of the two societies; it deviated so widely from the official Soviet line that it was hard to know what to make of it. Its effect everywhere had been powerful, in both East and West. There is no doubt that its effect on Jerrold was equally electrifying, for when, years later, near the end of his life, Jerrold turned his attention fully to the question of coexistence of East and West in the presence of the hydrogen superbomb, the influence of the Sakharov document on his thinking was unmistakable.

Jerrold persuaded Rabi to address the seminar, which he did on several successive Tuesday evenings; Rabi some months earlier had given a colloquium at the Education Research Center on the subject of the Sakharov document. Now he spoke in a smaller, more intimate setting, drawing on his personal knowledge of Oppenheimer and providing the students with an insight into his very complex character that would have been hard to achieve in any other way. It was natural that Rabi would look for contrasts and parallels between Oppenheimer and Sakharov, although he had known nothing at all about the Soviet physicist until he read Sakharov's paper in the *New York Times*. He had said, in fact, in his earlier talk that he had been suspicious at first: "When I first read it, I thought it was a 'plant'—that no such man existed and that it was a production of some agency, either the CIA or the Russian equivalent. The reason was the extraordinarily frank and profound criticism of the management and administration of the Soviet Union. . . . A similar paper in the United States at the time of the late unlamented Senator Joe McCarthy would certainly have been considered subversive."[29]

Rabi's sessions with the Zacharias seminar were recorded, and a transcript has survived. The conversation continued for many hours, to judge by the bulk of the transcript, and it ranged over a variety of subjects; Oppenheimer and Sakharov were the starting points, but the discussion continually returned to the United States role in Vietnam. Clearly this is what was uppermost in the students' minds. The transcript is lengthy and rambling, as conversations can be, but it is a fascinating document, nevertheless, for it illustrates the gulf that occasionally revealed itself, Rabi and Jerrold on one side of it, the students on the other—a gulf that no one truly succeeded in bridging.

The question was this: what was the appropriate role for an individual who believed that United States involvement in Vietnam was wrong? There was no disagreement on the premise that the United States ought to get out of Vietnam, but the students brought to the question a sense of disillusionment that was sometimes startling to the older men.

When the nation had been faced with a crisis in 1940, neither Jerrold nor Rabi had hesitated to drop what they were doing and to place their talents and abilities at the service of the government. Later, during the period of the cold war, they did the same. Their pain at the treatment Oppenheimer had received was only in part because they knew and liked him; it was also because he had been so badly rewarded for doing what they firmly believed— what they knew—was the right thing.

Jerrold spoke about his own relationship to government, recalling events that had happened in 1952, when some of the members of the President's Science Advisory Committee, frustrated over lack of access to the president, had wanted to resign:

I remember specifically . . . a meeting at the Institute for Advanced Study. I can find out exactly, because it was the day Ed Purcell got the Nobel Prize. We were all down at the Institute . . . and Oppenheimer read a letter from [James] Conant [President of Harvard University]. And Conant said "the time has come for all of us old fellows to get out." And he meant all of us. Get out of the government, out of the military and all those related things.

But the government was in a mess, and you just couldn't do it. . . . How do you separate yourself from the country, from the culture, from the government? from everything? You don't quit just because you disagree.[30]

Rabi, too, felt a deep obligation to the country and had said so many times. Now he spoke about the problems: "You take the United States, 200 million people . . . [composed] of . . . sectionalism, of religious groups, of ethnic groups, of groups that have different eating habits, different history. Boston as contrasted with El Paso. Entirely different. The miracle is that it's been kept together as a country. The first job of the President of the United States is to preserve the union. It's worth a heck of a lot of everything else." And then, in response to a direct challenge about the war, he added: "If we had to have the Vietnam war to preserve the union, I'd be for the Vietnam

war. Absolutely . . . but it doesn't. It's done more to destroy the union."
Jerrold agreed; the war "anti-preserves it," he said.

But the students were unwilling to see it their way. The preservation
of the union was not an issue for them; they did not see it at risk, there were
more pressing issues, and they did not trust their government. They had
formed their beliefs in a different time from the older men, and they did not
share the same experience. The gulf that separated them was a gulf of
unshared years; Rabi was seventy-one, Jerrold was sixty-four, and the
students were not yet twenty. A student coming of age in 1969 would have
been born after World War II; he would not remember a time when the
nation was united by a common goal and would be less likely to appreciate
how the country suffered from being divided. He would have fresh in his
memory the brutal police reaction in Chicago to the antiwar demonstrations
outside the Democratic National Convention and would have no wider
context in which to set that tragic event. He would know about the bombing
of North Vietnam and would find no sense in it. He would find little reason
to accept the idea of a well-intentioned government that had somehow gone
astray. It was too easy to think of the government as a set of villains.

One of the students tried to state his own point of view to Rabi: "The
way we look at it, or the way I, a graduating senior, look at it is this: I graduate
next year and assuming I don't get into some kind of draft-deferrable job . . .
I'll go to Vietnam, you know; get in the army and go to Vietnam and perhaps
lose my life in a war that I cannot see any point in at all. . . . Something is
wrong where we have to go and just throw away our lives."

They argued not so much about ends as about means. Rabi and Jerrold
tried to encourage the students to seek political change by going through
the system—by marshaling votes, writing letters, forming political action
groups, making the system work. But the students—and they were bright,
promising, talented young men and women, the best in the nation—had little
or no faith in the system. It was a fundamental point of difference. It was
not that they did not understand each other, but they had come to the
discussion with such different assumptions that agreement was out of reach.
They might even in some cases hold similar views about means, but then
they disagreed about urgency. Jerrold put in a plea for education as a means
of getting the country on the right track: "I've been working on education,"
Jerrold said, "for a long time. It's now twelve and a half years solid. [But]
changing education, even with a small army, I think is going to take another
two generations." A student responded, "But it may be too late by then,
Zach. That's what the problem is from our side."

At the end, Rabi spoke to the students with sympathy and understanding: "A whole generation is entirely dissatisfied with this age altogether . . .

everything about it is lousy. The air is polluted, the water tastes bad, the streets are dirty . . . basically, it's a lousy time to have been born. . . . There's this atomic bomb and you're facing instant destruction and it depends upon a very few individuals. There's very little rationality you can see. . . . That's one side. On the other side, never has there been so much opportunity . . . opportunities for action, opportunities to do things which were unimaginable before."

And one of the students responded, ending the recorded discussion: "But it seems to me there are some even more basic things. We've got to decide first what our values are."

It was hardly a conclusive ending or a satisfying one; to see the most dedicated and well-intentioned students be so doubtful of the good faith of their government must have been extremely painful. But the seminar had been useful. It had allowed the most free-ranging examination of ideas and expression of opinion; as an object lesson, at least, it was a good one.

Options and Alternatives

A few small seminars, however excellent and imaginative, cannot change the character of a place as complex as MIT. A large institution is as ponderous as a ship and has as much inertia; it changes direction slowly. Students at the outset of the 1970s still had little in the way of choice, whether in terms of subject matter or in the style of learning it. An MIT education tended still to be highly prescriptive for all except the most enterprising. Naturally, in any discipline there is a body of basic materials to be mastered. It was Jerrold's argument that there ought to be a variety of ways to master it; the ways of learning ought to be as varied as the people doing the learning.

A civil engineering graduate of MIT, Richard L. Meehan, class of 1960, wrote about his own experiences as a freshman.[31] He contrasted the requirements that had been imposed on him with those that had been imposed by MIT on his father, fully thirty years earlier. According to the 1956 course catalog, the requirements were as follows: "All freshman engineering majors will concentrate on fundamentals: 70 percent of their forty-eight hour week will be spent on calculus, physics, and chemistry. Required in addition will be a general Western culture humanities course, a course of military science, plus an elective." Examining the course catalog for 1926, the year that his father had been a senior engineering student at MIT, Meehan wrote: "The 1926 freshman year requires fifty hours of exercises and preparation, 62 percent spent on basics—a lesser percentage than in 1956. English and history, which is limited to European history since the dawn of the industrial revolution, is required, as are descriptive geometry, mechanical drawing, and military science."

About the only thing that had changed over three decades was the tuition fee. But a closer look would have shown that the technical material that needed to be mastered had grown enormously, and it continued to expand vigorously. Nevertheless, the form of education and the choices available within it had remained unchanged for a very long time.

It was exactly this rigidity that Jerrold had sought to abolish when he had asked Julius Stratton to establish the Committee on Curriculum Content and Planning back in 1961, and some progress had been made at MIT as a result. One consequence of this was a broadening of the task of the Science Teaching Center; one of the first steps taken by Howard Johnson, Stratton's successor as MIT president, was to redefine the center's purpose and to rename it the Education Research Center (ERC) to signify its broader scope: to include all of the schools of the institute instead of only the School of Science.

From 1964 to 1968 it had been headed by Robert Hulsizer, a nuclear physicist from Illinois who had earlier been part of the PSSC effort. In 1968 Hulsizer decided to return to physics research, and Jerrold picked up the reins again. He was after a fresh approach, something that would allow a kind of flexibility that would be new at MIT. He was full of enthusiasm, ideas, and proposals; in May 1967, he wrote a sixteen-page memorandum to the Committee on Educational Policy, offering a lengthy list of iconoclastic suggestions—some startling, some not, but all interesting.[32] Characteristically, he also offered a practical suggestion for how to begin.

The memorandum contains some of Jerrold's most fundamental beliefs about education. They are expressed in his own forthright style, describing what he had come to believe—for example: "I believe that it is possible for a human being, even though he is not a mathematician, to understand differential equations right down to the ground."

The very term *differential equations* can strike terror into the heart of a nonmathematically oriented college student, not to mention the population at large; how can he have believed it to be a subject understandable, as he said, to any human being? Since he spent much of the rest of his life trying to do something about what he called "mathophobia," writing about it and looking for cures for it, it is possible to have a pretty good idea of what he meant. He would have explained it with an example; this was one of his favorites: "If you can understand the meaning of this sentence: 'The price of coffee increased more this week than it did last week.'—then you understand the meaning of a second derivative. And if you understand second derivative, then you can understand everything." And so, he would argue later, differential equations lie within the grasp of anyone who is competent to shop in a supermarket. But in 1967 he was writing specifically about MIT students, and the proposition applied to them required somewhat less justification.

He continued, in the memorandum, to make an important point about mathematical rigor: "It may well be that [the] student is not impressed or attracted by the movement toward increasingly sophisticated rigor. To me, rigor is a matter of taste, and one can have as much of it as he wishes. But the usual complaint about rigor misses the point. Many students, especially the experimenters, live more with nature than with symbols. No college mathematics test that I have seen has helped match the learning, doing and thinking styles of naturalists to the abstractions which seem so homey to the theorists."

Almost nothing Jerrold ever wrote comes as close to his intellectual core as that. The division of the world into naturalists (of which Jerrold was undoubtedly one), more at home with nonverbal experience, and theorists, more at home with symbols and words, echoes in a curious way a speech made by the English writer Aldous Huxley at the MIT Centennial in April 1961, subsequently published under the title "Education on the Nonverbal Level." In it, Huxley spoke about the problem of excessive scientific and technical specialization:

At the Massachusetts Institute of Technology and in other schools where similar problems have arisen, the answer to this question has found expression in a renewed interest in the humanities. . . . All this is excellent as far as it goes. But does it go far enough? . . .

Science is the reduction of the bewildering diversity of unique events to manageable uniformity within one of a number of symbol systems, and technology is the art of using these symbol systems so as to control and organize unique events. Scientific observation is always a viewing of things through the refracting medium of a symbol system. . . . Education in science and technology is essentially education on the symbolic level.

Turning to the humanities, what do we find? Courses in philosophy, literature, history and social studies are exclusively verbal. . . . Training in the sciences is largely on the symbolic level; training in the liberal arts is wholly and all the time on that level. . . . But this world of symbols is only one of the worlds in which human beings do their living and their learning.[33]

Recognizing Jerrold's underlying beliefs about the many ways different people learn, it is not surprising to find in his memorandum further appeals to enrich the experiential side of education: "I believe that laboratories should be available in every department—in history, social sciences, and human behavior, as well as in physics, chemistry, biology, metallurgy and so on."

He had reason to believe that such laboratories could be made to work; more than that, he knew that they would be exciting and that the design of both the laboratories and the curricula that would surround them would attract some of the best talent in those fields. It had already happened that way at the precollege level in the ESI social studies program. Just as in physics, there was something to be learned at the university level from initiatives in secondary and primary schools.

The entire memorandum is a prescient appeal to recognize individual differences, to get away from the lockstep into which MIT, and other universities, tended to force its students. It was prescient because it predated the days of student demands and protests. Jerrold wrote:

Students arriving at MIT have always been different from one another, and this disparity is increasing rather than diminishing. Possibly this is because of the increasing inequality of the schools they have attended, or it may be because our sampling procedures are changing. Whatever the cause, I believe that the years at MIT must continue to allow each student to develop in his own way.

I believe that every student should take increasing responsibility for his education. Student responsibility should be greater than is now usual, and the responsibility should grow as the student grows, until as an adult he can assume his full role. . . .

MIT should consist of a very rich intellectual and artistic mix. In it there should be stuff of the highest quality—engineering, science, philosophy of science, Freud, law, history, literary art, and so on. . . .

There should be many paths through MIT. There should be variety in course offerings and we should pay particular attention to those parts of MIT which suffer at our present regimentation.

He had much to say about the teaching of small seminars as well as large lecture courses; about the apprentice-type relationships that he hoped could be fostered between students and faculty; about examinations, grades, and alternative means of evaluation. Finally, he called for an experiment to be made. It was against his nature to make hypotheses without suggesting an experimental test; he ended his long memorandum with the recommendation that the experiment be tried with 10 percent of the student body. "Take this ten percent of the students and handle them in one of these radically different ways," he said. "See what happens."

11

A Partnership Ends

Development

As far as I can see, the only meaningful recommendations that one can make to the federal government have to do with money and the stimulus that money can provide. And so I repeat that ten percent of gross should go into innovation in education. For *every* program this should be so, and it should be monitored somehow.

—J. R. Zacharias (1969)[1]

The President's Task Force on Education, of which Jerrold had been a member in 1965, signaled a sharp change in the way the government interacted with the educational community. Funds for experiments in education became hard to obtain, and with that change came a host of problems.

What happened was this: prior to 1965, federal support to education had been provided mainly through the National Science Foundation. The NSF was interested in curriculum reform, and its interest had been reinforced by the success of PSSC and related programs. The foundation therefore tended to concentrate its support on programs aimed along similar lines, including, to some extent—Jerrold would have said not nearly enough— the training of teachers to use the new programs. The emphasis had been to pull the system up from the top. The criteria applied by NSF were quality and effectiveness.

After 1965, the emphasis changed. The National Defense Education Act provided for large sums to be distributed directly to state and local school systems, with almost no strings attached other than the requirement that the funds be spent for school purposes. School systems were given an almost free hand in deciding how to spend the money, and the funds were distributed more according to political needs than to any other criterion.

The sums involved were dramatic, for President Johnson, anticipating a substantial political benefit, allocated $1 billion for these purposes in the first year following the act, twice what Congress had recommended. The funds were to be distributed regionally, on the basis of a formula. The National Institute of Education was created as a new agency responsible for administering a major portion of these funds to a group of regional laboratories, themselves newly established to receive this money. The NSF, always under severe pressure from Congress about the way in which it allocated its funds, substantially reduced its role in educational development.

The effects were felt very quickly in all parts of the educational system. For Jerrold, in particular, and for those allied with him, raising money for curriculum development, or, in fact, for any other experimental effort that broke with tradition, became difficult, for the funds now were largely in the hands of the conservators of tradition: the school boards and the school administrators, for whom change was often threatening. The days of making common purpose with a sympathetic and helpful NSF were over.

Adjustments had to be made. Educational Services, Inc. found it in its interest to merge with the Institute for Educational Innovation, which had been formed to establish the New England regional laboratory. The enlarged corporation that resulted received the new name of Education Development Center, Inc. (EDC). The new company was much bigger than the old one and quickly developed new aims, matching the increased social consciousness of the time. Its management style necessarily changed; no longer able to continue its unstructured, free-wheeling style, it instituted a Table of Organization, installed its first president, and sought permanence.[2] It could no longer function intellectually in quite the way it had done in the past. Formerly, it had helped to define the needs of education by pointing out unrecognized problems and proposing often unorthodox solutions; now it was necessary for the corporation to seek funds where funds were available, for new purposes and goals it had not helped to determine. Curriculum development did not disappear but played a sharply diminished role as EDC was obliged to turn its attention to projects thought to be more socially relevant.

At MIT, the newly renamed Education Research Center also had difficulty obtaining federal funding for its purposes. For a time, some supplementary funds could be obtained, by dint of hard solicitation, from private foundations, and the shortfall could be made up by MIT itself, but that was not a situation that could survive indefinitely.

The constriction in the flow of federal funding began at a time when the ERC was attempting to redefine its role within the larger context of

MIT. In 1968, Jerrold had returned to the directorship of the center; by that time, and with his encouragement, its programs had become adventuresome, highly experimental, and highly promising. By 1969, it was possible for the center to advertise itself to the incoming MIT students as follows:

The Education Research Center at MIT has begun a major piece of educational reform. The effort could, we believe, have an impact on many aspects of post-secondary education in this country. . . .

A fraction of the MIT freshman class in September, 1969, will enter an educational program that will question all of the assumptions of form and strategy: credentials, certification, style, scope, topic, pace, technique and organization—in short, most of the traditional constraints of American higher education.[3]

For a man then only a year away from MIT's mandatory retirement age, it was an ambitious venture. To the extent that it would be possible, the Education Research Center planned to put to the test many of Jerrold's most cherished ideas about individuality and flexibility in education. Students would be offered choices. For example, they might enter concentrated study programs, devoting all their time for a limited period—a month—to a single topic normally the subject of a one-semester course. They might, as sophomores, enter the Unified Science Study Program (USSP), which would allow them to build their academic programs around self-selected topics, with suitable guidance. They might try out the so-called Personalized System of Instruction, a form of self-paced teaching, learning at their own individual rates, rather than at an arbitrary rate set by others.

Of course, there would be problems. There would be schemes that would not work and students who might not adjust. It would take feedback, to use the engineers' term, to fine-tune the program as it went along, making adjustments to the varying needs of the students. Jerrold was confident that the outcome would be a better trained, better educated student, whether a scientist, engineer, or philosopher.

Perhaps if the problems had been only programmatic, that would have eventually been true, but there were other problems that in the end could not be dealt with.

Proposal

How and what to teach? These are hardly new questions, especially for an institution that has been the intellectual leader in its field for most of its history. Yet today these questions are regarded as more urgent by students and faculty alike than they have been for many decades.

—J. B. Wiesner (1972)[4]

In spite of the difficulty in obtaining federal funds for innovation, Jerrold remained convinced that good ideas would ultimately attract support, as they had in the past. He believed that MIT had an urgent, immediate need for the kind of research and development in education that ERC could provide but that it would require substantial involvement of MIT faculty. Finally, he believed that MIT could serve as an example to the nation's other colleges and universities, that it ought to be concerned as an institution with educational innovation, and that it should provide both training and encouragement for those of its students who might wish to become teachers. MIT-trained teachers, brought to maturity in the innovative atmosphere surrounding ERC's programs, would bear little resemblance, he felt, to the product of the teacher's colleges around the country, for which he had small regard.

The problem was that the Educational Research Center had never become a fully integrated part of MIT. The condition of the center was ill defined, Jerrold said; at the core of its difficulties was its lack of the right to make academic appointments. The staff of the ERC had become substantial by the time Jerrold resumed the directorship in 1968, and it continued to grow as new projects were started. By 1972, the center had a professional staff of about forty, including a few part-time faculty from various academic departments. But without the ability to make faculty appointments in its own right, the center was dependent on the sufferance of the departments, especially since it could not pay for the time of those faculty members out of its own budget. It was not to be expected, therefore, that significant leadership in ERC projects would emerge from members of the MIT faculty, and with one or two notable exceptions, such as the Concentrated Study Program introduced by John King, it did not. Jerrold had been able to obtain permanent faculty appointments for only two members of the center: Anthony P. French became a tenured member of the physics department, and Judah L. Schwartz received a faculty appointment in the School of Engineering. The center's difficulty was well exemplified by the oddity of Schwartz's appointment, which was not to any specific department but to the School of Engineering at large. To this day, by his own account, he remains an administrative anomaly.

Those of the center's staff who were not regular members of the faculty had few of the benefits faculty members hold valuable. More than academic prestige was at stake; job security at a university consists of academic tenure, and staff members were not eligible. Their jobs depended for the most part on soft money, that is, funds obtained through external grants from government or private foundations or from the hard-pressed general funds of MIT itself—and there was not much to be expected in this latter category. As nonmembers of the faculty, ERC staff were excluded from participation in faculty meetings and from service on faculty committees; they could not vote degrees or on degree requirements; they could not serve on the Committee on Education Policy, for example, which was particularly frustrating to people whose entire expertise was in education. The result, as one ERC member described it, was that the center's staff felt themselves severely isolated from the institute; they were not full members of the academy and could participate in its intellectual life to only a small degree.

Jerrold hoped to change this by altering the status of the center. In 1969, with Tony French, he addressed a proposal to Jerome Wiesner, who was then MIT provost, recommending that the center be upgraded to a school; like the School of Engineering or the School of Management, this would be the MIT School of Interdisciplinary Studies.

The new school, as they visualized it, would have its own dean and academic departments; they hoped to see included a department of undergraduate interdisciplinary studies and a graduate education department. The school would have its own faculty, preferably jointly with other departments; it would mount its own degree programs and award its own degrees. Jerrold had often expressed his worry that students at MIT became imprisoned by the demands of professionalism imposed too early. He and French hoped that this might be avoided by the emphasis on interdisciplinary work: "[The School's] interdisciplinary nature needs interdisciplinary students. These, thank heaven, are the undergraduates. By the time students have graduate degrees, they are already specialized and seek their kudos from their fellow prisoners."

They noted that there were wider MIT problems not addressed by their proposal. There was considerable faculty sentiment in favor of establishing a new MIT committee on education to examine the entire MIT educational enterprise. Believing that such a committee might favor their proposal, they joined in urging that it be appointed.

Student unrest was severe in 1969, and they believed that educational reform was urgently needed: "Until the students have access to a part of MIT which is devoted to working with them in non-traditional ways, the number of emergencies will increase. The students are trying to tell us something, and we need a School that is geared to listening to them."

The proposal was too ambitious for immediate action; it came at a time when the institute was under great internal stress, aware that something was seriously awry but uncertain of how to fix it. The Commission on MIT Education was appointed, as Jerrold and others had urged; it was the third such wide-ranging study group since MIT entered the postwar era in 1945.[5] The new commission, generally known as the Hoffman commission after its chairman, reported to the MIT community in November 1970. By that time, the MIT president, Howard Johnson, had announced his retirement, and a vigorous search was underway for his successor.

Jerome Wiesner became president of MIT in 1971. The members of the MIT Corporation must surely have had a moment's doubt; Wiesner was Jewish, and the time had not yet arrived when a Jew could become president of a major American university without comment, remark, or hesitation. Antisemitism was no longer as much of an issue in the academic world as it had been before World War II, but its echoes lingered still. In a talk he gave to students in 1971, Jerrold must have had Wiesner's candidacy in mind when he said: "It [the antisemitism] is ameliorated now, not fixed. And don't think it is, because there are very few Jewish university presidents. It's not over. There's just one Jew on the MIT Corporation, out of forty. But it was so much worse."[6]

Jerrold took it upon himself to visit as many members of the corporation as he could, on behalf of Wiesner; he reached nearly all of them. In the same quiet and effective way in which he had once recruited physicists, he asked each corporation member to consider a simple question: "Who have you got better?"

Honest judgment carried the day, and Wiesner, having served as MIT provost for five years, became its president. MIT moved into a new period, with a leadership prepared to bring an engineer's pragmatic approach to the solution of the institute's problems. Certainly some of the most pressing issues Wiesner faced were contained in the report of the Hoffman commission.

Much of what Jerrold believed to be wrong with MIT education was echoed in that report; the commission had arrived at its conclusions independently. Speaking of the frustrations that students feel, the report stated:

Students come to the university young and skeptical, but brimming with idealism and enthusiasm. Too often what happens to them is disillusioning. Reared by their culture to pride themselves on their individual talents and achievements, many find that they are confronted with stereotyped requirements drawn up for a mythical average student and administered by impersonal systems of schooling and certification. Anxious to learn how to do good, they are taught how to do well. . . . Open to the force of constructive example, they find that scholars in the separate fields and schools have little to say to each other and rarely if ever come together, as the name "university" might imply they should."[7]

The commission report supported many of Jerrold's hopes. For example, it called for "new approaches to education in which imaginative ways are found to integrate scientific, social and humanistic studies. . . . MIT can meet these challenges forcefully by restoring education in its broadest sense, especially undergraduate education, to the center of its institutional commitment. . . . By inviting faculty members to develop imaginative new programs and by rewarding them for their efforts, we hope to stimulate experiments that may have far-reaching consequences not only at MIT but for teaching and learning elsewhere as well."

The commission was specific in describing what it wanted; its examples were all drawn from the programs of the Education Research Center. In order to enhance and institutionalize such activity, the commission recommended the formation of a new administrative structure, to be known as the First Division, which would have full responsibility for the first two years of the MIT students' academic lives. Within the First Division, there should be an Experimental Section, with responsibility for initiating and evaluating major experimental educational programs.

It was almost what Jerrold had wanted, although it had not gone quite far enough. In early 1971, he addressed another proposal to Wiesner; this time, as recommended by the Hoffman commission, he called for the establishment of a full-fledged academic Division for Education Research, "to create, within MIT, a single organization that can provide initiative and leadership in exploring the assumptions, content and processes of education, and that can serve the needs of the growing numbers of MIT students who are deeply interested in education and who see themselves as future teachers as well as those interested in research."[8]

This was the first time Jerrold had formally committed himself to the goal of training potential teachers at MIT; he now proposed ways of extending the Ph.D. programs for interested students. Potential candidates for such a program had already appeared, unsolicited, at the Education Research Center, he said; he was confident that there would be many more applicants to an established program.

There were those who saw a contradiction in this with Jerrold's well-known disdain for schools of education. Would not this amount to another school of education? It sure would, said Jerrold; but it would be "the goddamnedest school of education you ever saw."

Division

Now, I also believe down to my shoes that the laboratory for an educator is a laboratory in which the molecules are people, pupils, students, other teachers, and so on, just as the laboratory for a professional physicist contains molecules one can buy in large quantity.

—J. R. Zacharias (1969)[9]

Well, who in his right mind would stop interrogating molecules and start interrogating people? You know, a molecule may not talk, but it never talks back. It never does the opposite of what you ask it to do just because you asked it.

—J. R. Zacharias (1977)[10]

Within months of assuming the MIT presidency, Wiesner was ready to move forward on the proposal for an Education Division. A division would be a new structure for MIT, equivalent in standing to one of its schools, perhaps, but surely not the same, for the division would be a single entity designed to interact strongly with existing departments outside of itself, whereas schools are collections of relatively autonomous departments having substantial interests in common. There were many questions to resolve, many problems to work out. While it would have been within Wiesner's authority simply to recommend the establishment of a division to the MIT Corporation, and within the corporation's authority to establish it, the academic programs of the division would require the approval of the MIT faculty, and it would have been pointless to proceed without substantial faculty agreement. A steering committee was therefore established at the beginning of the summer of 1972, with William Ted Martin of the mathematics department as its chairman. In all, the membership consisted of three mathematicians, one of whom, Seymour Papert, was also the co-director of the Artificial Intelligence Laboratory; two psychologists, one of whom was from MIT's Sloan School of Management; a psychiatrist, Benson Snyder, who had served as MIT dean of institute relations; a professor of electrical engineering who was the director of the MIT Center for Advanced Engineering Studies; a professor of urban studies; and a physicist. Wiesner expressed his keen interest in the possibilities of a division by attending many of the committee's meetings; his provost, Walter Rosenblith, came to most of them. The committee was supplied with a staff person from Wiesner's office, and its mandate was to produce a design for the division—its structure, its programs, and its function—in time for it to begin operation in the fall of 1973.[12]

Jerrold was the physicist member of the committee. At sixty-seven, he was now past the MIT mandatory retirement age and had formally been retired, but in fact, his retirement had meant little. His title had changed—he was now Institute Professor Emeritus and he had gone on half-pay—but he continued as director of ERC, and ERC continued to be the principal active locus of educational innovation and experiment at MIT. By then Jerrold had already devoted well over a decade and a half of intense work to the problems of education; there was no one at MIT, or for that matter anywhere else, who could match his experience. On the nine-man committee itself, only Ted Martin had ever been engaged in educational development, and his involvement had been limited almost entirely to school mathematics in Africa; it would have been unthinkable for Wiesner not to draw on Jerrold's experience by appointing him to the steering committee.

There was trouble from the outset. It is interesting to contrast the performance of the steering committee with that of the CCCP, nearly ten years earlier. Under Jerrold's guidance, the members of the CCCP had spent nearly a year coming to a deep understanding of their problem; they had familiarized themselves through intense, full-time study with the methods, procedures, and substance of all the areas they intended to deal with. Only then, when they had come to know not only the problems but also each other, did they begin to decide what to do. When they did, they found that there was little mutual misunderstanding and little conflict. The steering committee for the Education Division, on the other hand, chose to consider the Ph.D. program immediately, plunging right in to the hardest part, before its members had come to know each other. Their discussions were immediately marked by mistrust, hostility, and what in hindsight appears to have been considerable false supposition. Questions of turf arose but were not fully articulated. It was clear, nevertheless, that the division was thought of as a major prize. Many members of the committee were leery of Jerrold; he had a forceful personality and an awesome reputation, and he had the Education Research Center at his back. They resolved almost at the outset not to allow him to take over the committee, as they feared he might do, and they resolved as well not to allow the ERC to become the core of the division. The prize was not to be Jerrold's.

Thus, what should have proved a valuable asset—Jerrold's experience—became instead a matter of serious dissension. By starting with an argument about how to construct a graduate program, which would define the academic nature of the entire division, the committee unwittingly guaranteed that every conceivable difference of opinion about philosophy, method, style, content, and approach would promptly come to the surface. The committee became irretrievably polarized, almost from its first meeting.

Jerrold was not the only committee member with a forceful personality. Seymour Papert, in particular, was equally forceful, and the committee meetings quickly became head-to-head conflicts over virtually every issue: the nature of research in education and the nature of educational development itself. Jerrold's view was that education is a practical subject; one learns to be a teacher mainly by teaching "real stuff," as he put it, and the attempt to teach others is itself a splendid way to learn content. In order to make such an experience practical, however, he claimed that it is necessary first to get the subject matter straight—in science, to get the "story line" correct and to do the equivalent in other fields. Given mastery of the subject matter, which must come first, teaching has many aspects of a craft, best acquired through apprenticeship.

Papert, and eventually the rest of the committee, disagreed; the real problems, according to their view, lay in the understanding of the cognitive processes of the student. Papert's own experiences, first as an exponent of the ideas of Piaget, then as the designer of the highly successful interactive computer language for children, LOGO, led him to push hard for research in what the committee called developmental epistomology, that is, studies in areas such as psycholinguistics, history of scientific thought, and cognitive psychology. These were subjects with acknowledged disciplinary backgrounds; the committee believed that they would be considered more respectable by the MIT faculty at large than the hands-on approach to curriculum development favored at ERC.

It is evident that the committee, unfamiliar with Jerrold's outlook and sometimes oblique way of expressing himself, did not understand very well what Jerrold meant by apprenticeship in graduate education. The sort of education that had produced more than thirty Ph.D.s in his molecular beams laboratory was clearly a kind of apprenticeship education, in Jerrold's mind. He had had many apprentices who were now successful physicists, and he had been an apprentice himself in his earlier days; it had been almost the only mode of advanced physics education in the days before World War II. It would have advantages now as well, Jerrold claimed. He particularly admired a program that had been established at Western Reserve Medical School by its dean, Dr. Douglas Bond, Jerrold's sometime sailing companion. Jerrold wrote about it:

Every entering medical student is assigned to the case of a family in which the wife is already pregnant. He follows the case for about two years. (Two years seems to be about enough. They have tried four years, and that seems to be too long.) The student follows the expectant mother to the obstetrician, to the hospital, through post-natal care, and so on.

At the same time, students are assigned in groups of eight or nine to a discussion leader. These groups discuss the cases which the students have been following. When the students meet with difficulties . . . about which the physician-discussion leader is not an expert, he naturally calls in other physicians as appropriate. . . .

In any event, the freshman medical student . . . is right there, learning what it means to be a doctor, and learning why he really should know something about blood chemistry, microbiology and neurophysiology.[13]

Jerrold thought an apprenticeship of that sort would be just the right kind of graduate education for a teacher, provided that the apprentice teacher, like the physician and the experimental physicist, would also be given the opportunity—the obligation, he said—to get the subject matter right, with whatever learning aids could be made available. It was here that the division could make its contribution: by inventing and modifying curricula until the subject matter was straight, understandable, and therefore teachable and by developing an ever better roster of learning aids—films, books, experiments, and whatever else might prove useful—for teachers and their pupils. Innovation in teaching methods would follow similar lines; the laboratory would be the classroom or the corridor or the museum— wherever learning might take place.

But as summer progressed into fall and the committee continued to wrestle with the relationship of the division to the Education Research Center, it became clear that the committee did not want that sort of work to be even a part of the division's function. The members saw little value in a design that would produce either teachers or innovative curricular approaches, and they turned away from such activities. What they wanted was the research and scholarship in which they were generally interested and which they imagined would confer respectability on the enterprise; in the committee view, neither curriculum development nor teacher training met that description. Those subjects had been spoiled, they felt, tainted by belonging to the standard teachers' colleges, and of low prestige. This, at any rate, was the widespread if somewhat parochial view at MIT of the field of education, and the committee members had no greater experience of the educational community than did the vast majority of their faculty colleagues. Barbara Nelson, the staff person assigned to the committee, commented afterward: "I remember being absolutely dumbfounded at the time when I suddenly realized that these people had no intention of looking around the country or the world to figure out what was going on in education. I was astounded! Because I thought that's what the task was."

To Jerrold's dismay, it was held that subject matter and curriculum were at best to be secondary to such erudite concerns as the theory of cognitive structures or the social context of learning, and at worst not present at all. As for educational innovation, the committee said that there should always be a home for it at MIT—somewhere, but not in the division. The division would not tackle the education of MIT students in the sciences or in engineering, and for any such activities to be connected to the division, they would have to have a research component related to one of the primary divisional interests.

The issue was certainly a matter of substance, and perhaps deserved argument on purely intellectual grounds. But the debate was heated and often became unpleasant and personal; individual conflicts and dislikes became mixed with concerns that should have remained professional. According to the official historian of the division, Jerrold found himself obliged to defend what he knew to be highly successful curriculum development programs, such as PSSC and ESS, against charges that they had failed or had been insufficiently radical.[14] They had "failed" to take sufficiently into account cognitive differences among children, or they had "failed" to take into account the different social contexts in which the programs would be used. Most of all, the charge was laid that programs like PSSC and ESS had not been tied to any educational hypothesis and therefore had no intellectual underpinning.

Before long Jerrold had had enough. He simply stopped coming to meetings. The committee continued to debate what the division's core programs should be. They decided that there should be four, but they could in fact suggest only one; another was added later. As Wiesner subsequently reported to the Corporation, these were:

"(1) a style of education research with an emphasis on knowledge structures and individual learning of the kind called information processing theories of cognition, and (2) a style of research with an emphasis on institutional processes and the concept of public learning."[15] To deal with the problem of ERC, the steering committee formed a subcommittee to develop recommendations "regarding which parts of ERC it would like to invite to join the Division." Jerrold was out of the country during much of the period that this discussion took place. It is not likely that he would have participated in the dismemberment of ERC had he been available.

In the end, the committee made its recommendations within the time frame established, so that the Division for Study and Research in Education (DSRE) began operations in the fall of 1973. Its first director was Ted Martin; Papert accepted a three-year appointment in the division as professor of education, made possible by a generous gift from an MIT benefactor. In 1975, Martin retired and was succeeded as director by Benson Snyder, who had also been a member of the steering committee.

In setting up the division, the committee had managed to agree mostly on what it wanted not to do, and its recommendation not to include any form of curriculum development carried the day. There was to be no reform of MIT curricula by the division; neither would there be programs designed to train the uniquely qualified graduate teachers whom Jerrold was certain MIT could produce. There would be no teaching laboratory, no place where Jerrold's "molecules" could be studied.

Wiesner, a long-time admirer of the field of artificial intelligence, supported the recommendations warmly; Jerrold was shocked and appalled.[16]

He had not expected that. Inevitably, a temporary coldness developed between the two men who had been so close for so long; they had been more like brothers than friends, and they now became estranged from each other. Jerrold felt badly let down by what had happened.

For a brief time he managed to keep alive the hope that the Education Research Center could survive, but since the division did not want its programs, and in the face of competition from the division for institute funds, it could not. Arrangements were made for some of the professional staff to transfer to other MIT programs; Judah Schwartz, for example, became a part-time member of the division. Others left for employment elsewhere. The Education Research Center ceased to exist in 1973.

The demise of the Education Research Center meant, for Jerrold, that there was no longer a place for him at MIT—no job for him to do, no active position for him to hold. His official relationship to the physics department had ended when he was made an institute professor in 1966, but he had actually stopped doing physics ten years before that, and his connections to the department were almost entirely social. He had many individual friends, of course, long-time colleagues and companions in arms, but for the first time in nearly thirty years, since he had come to MIT right after World War II, the institute was not prepared to make use of his abilities and experience. In May 1973, the remaining staff of ERC presented him with a bound volume of the occasional papers of the center; then, without further ceremony or ritual, he quietly left MIT. No formal notice was taken.

The Division of Study and Research in Education was not a success. In 1982, shortly after Wiesner retired from the MIT presidency, the division was dismantled, barely nine years after its establishment. It had never developed a clear focus or unambiguous purpose. According to Silvia Sutton, the division's historian, Papert's style remained far too idiosyncratic for him to become the Division's intellectual leader. She wrote:

He did not keep track of his graduate students, or conduct standard courses, or show up for meetings, or—on occasion—follow through on grant and contract obligations. Accordingly, "intellectual" leadership came into conflict with "administrative" leadership, argument followed, and personal relations deteriorated; the effects were demoralizing for everyone in DSRE. Papert became more and more of an absence in the Division, creating administrative headaches from afar. (I persist in my admiration for Papert. I liken him to a chess master who begins a series of brilliant games, stimulates his opponents, but suddenly loses patience and sweeps out of the room, knocking over a few boards in his path. The Institute must make a place for people like him, but it should be a place where they can exercise creative intellects without doing too much damage.)[17]

In its nine years, the division did not succeed in establishing its own graduate programs, obtained relatively little in the way of outside financial support, had no impact on MIT's programs, and never won the confidence and support of the MIT faculty. Whatever opportunity has been represented by the division was lost.

Numbers

Elephants, Kisses, and Grains of Wheat

Long ago in a faraway land a man invented a game called chess. "I am delighted," said the king after he had played and won a few games. "Name your reward. Whatever you desire shall be granted."

The king's advisors looked alarmed.

"Your pleasure is my highest reward, your Majesty," said the Inventor. "But if you so wish I would like some wheat. Let me have enough wheat to put two grains on the first square of the chessboard, four grains on the next square, eight on the next, and so forth, doubling the amount each time for all 64 squares."

"Is that all you want?" said the king. "A little wheat? So be it." The advisors sighed with relief.

—From "The Reward" (1974)[1]

What was it about mathematics, Jerrold had long wondered, that confused, frightened, and paralyzed so many people? The fear of mathematics—or mathophobia, as he called it—seemed irrational, yet it interfered with many people's understanding of the modern world. It kept them from being fully engaged and responsible for their own lives. What could be done about it?

Characteristically, he drew up lists, many of them, with titles like "Things About Math That Drive People Up the Wall" and "Possible Causes of Math Anxiety." How does one explain to a child, he demanded, why (minus) × (minus) = (plus)? When is it all right to neglect something, and when is. it wrong? Can you ever use approximate answers in arithmetic? When is a theorem obvious? What is set theory for? Why do we use emotion-loaded words like *complex, imaginary, irrational*? What is the difference between zero and nothing? When can you add dissimilar things: two elephants, three kisses, a lungful of air, a yard of string, a lover's promises? You can multiply five feet by two feet and get ten square feet. But is the reason you cannot multiply five apples by two apples simply that there are no square apples?

Mathematicians themselves were hardly likely to be helpful in curing mathophobia; Jerrold believed that the pitfalls of mathematics were often simply invisible to mathematicians. The New Math, developed in the 1950s and 1960s, had been largely oblivious to the confusion and anxiety it frequently produced in a child's mind and had only made things worse. Jerrold wrote, "Mathematical ideas are so easy, so natural, so apparent for some people—possibly 200 thousand out of 200 million (one can count all of the mathematicians). It is safe to say that barely one tenth of one percent of the population has neither been frightened nor snowed by a shorthand that should be easy for surely several hundred times that number."[2] Certainly something needed to be done; a nation could not go forward with the majority of its citizens unable to participate fully either in the obligations to or the benefits of the society in which they live:

A citizen [Jerrold wrote] cannot take advantage of and contribute to our social institutions without an understanding of science and mathematics. He or she must make a living using the tools and facts of science and technology. . . . The citizens of today's world must use concepts and facts—computations, logical and quantitative reasoning, knowledge of their own bodies, understanding of causal relationships in the natural world. Without these, individuals are powerless... Without understanding the arguments of the issues, a citizen would have to vote by following a leader blindly.[3]

Nor was help to be expected, said Jerrold, from the schools of education; there was no new generation of young teachers on the horizon, armed with new skills and new insights. There had been an opportunity to create a school at MIT that would produce such teachers, but the opportunity had not been taken. Jerrold commented on that, too, a decade later, in a 1982 interview: "Ed Schools don't really exist. It is charitable to say nobody has invented an Ed School yet. It would be good if there was one. We look forward to it. . . .

I mean, there's one up the street with a great reputation. Doesn't do anything about mathematics at all. At any level. As if you could teach anything at all about science without getting your head straight about mathematics. . . .

We tried to start a different kind of thing at MIT and it flunked."[4]

Nevertheless, as far as Jerrold was concerned, it was still a job very much worth doing. He had already begun to think about it before he left MIT, and he was prepared to go on with it. He was certainly not ready to retire; it seems completely unlikely that he would have known how to tolerate the inactivity or the aimlessness that our society often manages to impose on its elder members. On the contrary, for the rest of Jerrold's life there would not be a day, a moment, when he did not have some project under way.

He moved his base of operations to the Education Development Center, although he still kept an office at MIT; some of the ERC staff moved with him, bringing their projects along. The shift had considerable symbolic value, had anybody taken notice, for EDC was physically separate from the institute campus, located some five or six miles away in a reconditioned redbrick factory building in the town of Newton, Massachusetts, on the other side of the Charles River from MIT. Jerrold was now truly professor emeritus.

He had always maintained a certain level of activity at EDC and had even served for a short while as its acting president after Arthur Singer left for the Sloan Foundation in 1968. A special formal title had been invented for him there; he was Jerrold Zacharias, founding trustee. He had no need for a transition period or for any hiatus at all. The daily conferences, arguments, planning sessions and discussions continued; only the location changed. Jerrold simply went on generating ideas and proposals as he always had. It was the only way of life he knew or cared for.

The Prime Minister held up a map. "Your Majesty," he said, "the entire kingdom could not grow enough wheat in a hundred years to meet this man's request. That many sacks of wheat would cover the entire kingdom many times over!"

"How can that be?" demanded the king. "It seems like such a simple request. Two grains, four grains, doubling each time . . . anyway, how much wheat is it?"

"Two to the sixty-fourth is 16×10^{18} and the total is double that or 32×10^{18} or 32,000,000,000,000,000,000." said the Inventor.

"Approximately," said the king.[5]

Much Ado about Nothing

As Jerrold approached the biblical age of seventy, he continued to carry with him a certain lightheartedness, a willingness to find fun in most situations, that others found very attractive. He was ready to make a friendly bet about almost anything interesting: the outcome of an election, for example, or the future of fusion power in his lifetime. The standard stake was a magnum of champagne (which might escalate to a case if the other fellow could afford it.) On other occasions, he might offer champagne as a prize. These were some of the ways he found to engage others, sometimes eliciting unexpected whimsy from those who were notoriously dour and sober. But those who were involved in these small games remember them with affection. They were an important aspect of Jerrold's personality, and often they are among the most vivid things that people recall about him.

The problem of Zero was one example of how people might be drawn in, challenged by Jerrold's sense of fun. The correspondence that follows is self-explanatory.

From Dr. Douglas Bond, M.D., president, Grant Foundation, New York, and professor of psychiatry, University Hospitals of Cleveland:

October 8, 1974

Dear Jerrold,

What they never told me in school was that there were two powerful numbers—zero and infinity—so that whatever you do with those numbers ends up as those numbers. There is another problem about zero. Zero is nothing, but it really isn't nothing. It makes no sense that you have something and you multiply it by nothing and you end up with nothing. It is clear if you have nothing and you multiply it by a number, it is nothing, but it never makes any sense to a child of my age that the multiplication of another number by zero is nothing.

You see I do think of you.

Sincerely,
Doug

P.S. If you have nothing and you take something away from it, you've got nothing. On the other hand, if you take one away from zero, you have -1.

$$0 + 5 = 5$$

$$0 \times 5 = 0$$

and yet multiplication is addition.

October 10, 1974

Dear Jerrold,

I called back but I couldn't get you. . . .

I had been under the impression that the Dodo was extinct. However, after I left you I went to the MGH [Massachusetts General Hospital] and, lo and behold, I found many. I was fortunate also to take the plane to Cleveland and sit behind one of them and Alice. I didn't have my recorder, but . . . it went like the enclosed.

Sincerely,
Doug

Enclosure:

"Why," said Alice, "is zero nothing?"

"Don't be stupid," said the Dodo. "Sometimes it is and sometimes it isn't."

"When is it?" said Alice.

"When you have 5 and you take away 0 you have 5 left because 0 is nothing. If you have 5 and add 0 to it you have 5 because 0 is nothing. Anybody knows that," said the Dodo.

"My, what a weak number 0 is," said Alice. "Or is it a number?"

"It's all very clear," said the Dodo. "If you are playing cricket and you have no runs you get a zero, or a l'oeuf in tennis, or a goose egg in baseball. They are all the same, of course, but in such instances zero is nothing because there is nothing below it. That's not true at other times because zero can be in the middle and then zero is something."

"It sounds a little confusing," said Alice.

"Think, girl, think," said the Dodo. "If you have 5 and you multiply it by zero (nothing) that just tells you how many times you add and you get zero."

"But," Alice said, feeling a little annoyed at the Dodo's attitude. "If I have 5 and I multiply by nothing, I don't see why I don't have 5. If I have 0 (nothing) and I multiply by 5 then I would have nothing."

"There," said the Dodo. "You've just shown what a silly girl you are. You think that multiplying by nothing (0) is the same as not multiplying by anything and it is entirely different."

"You aren't very clear," said Alice.

"You are quite cloudy," said the Dodo. "Anyone who thinks can see that 0 times anything is 0."

"My, what a powerful number," said Alice. "Does $0 \times \infty = 0$?"

"Of course," said the Dodo.

"Oh," said Alice. "I think I see what you mean. The thing you multiply by tells you how many times you add the thing multiplied. So $5 \times 3 = 3 + 3 + 3 + 3 + 3 = 15$ and $3 \times 5 = 5 + 5 + 5$ and $0 \times 5 = $ (no fives) $= 0$."

"My child," said the Dodo, "you are not as dumb as you look."

"Well, if that's what you meant," said Alice, "why didn't you say so?"

October 15, 1974

To: Project ONE and Project TORQUE
From: J. R. Zacharias

The attached correspondence from Douglas Bond requires an equally crisp answer. This need provides a fine occasion to offer a prize (I have ready a bottle of Moet-Chandon Dom Perignon 1964) for the best one submitted. . . . Bond has agreed to serve with me as one of the co-judges.

From Sheldon H. White, Professor of Psychology, Harvard University:

November 6, 1974
Dear Jerrold,

It seems to me that a psychiatrist should have the heart to appreciate the true sadness of Zero, which is not very powerful at all. It is a rather sad creature, a Place that yearns to be a Number.

Zero is a place where numbers go, like a kind of waiting room. We put Zero wherever we expect that a number might come someday and what Zero always says, wherever it is, is "Watch this place. There's nothing here at the moment, but a Number could be here." It's all a little upsetting to Zero. In the first place, Zero is intensely aware that if there were a Number to go in the place, Zero wouldn't be there. The Number would be there in its place. And, furthermore, it's not really serious work to sit around holding places. Zero would like to get into the action.

I suppose we've all felt a little sorry for Zero so we've dressed it like a Number and treated it like a Number . . . sort of a numerical affirmative action program. That's a bit of backhanded sympathy, because things keep going wrong for Zero when we all pretend this way:

• When Zero tries to buddy up to a number, it gets rudely shoved aside. Once, for example, when $0 + 5$, the outcome was 5. Then there was the case $25 - 0$ and the outcome was 25. A few years ago, there was the rather sad case of 6,352,784 $+ 0 - 0$. Zero did everything it could to get together in this case, with no effect on the Number. But I could go on and on.
• Frustrated in its efforts to be friends with other numbers, Zero tried bossing them around. This leads to another kind of painful outcome. The other Numbers just go away. Once, when 0×25 the outcome was 0. Another time when $0 \times .00018$— just a tiny little number, maybe something will happen?—the outcome was still 0. . . .

Can we help poor Zero with his plight? I believe we can. In the first place we should say to Zero gently but firmly. "Come, come. You'll have to face it. You'll have to adjust. You are not a Number but a Place. Places are good, and important, and they do useful work." Not only should Zero come to believe that, but we, too, who deal with Zero every day, will have to accept it. We will have to understand that Zero is not a Number but a Place, and Places are perfectly fine. In fact, quite powerful in their own way. Zeroes can tell us how big a Number is, or how small. Just watch the gang of Zeroes the Number has with it. Or, when we are getting rid of things, a Zero always tells us we have gotten rid of everything.

I might say that one of the main sources of Zero's present plight is that he has been dealt with only by Mathematicians and Physicists and people like that just don't understand Mental Health. Put Zero in a series of Encounter Groups with other Numbers and Functions. Let him face the fact that he is not a Number; let him find his place as a Place. Charge him exorbitant fees until he comes to realize the fiscal value of Zeroness. You would shortly see him stop dithering around after Numerosity and settling down to a round, comfortable Placidity.

With best regards,
Shep

P.S. I showed this letter to a Zero friend of mine and, true to his Numerophilia, he said, "Very smart. Very smart. But listen here, smart guy. You can eat 1 banana very nicely. But you try eating 10 bananas or 100 bananas and you'll find out I'm a Number, all right." I gave it to him right between the eyes. I said, "Zero, baby, without a number somewhere around you you're nothing and always will be." It was not a kind thing to say but I hope it helped.

From Gerald S. Lesser, Harvard University Graduate School of Education:

November 8, 1974

Dear Jerrold,

Here is my entry for your Douglas Bond contest:
"Beyond your questions lies an even more profound truth: GORNICHT MIT GORNICHT IZ ALS GORNICHT."

Regards,
Gerry

From Erik L. Mollo-Christensen, professor of meteorology, MIT:

December 1, 1974

Dear Jerrold:

I suggest that an attempt to reform numerical practices at the University Hospitals of Cleveland, and also to change the way physicians feel about numbers, would be a project beyond the scope of our resources.

—E. M-C.

From a comic strip by Bud Blake, October 18, 1974, submitted by Sheryl Slone, College of Education, University of Illinois at Chicago Circle:

First child: What is zero multiplied by five, Hugo?
Hugo: Five.
First child: Incorrect. Zero multiplied by anything remains zero.
Hugo: Then why bother multiplying it?

Regrettably, the name of the winner of the bottle of Dom Perignon was not recorded.

Infinity Factory

I got those negative number blues;
I got less than nothing to lose.
Some folks think that zero
Is as low as you can go.
But I learned the hard way
That it just ain't so.
'Cause those negative numbers
Go right on down the line,
Past zero on forever.
Into infinite time.
I got those negative number blues;
I got less than nothing to lose.

—Song from "Infinity Factory"[6]

Jerrold's concern with mathophobia considerably predated the closing of the Education Research Center at MIT and his transfer full time to EDC. One of the more central projects at ERC had been the development of a new kind of calculus course, one that would utilize a hands-on approach. The course would have a laboratory where students could carry out physical procedures involving differentiation, integration, and the solution of differential and even integral equations. The project was headed by William U. Walton and Harry Schey. Walton was a veteran of the African Primary Science Program and had lived and worked for two years in Kenya. He had first taught in a small Minnesota college and then had taught PSSC physics at Webster College, in the outskirts of St. Louis, where he had been discovered by Jerrold. Walton was a good example of Jerrold's readiness to embrace the nontraditional, for he had no advanced degree and lacked advanced training. But he was endlessly inventive and brought to the project skills and insights that advanced training sometimes seems to inhibit.

Schey had a more thorough training, at least in the conventional sense. He had been a research physicist at Livermore Laboratory, where he worked with Judah Schwartz, helping to develop computer-generated films illustrating quantum mechanical phenomena. He was responsible for the theoretical aspects of the laboratory calculus course. When Jerrold moved to EDC, Walton, Schey, and the calculus project all moved over with him.

Another unconventional type who moved to EDC from the Education Research Center was Mitchell Lazarus. Lazarus's background was unusual: he had received a master's degree in electrical engineering but had then changed fields. When Jerrold hired him in the spring of 1971, his Ph.D. in experimental psychology from MIT was just one year old; Lazarus felt that his encounter with Jerrold had saved him from an unrewarding teaching career at some lesser institution where he would probably not have been able to carry out any research or do anything interesting. He preferred the uncertainty of an interesting job on soft money to the secure but almost certain boredom of the job for which he had otherwise been headed. Always idiosyncratic, he subsequently became a successful attorney in Washington, D.C.

There were others, like Judah Schwartz and Edwin Taylor, who retained positions at MIT—Schwartz as a tenured professor in the School of Engineering, Taylor as a senior research scientist in physics—but continued their involvement with Jerrold's enterprises to greater or lesser degree, as they were able. Compared to the old PSSC, the group was small and lacked both seniority and visibility, but it was nevertheless a substantial resource on which Jerrold could draw for advice, counsel, and the kind of give-and-take argument on which he thrived.

It was not long before the next project got under way; called Project ONE, it embodied some of Jerrold's most cherished notions about teaching mathematics. He had, in time-honored fashion, visited the NSF seeking support for some project—which one is now lost to memory, for he was unsuccessful. Mitch Lazarus recalled that Jerrold told him that he had asked himself, in some exasperation, "Well, what do they want?" and had decided that it might be a kind of math Sesame Street, aimed at minority children. It was worth thinking about, said Jerrold. "What are the kids doing?" he asked. "They're watching TV, so if you want to reach them, go where they are."

Lazarus recalled that they spoke to hundreds of people about how to create such a project. They did not expect that it would be easy. Jerrold saw it as a chance to do something effective about mathematics teaching, for, as Lazarus said later, "We were able to bypass the school boards, the publishers, the bad teachers and the parents who didn't care. The stuff would go straight from us to the kids. There was something enormously seductive about it, but it was enormously difficult."

Clearly Jerrold still knew how to get things moving; they were able to obtain start-up funds from the NSF and from four private foundations.[17] The project wound up with the usual unwieldy but informative subtitle: "A Program in Mathematics, Drawing upon Science, Technology and the Arts, and Using Television and Manipulable Materials." It was aimed primarily at minority children of ages eight through eleven, and it was to consist of sixty-five half-hour television programs. On June 28, 1973, after a year of planning, the U.S. Office of Education announced a grant to EDC of $4,023,027 to fund the project; Jerrold was named project director. It was not a large sum, averaging out to just over $60,000 per program; commercial television spends that much on a two-minute commercial.

It was, as Lazarus predicted, very difficult. Jerrold had clear and unequivocal ideas about the teaching of mathematics, and he tended to have equally definite ideas about what minority children needed, based on his experiences with the Negro Colleges Program nearly a decade earlier, and on his work with the Bank Street conference, "Education of the Deprived and Segregated."[8] Agreements were not always achieved to the satisfaction of the minority staff members on the project, and there was often considerable friction. Pressure at times was intense, especially because there was close and continuing oversight from the federal government, itself deeply concerned at the time with minority matters. There were resignations and reassignments; the project experienced a kind of turbulence that became, in Lazarus's words, "part of the folk legend of the industry."

Nevertheless, the project, together with its subsequent renewals over a four-year period, eventually produced ninety-five programs under the title "Infinity Factory." *Time* magazine described them:

The show . . . tears along at a breakneck pace to the beat of finger-snapping rock music. . . . Episodes take place . . . in neighborhood settings. Because black and Hispanic children are a special concern, many of the shows are filmed in black areas or in the barrio."[9]

The programs were shown by the Public Broadcasting System in approximately 200 television regions, several times a week. The results were highly creditable, especially given the difficult circumstances. In each of the 200 markets, without exception, "Sesame Street" ranked first, and "The Electric Company," another children's television series, ranked second in popularity. In about half of those regions, "Infinity Factory" came in third, "Masterpiece Theatre" fourth; in the other half, it was the other way round. Teachers, according to the evaluations that later were provided, were highly pleased with the programs and with the program guides and other teacher aids that were made available to them prior to each broadcast.[10] Although the other two children's programs provided some educational segments, "Infinity Factory" was the only one with education as its principal purpose, which made its success all the more remarkable. As *Time* magazine said, there was nothing else like it anywhere.

Jerrold had his seventieth birthday while the programs were being broadcast; the project, contentious and difficult as it had been, was not a bad achievement to contemplate as he reached the canonical age of wisdom. There was considerable satisfaction in it, for it had been a struggle, but there was a great deal of frustration as well, for the struggle was not over. Lazarus gave voice to the problem:

Jerrold carried a pocket knife and found it very useful. He had a little magnifying glass in his pocket and he found it very useful. And he carried his math skills around with him and found them very useful and he wanted to sell everybody else on the pocket knife, the magnifying glass and the mathematics. There was always a lot of tension between Jerrold and traditional math educators for whom mathematics was an end in itself.

You didn't learn math because it was useful for something else; you learned math because it was good for you. The teachers felt that way—the mathematicians felt that way, Lord knows—the parents felt that way. So Jerrold had an awful uphill fight, trying to sell this idea. . . . This idea has not yet caught hold in American education. It will, someday.

But beyond such problems, Infinity Factory confirmed that there remained yet another massive obstacle on the road to good education: the standardized examination. All the effort that could be mustered—all the ingenuity, skill, cleverness, and good taste—would be useless if the students still had to pass traditional examinations that stressed the wrong aspects of learning. The examination was still the tail that wagged the dog.

The Thunderer

I get people angry. I don't know why; I don't know how.

I asked Jerry Wiesner about it. . . . I said, "Jerry, look, you get away with it and I don't. What's the difference?" He said, "I mumble. I don't talk loud enough. It's impossible for me to do it, so people don't quite hear."

—J. R. Zacharias (1976)[11]

Jerrold didn't mumble. When someone or something bothered him, he was inclined to come right out and say so, and what he found when he looked into the question of testing bothered him a lot. He had never been especially tolerant of anything he considered second rate, but he felt that the testing industry was not merely second rate; it had actually inflicted considerable damage. He lashed out at it: "I feel emotionally toward the testing industry as I would toward any other merchant of death. I feel that way because of what they do to the kids. I'm not saying they murder every child—only 20 percent of them. Testing has distorted their ambitions, distorted their careers. Ninety-five percent of the American population has taken an ability test. It's not something that should be put into the hands of commercial enterprises."[12] In 1975, he had written, "I have often referred to tests as the Gestapo of education systems. Uniformity and rigidity require enforcement, so I have chosen a most denigrating title for the enforcement agency. Its hallmark is arbitrariness, secrecy, intolerance and cruelty."[13] And in a 1976 speech given in Dallas before the National Association of Elementary School Principals, he said: "For the last couple of years I've been studying the tests as they come out of the industry. . . . Achievement tests are a combination of lack of understanding and simple wrong-headedness and they're ruinous to the kids. There is something called copyright law, which serves as the guardian to keep people, such as me, say, from publishing the achievement tests. In other words, I could manage, somehow or other, to buy the achievement tests, publish them in *the New York Times* . . . and then be sued. Now those tests control what goes into the textbooks, and [what's in the textbooks] is what goes on the tests. It's what I call a vicious circle. It isn't vicious because the people are vicious. I've never said that. They're not vicious. They're just stupid.[14]

It was perhaps a bit disingenuous for Jerrold to say that he didn't know how or why he made people angry. If he had simply read his own prose, he would have known clearly enough. But he was angry and indignant himself, as his language reveals; he was angry because the tests served as cruel barriers, discriminating against the unusual, filtering out everything except conformity, unfair to most of those who were tested. He was not the first in the field on this issue; the mathematical physicist Banesh Hoffman had sounded the alarm years earlier, in 1962, in his book *The Tyranny of Testing* in which he had pointed out, with many cogent examples, the failure of standardized tests.[15] "Hoffman's book," Jerrold wrote, "should have spearheaded an immediate, major, indignant revolt against the misuse of 'standardized,' machine scored aptitude and achievement tests. It did not. The standardized achievement test business flourishes and we have continued on with the insidious undertaking of trying to measure 'intelligence.'"[16]

The achievement tests were culture bound; children of other cultures, minority children, inner-city children, the "deprived and the segregated"— all were placed at a permanent disadvantage by a system that pretended to measure something with precision that was far more complex than the would-be measurers understood. Jerrold wrote:

The worst error in the whole business lies in attempting to put people, of whatever age or station, into a single, ordered line of "intelligence" or "achievement" like numbers along a measuring tape: 86 comes after 85 and before 93. Everyone knows that people are complex—talented in some ways, clumsy in others; educated in some ways, ignorant in others; calm, careful, persistent and patient in some ways; impulsive, careless or lazy in others. . . . The traits tall, handsome and rich are not along the same sets of scales as affectionate, impetuous or bossy.

As an old professional measurer (by virtue of being an experimental physicist) I can say categorically that it makes no sense to try to represent a multidimensional space with an array of numbers ranged along one line. This does not mean it is impossible to cook up a scheme that tries to do it; it's just that the scheme won't make any sense. It's possible to strike an average of a column of figures in a telephone directory, but one would never try to dial it.[17]

Probably Jerrold's indignation was heightened by what he saw as the pretense of precise measurement. As he pointed out, he knew something about measurement—enough to recognize when it was a sham. Every advertiser of soap or patent medicine knows the psychological impact of precise but meaningless numbers—for example, "99 and 44/100 percent pure," or "four doctors out of five say . . ." Precision can serve as a mask for falsity; it looks like truth. Jerrold expressed something of this view in another, more serious exchange of letters in with Sheldon H. White, the Harvard psychologist, in 1978:[18]

7/11/78

Dear Shep,

"When you can measure what you are speaking about and express it in numbers, you know something about it, but when you cannot measure it, when you cannot express it in numbers, your knowledge is of a meager, unsatisfactory kind."

These words of Lord Kelvin, a noted British physicist (1824–1907), have been a major source of misunderstanding and of misdirection of attempts to improve the schooling of children in many parts of the world. Even if Kelvin had interpolated a phrase making the quote read—"but when you cannot measure it, when you cannot express it in numbers, *when you cannot express it in words, phrases or sentences*, your knowledge is of a meager and unsatisfactory kind," he would still have been a source of much misfortune, much mischief.

Jerrold

Dear Jerrold:

I thought that an appropriate counter to Lord Kelvin would be a famous quote from Daniel Yankelovich. The first time I saw it, I was so delighted with it I copied it out, but I have since seen it in several places. On the chance that you haven't seen it, it goes as follows: "The first step is to measure whatever can be easily measured. This is OK as far as it goes. The second step is to disregard that which can't be measured or give it an arbitrary quantitative value. This is artificial and misleading. The third step is to presume that what can't be measured easily isn't very important. This is blindness. The fourth step is to say that what can't be easily measured really doesn't exist. This is suicide."

I first found the quote in Adam Smith's *Supermoney*. Then someone used it in *Science* to comment on a Vietnam war run by body counts. I believe this quote should be read three times . . . by a minister, priest and rabbi . . . before each annual meeting of the American Psychological Association.

Shep

Jerrold did not claim that education should do without tests of some kind. Children, teachers, schools, principals, school systems, curricula—all need some kind of evaluation, both to improve the overall system through feedback and to enable decision-making processes to take place. But in his view, IQ tests measured nothing useful and constituted a dishonest distortion of the evaluation of children. Philip Morrison agreed; he pointed out that IQ scores fall nicely along the bell-shaped curve called the normal distribution, which may perhaps give confidence to statisticians that the data are unbiased. But Morrison did not stop there; after first displaying a variety of measurement curves—some normally distributed, some not (the heights of 20-year-old men are normally distributed, for example; the distribution of income is not)—he went on to say:

It should be remarked that each and every one of these results follows from a careful but rather simple measurement. We know what we are measuring in each case, and the means of measurement has been developed with little reference to the distribution that results. There is one grand exception: the tests that lead to the IQ. With perhaps one exception . . . these tests have been composed of items selected after trial for observed conformity with the normal distribution. Items that showed little correlation with the overall expectations, or with the results of previous tests of the kind, have been systematically excluded.

. . . The mind is a complicated animal and the IQ "measures" it in a way no one can describe. The justifications are after the fact and profoundly statistical, and they stand under the heavy indictment of circularity.[19]

When Jerrold was invited to give his speech to the elementary school principals gathered in Dallas in 1976, the title suggested to him was "Curriculum Reform and What Went Wrong." He changed it to "Curriculum Reform: What Went Right, What Went Wrong and What Should We Do Now?" The new title, especially the last part, represented the difference between merely criticizing—however valid the criticism might be—and being an active agent for change. Jerrold could and often did make people angry by expressing his opinions, but he could not be accused of leaving the work for someone else. He had a little motto—one of the dozen or so that constantly reappear throughout his notebooks and papers—"If not, what?" His idea of useful criticism required offering an alternative.

He asked himself the question publicly, in the course of the Dallas speech. "If you ask me," he said, 'Do you know now what you would do, how you would make tests . . . ?' The answer is no, I don't know. . . . I've thought about it some. . . . You can't make tests unless you know a way to find out what a student does know and does want to do and is proud of. The present testing arrangements are just ordeals with pencil and paper to try to find out what the student can't do. Now I can't play the saxophone and I can't speak Russian and I can list more things that I can't do than I can do, by a lot."[20]

When the African Primary Science Program produced new materials in the 1960s, there had been scant thought of introducing standardized achievement examinations for the children. The variations in culture, language, and circumstance across that continent would have rendered such an idea laughable. Nevertheless, the program had to be evaluated on behalf of its American sponsors, at least, and for African school authorities as well, and that meant that the children had to be evaluated. The task fell to an educational psychologist, Eleanor Duckworth, herself a student of and sometime translator for the legendary Piaget. Later, when Jerrold wrote about testing, he found it useful to reproduce some of the twenty questions that had formed the basis of the Duckworth study:

Can the child show others what he has done so that they can understand him?
Does he puzzle over a problem and keep trying to find an answer, even when it is difficult?
Does he give his opinion when he does not agree with something that has been said?
Is he willing to change his mind about something, in view of new evidence?
Does he make things?
Does he feel free to say he doesn't know an answer?
Does he talk about his work at other times of the day?
Does he make comparisons between things that at first seem to be very different?
Does he start raising questions about common occurrences?
Does he ever repeat one experiment several times, to see if it always turns out the same?[21]

It is difficult to imagine how any of these desirable behaviors—or their absence—would be revealed in a standardized, multiple choice, computer-graded examination. But any good and reasonably observant teacher would be able to recognize these patterns in students and would seek to develop and encourage them. Jerrold felt it ought to be possible to design tests that would emphasize and reward such behavior patterns instead of rote performance or the ability to guess what the examiner may have meant by an ambiguous question.

Project TORQUE (Tests of Reasonable Quantitative Understanding of the Environment), funded by the Carnegie Corporation and the Ford Foundation, got underway at EDC in April 1974 as a spinoff of "Infinity Factory." Its purpose was to develop alternative assessment materials and techniques—alternative to the kinds of tests then in use, which Jerrold referred to as "desperately bad, bad for the children, bad for the schools, bad for the system."[22]

The project was restricted to mathematics, but it attempted to develop principles and techniques that would be far more widely applicable. It accepted that education needs tests but not for putting children in some kind of rank order; that, as Jerrold pointed out, is not a valid need, and in any case the tests do it very badly.[23] But they remained mindful that there is a valid purpose for tests. Lazarus and Schwartz, writing about the project, put it this way: "Any complicated system (and education is certainly that) must monitor its own consequences for the results it wants. Ideally, school achievement tests should work something like a furnace thermostat: telegraphing outcomes back into the system so it can adjust itself for the best possible effects."[24]

Project TORQUE ran at a time when there was a strong public outcry for programs labeled "back to basics" and "minimum competency testing." Jerrold felt the inadequacy of such demands; they were too simple and too crudely fashioned. Writing about the back-to-basics movement in 1978, he traced its origins and marveled at how it took off like the panic reaction that accompanies fire in a theater. The stimuli were various. A strange, widely circulated and poorly substantiated report, paid for by Educational Testing Service and formulated by a prestigious "blue-ribbon" committee chaired by Willard Wirtz (past secretary of labor) called public attention to what was called "A Decline in Scholastic Aptitude Test Scores."[25] "There were no minority opinions included in the Wirtz report. [Some educators], in journals with small and irrelevant circulation—it was like talking into a funnel connected by rubber tubing to the speaker's own ears—brought contradictory data to bear, but the public paid them no heed.

Testimony before a Senate subcommittee . . . placed the blame for a widespread lack of skills directly on the children. The hue and cry was taken up by other groups. . . . The solution was clearly [to be] 'Readin' and Ritin' and 'Rithmetic, Taught to the Tune of a Hickory Stick.' If you can't teach 'em, beat 'em."[26]

Project TORQUE had to make whatever headway it could against that powerful current of public demand. The ideas were sound and ambitious. Tests would seek to measure a child's performance against some previously set, publicly known criterion, but children would not seek to grade children against each other. The tests would not be secret; parents have a right to ask, "Is my child doing well enough?" and must therefore have easy access to information about what is "well enough." There should be, Jerrold argued, a publicly available test item bank, with perhaps thirty subject areas and four grade levels included, using some 300 or so questions per subject area. The test item bank would have $300 \times 30 \times 4$, or about 40,000 questions[27] (he thought the number would probably have to be substantially larger), available for general use, for anybody who wanted them. There should be booklets giving a variety of good responses; teachers should be trained to recognize good answers or sufficient answers instead of only the correct answer. Above all, the tests should be part of a teaching-learning process, discussable shortly after the event, and if the tests are to be used as part of a selection procedure, then the tests themselves, and not simply the disembodied scores, should be part of the student's dossier.

Project TORQUE designed and produced tests that met the new criteria. Using games, puzzles, and other appealing techniques, they showed that fear need not be a component of an examination, and they distinguished the valid purposes of an examination system from those that are invalid. Without compromising on quality, the project made use of the much underappreciated resource in the American education system: the caring attitude and the intelligence of the teachers. Project TORQUE represented nothing less than a bill of rights for schoolchildren, parents, and teachers alike; certainly long overdue, and unfortunately still nowhere implemented.

Symposium

On November 4–5, 1977, a symposium took place at Columbia University as a joint celebration of two happy events: the fiftieth anniverary of the opening of the Pupin Laboratories and the eightieth birthday of I. I. Rabi. It was a splendid occasion, appropriate to the remarkable distinction both names had brought to Columbia. Jerrold, himself now 73 years old, was one of a dozen notables invited to address the two-day symposium; naturally he took education for his topic.[28]

He began by paying tribute to Rabi, with whom his relationship continued to be close and warm. He took note of many of the superb achievements in which Rabi's influence could be detected, not restricting himself to physics. He spoke of the days in the laboratory, of molecular beam magnetic resonance, but he spoke also of some of Rabi's other activities, including what he called "Rabi's antimilitary-nonsense" accomplishments. And he pointed out that when he himself had set out to "do something" about science education twenty-one years earlier, there had been, not surprisingly, many friends of Rabi who went along with him. Several of them were on the program with him at the symposium: Killian, Wiesner, Land, Purcell, and Rabi himself. Jerrold described Rabi as a "nucleation site for nucleation sites," someone who, in addition to starting things himself, inspired others to go out and start things.

Jerrold's main theme was what needed still to be done: the headings were Teachers, Television, Textbooks, and Tests. He had tried his hand in all these areas and had a sense of what could be accomplished and how to go about it. But a new note appeared in his talk, one he had not sounded before: "Let me just say frankly that it is not enough for academia to stand aside and let early schooling be carried by people who are chiefly devoted to schooling before the college years." He would have more to say about this on later occasions; he would insist that education reform must be regarded as a continuing process, not an end, and he would insist that the professional working scientists were wrong to have left the problems to the educators, especially after having made such a good start. It is difficult to avoid the impression that he had taken a look around and had seen that of all those scientists who had started out with him two decades earlier to change education, he now stood virtually alone. Not completely alone, perhaps, but the exceptions were few. Phil Morrison, for example, had never given up his active interest and participation in early education, and Ed Purcell continued to be productive at the introductory college level in his own highly creative way. There were also younger scientists who had gone into education partly, as one of them said, because Jerrold had made it respectable, but they had little influence on the sources of support. "How do we go from a gleam in the eye to a program of national scope and importance?" Jerrold demanded.

Perhaps Jerrold ought to have been disheartened, but evidently he was not; three years later, Gerard Piel, the publisher of *Scientific American*, had occasion to refer to him in an article as "undiscourageable."[29] Jerrold finished his talk at the Rabi symposium with this remark: "One day when Jerry Kellogg and I were feeling defeatist about an experiment we were trying to do, Rabi said, 'You know, there is one important thing about a molecular beam experiment. You have got to want to do it.'" He left no doubt that as far as he was concerned, the education experiment was not yet finished, and, yes, he still wanted to do it.

In his response to the tributes paid to him, Rabi paid some tributes of his own. Speaking of Jerrold, he said

You may remember the grand old days of Sputnik, which burst upon the world, and suddenly we felt we were confronted with a nation of twelve-foot Russians. It was inconceivable that they would get ahead of us, but they did. . . . There was a wringing of hands until Jerrold Zacharias decided he was going to do something about it. . . . Jerrold applied the lesson [of big physics] to education. That is, he implied rather than said, that if you want to do something as big as education for the United States, it can only be done massively. You must think in large numbers—the books that have to be made, the teachers that have to be trained, and so on. . . . And the world of education has never been the same. . . . The main thing is to be enthusiastic, to feel you can do something. But do not underestimate the problem.

That was November 4, 1977. Five weeks later, Jerrold was seriously stricken; the coronary symptoms that had had him "popping nitro pills" ceased to be stable, and he experienced an alarming and acutely painful attack of angina. He was brought to Massachusetts General Hospital where he underwent triple bypass coronary artery surgery on December 10.

In spite of his undaunted, unflagging enthusiasm, education would have to wait a little while after all.

In the Public Interest

Recovery

There are few places as devoted to or as dependent on high technology as the intensive care unit of a major hospital. Oscilloscope screens—for a Radiation Lab alumnus, nostalgically reminiscent of early radar scopes—are placed around the room, monitoring heartbeat and breath: a distant early warning system responding to the body's vital signs. Jerrold's principle that "people have eyelids but not earlids" is much in evidence in such places, for the oscilloscope displays are often accompanied by repetitive beeping: any variation in sound immediately alerts the nurse to a problem. All that was missing from Jerrold's bedside was Sparky, the mechanical dog of sainted memory. But the devices that surrounded and sustained Jerrold as he emerged from his ordeal were clearly Sparky's electronic descendants.

Jerrold said later that the entire medical procedure had been a vivid reminder of the need for more widespread public understanding of science. He had thought about this many times before, but now, as he recovered from his surgery, the idea seemed to take on additional importance. He was fascinated with the technical aspects of his treatment, but he wondered how many people, otherwise well educated, really understood the sorts of things that happened to them in hospitals. How could they be brought to the sort of understanding that was necessary? How could science literacy be made a part of people's lives?

He realized anew, he said, that the battle over science education had to be fought not only in Washington but also at the level of the local school board, and that meant involving the public at large. Without some public understanding of science, there would be little local support for the teaching of it. People are generally comfortable only with what is familiar and comprehensible. Without some degree of familiarity with the germ theory of disease, for example, or even just with the notion of cause and effect, people would never feel truly at ease with the kinds of technological events that happened to them, not only in hospitals but in increasing numbers of other contexts as well. How could people make rational choices among the alternatives they faced? It was clear, he said in an article written a little over a year after his surgery, that "the citizen must use physical things—appliances, calculators and computers, automobiles, chemical products. And the citizens of today's world must use concepts and facts—computations, logical and quantitative reasoning, knowledge of their own bodies, understanding of causal relationships in the natural world. . . . Citizens must constantly cope with decisions and acts involving science."[1]

In accordance with standard medical practice at the time, he remained in the hospital for nearly a month; when he returned home, it was still another month before he was allowed up and around. He grew restless and impatient as he grew stronger, for he had little appetite for indolence; he had no hobbies except sailing, now forbidden; he had no sidelines and had never learned how to kill time. He read but was not a voracious reader; mystery stories, police procedurals, and the other standard fare of convalescents were not for him, and reading could not fill an entire day for him. His friends rallied round, naturally. There were visits, calls, and messages from scores of colleagues and other well wishers. He became fully and gladly reconciled with Jerry Wiesner; their earlier differences no longer seemed very significant in the overall scheme of things, not compared to a thirty-five-year friendship. Slowly health and strength returned; his recovery was complete. A year after his surgery he replied to a hospital follow-up questionnaire that he was as good as new.

In fact, free of the angina that had troubled him, he was in better shape than before. Those who knew him well thought they detected in him a kind of mellowing, a rounding of sharp corners. His mood was good; he resumed his habit of whistling Gershwin or Mozart in the corridors of EDC. But he had been interrupted in the full course of his continuing campaign to get education onto a better path. He remained as committed as ever to getting something done, and he picked up now where he had stopped, pointing out whenever he could—and as emphatically as he could—what might yet be accomplished if the proper support were available.

Jerrold's thinking on these matters continued to show a sense of scale that appeared mind-boggling to some of his listeners. In his talk at the Rabi symposium, he had talked of the need to reach some 300,000 to 500,000 talented but disadvantaged students each year. Such numbers seemed obvious to him, since they could be derived directly from demographic data: there would be roughly 40 million students between the ages of 8 and 18 in American schools by 1984, and 1 percent of 40 million is 400,000. He was thinking of the 1 or 2 percent of very bright children who fall through the cracks and are lost. Most often, these are minority children, who are served least well by the educational system—children from groups he described as "underrepresented in the major professions, especially in those professions that demand ease and familiarity with mathematics, the sciences and the technologies."[2] He produced a proposal at EDC, addressed to the National Institute of Education; under the title of Project M (short for Multi-Cultural Leadership Program) it called for a summer study to examine "improving the quality of education and creating broad access to educational opportunity for initially the top 1–2% and ultimately the top 10% of the gifted-but-educationally-disadvantaged students."[3] In a typical bravura display of exponential arithmetic, he offered the following assessment of growth:

Some 30 thousand dollars to initiate and bring into being a summer study which would require 300 thousand dollars for operation. Any resulting projects (e.g., research, tests and curriculum design, teacher training, etc.) might well require some 3 million dollars. The first year of operation might require 10 million dollars and in 5–10 years the size of the program might reach 300 million dollars or more per year. It may seem wrong to use dollar amounts to describe the growth of the program but the number and quality of the people involved in every aspect . . . will be determined by the availability of dollars.[4]

Three hundred million dollars per year sounds like an extraordinary sum of money, to be sure: it represents a prodigious rate of expenditure, and not even Jerrold knew how to spend at such a rate. It was more than twice as much as the entire NSF budget for science education had been in its best year.[5] But as Jerrold never ceased pointing out, education was a $100 billion business in 1977; it was big, he said: "It is big like the military. It is big like transportation." But then he added, perhaps realizing that "big" can also be alarming, "It is not as big as the energy budget." He was proposing an allocation of only three-tenths of a percent of the education total. It was what was required by the size of the problem. The $300 million would steer an expenditure more than 300 times larger, and any time you can do that, said Jerrold, "you are doing beautifully."[6]

He felt strongly enough about the need for action on this scale that he sought support for Project M directly from the White House; he brought it to the attention of President Carter's science adviser, Frank Press, a geophysicist and colleague from MIT. But the National Institute of Education was the agency that had to be convinced, and it was not prepared for innovation on such a scale. In fact, no one was. Writing in 1975, in a review of the accomplishments of PSSC and all the other ground-breaking programs that followed it, Gerard Piel, the respected publisher of *Scientific American*, wrote: "The inertia of the country's vast, pluralistic, independent, locally controlled school system, taken together with the high risk and intense competition in educational publishing, has tended to inhibit innovation and to promote uniformity at a safely mediocre level."[7]

This was a poor time from the political point of view to push radical new educational initiatives. Piel's comment was made just when those policies of the National Science Foundation that supported innovation and risk taking were under sharp attack by some members of Congress. The role of government in educational reform enterprises was being seriously questioned. In the conservative climate that prevailed, none of the funding agencies was likely to take on something that appeared big, chancy, and probably controversial. Without the spur of *Sputnik*, or something like it, the American will proved impossible to mobilize on these issues.

Foremost among the programs that had induced such congressional hostility to innovation was *Man: A Course Of Study (MACOS)*. Why it should have caused difficulties in 1975 was not clear; it was hardly a new program. It had begun in 1962 as a program in social studies sponsored by EDC (then ESI), under the leadership of the psychologist Jerome Bruner. The curriculum was built around the question "What is human about human beings?" and it was in essence a course in comparative anthropology; it examined the social behavior of salmon, of herring gulls, of free-ranging baboon troops, and of a tribe of Netsilik Eskimos. Much of the material that had been developed was of a remarkably high quality, and the program had proved to be very effective.

Nevertheless, it was clear that some congressmen objected to the content, especially the candor with which the Netsilik culture was portrayed. Congressman John B. Conlan of Arizona charged on the House floor that the curriculum offered "stories about Netsilik cannibalism, adultery, bestiality, female infanticide, murder, incest, wife-swapping, killing old people, and other shocking condoned practices."[8]

It was all the right-wing press needed. An article in a Washington newspaper, for example, was headlined "Teaching Fifth-Graders about Eskimo-Style Sex."[9]

Some in Congress demanded that both EDC and the NSF be held accountable; EDC found itself parrying charges of financial improprieties having to do with the publication and dissemination of MACOS. There had been no warning; a press release from the office of Congressman Olin E. Teague of Texas, chair of the House Committee on Science and

Technology, on May 8, 1975, simply announced "that financial irregularities may have occurred in connection with the implementation of project MACOS."[10] There was no explanation provided as to what irregularities "may have occurred." The foundation itself, over which Congress exercises budgetary control, was vaguely accused—more by suggestion than by direct charge—of failing to control the content of what it funded with the taxpayers' money. Quickly the issue became one of what constituted censorship in this case; many felt that there was an impossibly narrow line separating curriculum innovation from sponsorship of a federal curriculum. Both NSF and EDC had managed to walk that line successfully so far, but it was now proving very slippery.

Congressman Teague, evidently much exercised by the content of MACOS, appointed the Review Committee on which Piel served; that committee studied the problem but remained divided in its view. The matter thus continued to be controversial, although some of the worst harm was avoided. The more restrictive congressional motions for continuing close oversight of the NSF—including congressional veto power over individual grants—did not carry. But any interest on the part of the funding agencies for taking educational risks evaporated.

Where Have All the Scientists Gone?

When all was said and done, Jerrold said, it was the National Science Foundation that ought to have been most responsive to innovative suggestions and proposals in education. Other agencies had to be concerned with keeping the system going, but the NSF, especially because of its mandate for supporting research, should have been hospitable to new ideas. If it was not, it was because the NSF had never been an entirely free agent; in addition to being hammered at continually by congressmen and other politicians, the agency had the problem of resolving its own internal conflicts. Its policies had to deal with the needs of both science education and the nation's scientific research enterprise, and these needs were set in competition with each other for limited funds. Keeping the nation's research in a healthy state meant more than just providing laboratories and grants; it also meant ensuring a steady flow of well-trained scientists into the field.

It was puzzling to Jerrold why these needs were perceived to be so much in competition when in fact they were opposite sides of the same coin. NSF policies were set with the advice and guidance of the National Science Board, nearly all the members of which were or had been successful practicing scientists. Surely, he argued, they ought to know how new scientists are made. Testifying before Congress in 1980, he said: "What I'd like to know, and I don't know how to find out, why does the Science Board that governs the NSF budget, the directors and so on, continue to act as if scientists are generated in graduate school when every one of them who is a scientist knows perfectly well that he caught fire when he was a kid?"[11] For the NSF to continue to emphasize graduate student support made sense only if there was comparable emphasis on the paths leading to graduate school, and those paths began pretty far back. In his own case, he claimed , he could trace it back to about the age of four.[12]

He knew about the budget constraints, of course, but, as he had written in answer to an NSF questionnaire ten years earlier, in 1969:

Question: If you believe NSF should continue to make a substantial commitment at the pre-college level, what should NSF's future role be?
JRZ: When you have to do everything, priorities don't mean a lot. It is taste and choice that matter. NSF has to support science education hard, well, and in multiplicity, but not frightened of duplication. . . . Science and technology impinge on almost every aspect of the human condition and will continue to do so. You cannot afford *not* to do any piece of it [education] and the NSF cannot slough off to any other agencies in the government any aspect of education in the sciences and technologies.[13]

That is the attitude that he felt should infuse the NSF and everyone who had anything at all to do with education. The immediate problem was to get the NSF to see it that way. Who was going to lobby the NSF, and who would persuade the Science Board that a change in priority was in order? More important, perhaps, who could help establish the necessary atmosphere of receptiveness where it counted, which was at the local level?

He felt keenly the absence of his fellow scientific professionals. Writing in 1980, he offered these ideas in a piece entitled "The Case of the Missing Scientists":

The question, then, is: why with several million adult Americans educated in the sciences and the applied sciences have the schools largely given up trying to incite interest in technological subjects? Ten percent of the schools may believe that they are effective, but data . . . indicate that effective science learning is scarce indeed.
    Whenever the schools are not doing what someone thinks they should be doing, it is always tempting to blame several groups of people—the teachers, the publishers, the students, the principals, the superintendents, the school boards, the PTAs, the state departments of education, the testing companies, the people who use the tests, the psychometricians and the schools of education. Personally, I have been tempted to blame the Education Directorate of the National Science Foundation, which pulled the plug on the funding of the science education reform movement in the middle 1960s. But that's not where the troubles really lie. . . .

We have enough good science material to make a start. . . . We have enough experienced teachers to provide teachers of teachers. We have enough first-class laboratory equipment that is not expensive. . . . So the problem must lie instead with some group not named in the previous list. And, indeed, I believe that is the case.

That group is the scientists and technologists themselves who pay absolutely no attention to the major problems of school science—problems that consist of the lack of enough classroom time devoted to science, inside and outside of school, to permit the existence of really interesting programs. If every school board contained a near majority of scientists and technologists who could bolster each other's support for "science time in the schools" then something might happen.[14]

It was a very large if. Realistically, he knew that the only thing that could draw scientists away from their laboratories and back into education would be the expectation that their efforts might make a difference. Unless the resources of government were to be available, it was hard for anyone to see how a significant impact could be made. "Without money," Jerrold said, "there is no way to keep the entrepreneurship going. [In the 1950s and 1960s] there was enough money so that we could recruit busy people and assure them that if they got into this, something would come of it."[15]

But leadership was still necessary, and he saw none. "The Federal government is no source of educational reform; it's just a check-writing operation." Jerrold still held to the grand vision; he still thought in terms of restarting the reform movement. All the ideas were there; what was needed, he said, was "to build a political base for science first and then attempt to get it back into the schools."[16]

He thought he saw a way to do it: the nation had found itself in serious trouble because it had not understood and still did not understand the energy problem. In 1980 it was still very much in people's minds; energy, he hoped, might provide the boost for education that *Sputnik* had provided twenty-five years earlier. "I'm still crazy enough to try something," he said.

For an Educated Public

The 1970s had not been a particularly good decade for Jerrold. There had been some accomplishments, such as Project ONE and Project TORQUE, but he had not been highly satisfied with them. In the former, he felt it had been somewhat wide of the mark, and in the latter that they had not been able to go far enough. In retrospect, one can sense an uncertainty of direction—not so much because Jerrold was uncertain of what to do but because there was a pervasive sense of uncertainty, unreliability, and inconstancy about the agencies and institutions he might use to get anything done.

The network of colleagues on whom he depended was aging; its members were weary and losing influence. He had depended on MIT and had been disappointed; the NSF had backed off, he felt, under political pressure.

As always, Jerrold's life seemed to take on the flavor of what was going on elsewhere, and if the 1970s had not been a particularly good time for him, they had been far worse for the country as a whole. As the decade opened, halfway into Richard Nixon's first term, the United States was still deeply entangled in Vietnam, still vainly trying to extricate itself by renewing the bombing of the north. There was progress of a sort, nevertheless, and by early 1973 a face-saving Vietnam peace agreement had been signed in Paris, and the last U.S. troops had left South Vietnam. But later in that same year, the vice-president of the United States, Spiro Agnew, had resigned after pleading nolo contendere to tax evasion charges, and he was followed into disgrace by President Nixon himself, brought down less than a year later by the scandal of Watergate and threatened with impeachment. There had been a gross failure at the highest levels of government; as Jerrold said later, there was "no shred of leadership" anywhere—not then or at any later time in the decade.[17] He had been a member of a group of scientists that had formed to help elect George McGovern, the Democratic candidate in 1972, but McGovern had been overwhelmingly defeated by Nixon. The catastrophe that followed was not to be avoided.

The Watergate affair, with its break-ins, wiretaps, deceptions, secret funds, and, above all, the arrogance of those implicated, brought sharp recollections to Jerrold of the McCarthy period and of all that he hated, and he was gratified that justice had won out. Dinner guests at the Zacharias home in the years following 1974 were surprised and often pleased that meals always began with a formal toast to John J. Sirica, "the man who saved our democracy."[18] It had been, Jerrold said, a very close call, and the cost had been a serious loss of faith in government on the part of the people and the development of a pervasive cynicism.

In October 1973, the Arab oil-producing nations imposed a total ban on petroleum exports to the United States in order to press the United States and its allies to end military support for Israel. The event took place in the middle of the chaos brought about by Watergate, Vice-President Agnew's resignation, and the beginning of talk about impeachment of the president. The nation was in a poor position to react to an unanticipated energy crisis, a fourfold increase in oil prices, and double-digit inflation. Nixon called for a drive for the nation to become energy self-sufficient within a small number of years, but it was manifestly impossible. The graph was going the wrong way: the United States had been a net importer of energy since the early

1950s, and not only did the rate of consumption exceed the rate of domestic production, but the consumption rate was increasing faster than the production rate. As Jerrold had pointed out years earlier, in a different but comparable context, "It's the second derivative that will kill you."[19]

But Richard Nixon did not appear to understand this, nor did Gerald Ford, his successor in the White House.[20] It had been painfully evident in the long, angry gasoline lines of the summer of 1974 that the American public did not understand it at all. And that, said Jerrold, was a job for education. Testifying before a congressional committee in 1980 (on the NSF budget appropriation), he said

The energy issues that the public faces—I would love to recite on energy now; I have been working on it for six months to try to see if we can get it understandable— you can make it clear but you have to understand the numbers. You have to understand the graphs. You have to be what I call "quantitatively literate." . . .

Now here is a simple case of something which is so needed by the public: to be able to plot graphs, to be able to handle those graphs and handle the quantities easily, to understand the nature of acceleration—accelerating rates; namely, the inflation rate isn't steady, it is going up. . . .

So let me say it another funny way.

A few months ago on the editorial page of the New York Times, in the editorial column, there was a graph. I forget now what it was about. And I wrote the editor a letter of congratulations for putting a graph in an editorial column. It is the only way you could possibly have understood any of this, the only way you will understand the energy business, the only way you will understand most of what has to happen.[21]

Jerrold entertained hopes that the energy crisis might be sufficient to mobilize government to act on issues of science education. He hoped that Gerald Ford's successor in the White House, Jimmy Carter, trained as a nuclear engineer, might be able to move the nation along some sensible track. But in fact, lacking leadership, the nation as a whole was not mobilized to do anything at all and took only the most reluctant steps, dictated by shortages and pocketbook concerns—rarely by principle or policy—in the direction of energy conservation and energy efficiency. The problem of energy did not prick the nation's self-esteem as *Sputnik* had done. Perhaps, after Vietnam, Watergate, and national helplessness over the detention of hostages in Iran, there was little self-esteem to be engaged.

Ronald Reagan was elected president in November 1980. It was not long before the United States adopted a new, threatening, truculent posture: missile-rattling, some called it. Jerrold, in spite of his perennial distrust of the Soviet Union—or perhaps because of it—was alarmed. More urgent than the energy problem was the need to understand the numbers associated with nuclear weapons.

Time was running out on Jerrold; he knew he had slowed down. He had to pick his projects now with great care; he had to choose where he might still make a difference. He found himself back where he had been in 1950, after Hartwell: trying to get his head straight about the bomb.

Nuclear Dilemma

In 1963, in an act that was clearly in their mutual self-interest, the United States and the Soviet Union entered into a limited test ban treaty, ending there and then their atmospheric nuclear testing programs but leaving their underground testing programs intact.[22] The negotiating process had taken just eleven days. Jerrold liked to point out that it proved that when the nations wanted to agree on something, they could do it; if there was the will to reach agreement, the negotiators could always work out the details.

As the Reagan administration gathered full momentum in 1982 and 1983, there did not seem to be much will on either side to reach agreement. On the contrary, both American and Soviet foreign policy began to sound as though nuclear war was a genuine possibility, to be avoided only through strength and mutual menace. Each side made arms control proposals that the other could not accept, and each accused the other of violating past agreements.[23] The United States prepared to deploy nuclear-armed Pershing II missiles and long-range cruise missiles in West Germany to counter the Soviet SS-20s; at home the nuclear-freeze movement gathered strength.

The confrontation was deeply troubling. Jerrold pointed out that there was almost certainly no one remaining at the decision-making levels of either government who had ever witnessed a nuclear explosion or had sensed at first hand their enormity. There seemed to be no one at the policy level who understood anything about the catastrophic effects of nuclear explosions—what Jerrold had described on Project Lincoln thirty years earlier as "the horrors."

Jerrold was sure that Caspar Weinberger, for example, did not understand about nuclear bombs in any realistic way. Weinberger was the secretary of defense, charged in the Reagan period with drawing up a defense plan for the United States in the event of a nuclear conflict. Like many others who had thought much about the matter, Jerrold had long ago come to the conclusion that there could be no such thing as a defense plan; the use of nuclear bombs would be tantamount to suicide. Careful estimates had shown that even an attempt at a so-called surgical strike by the Soviet Union—that is, a strike directed solely at American missile emplacements rather than at population centers—would lead to tens of millions of casualties.[24] Only a madman, said Jerrold, would imagine such a blow to be bearable.[25]

Nevertheless, according to newspaper reports, the guidance paper that was offered by the Department of Defense in May 1982, over Weinberger's signature, included these points: "The armed forces [be] ordered to prepare for nuclear counterattacks against the Soviet Union 'over a protracted period' . . . [and] . . . that American nuclear forces 'must prevail and be able to force the Soviet Union to seek earliest termination of hostilities on terms favorable to the United States.'"[26]

The paper, officially called *Fiscal Year 1984–1988 Defense Guidance*, laid out the military thinking of the Reagan administration, calling for new weapons, including "prototype development of space-based weapons systems" and the deployment of the massive Trident II submarine-launched missile, among other recommendations. But it was the concept of a protracted nuclear war that became the focus of protest, in the Congress as well as in the press; the term had been used more than once and was not inadvertent. The idea of "termination of hostilities on terms favorable to the United States" suggested to many that the United States might be contemplating the possibility of winning a nuclear war. Secretary Weinberger denied that was in his mind.

In fact, the idea of preparing for a limited, protracted nuclear war as part of American defense policy did not originate in the Reagan administration but had already been endorsed by the previous administration. According to a 1980 newspaper report, "Mr. Carter . . . is said to have signed a document known as Presidential Directive 59, which asserts that the best way to prevent a major conflict with Moscow is to be capable of waging a prolonged but limited nuclear war. Government specialists maintain that the Soviet Union has long been committed to acquiring such a nuclear capability."[27]

On June 25, 1982, Secretary Weinberger gave a speech to an audience of 2,500 people at the graduation ceremonies of the Naval War College; the graduating class of the War College consists of senior officers of the navy and the marine Corps. Weinberger continued to maintain the same position:

We do not believe a nuclear war is "winnable." However, successful deterrence does require responsible and effective contingency plans should deterrence fail and we are attacked. In those plans we are *not* planning to lose. . . . We see nuclear weapons only as a way of discouraging the Soviets from thinking that they could ever profit from attacking the United States or its allies. . . . Our goal is to make certain that the Soviet Union never reaches the *conclusion* that it can gain from a nuclear attack of whatever duration. . . .

That is exactly why we must have strategic forces that are capable of surviving Soviet strikes over an extended, that is to say, protracted period. . . . In short, we cannot afford to place ourselves in the position where the vulnerability of our deterrent would force the President to choose between using our strategic response before it was destroyed or surrendering.

Those who object to a policy that would strengthen our deterrent, then, would force us into a more dangerous, hair-triggered posture.[28]

Weinberger's statements, perhaps more clearly than intended, had exposed the essential paradox inherent in the idea of deterrence. U.S. policy regarding nuclear weapons had taken various forms over the years. Early on, under Eisenhower and Dulles, it had been based on the threat of massive retaliation; then, under Kennedy (and Johnson), guided by McNamara, the concepts had been mutually assured destruction (MAD), damage limitation and flexible response—policies that have remained central throughout the 1980s. The Nixon-Kissinger policy had added détente: good behavior in return for good behavior, restraint for restraint. In the Carter years, the idea of counterforce with its notion of surgical strikes was emphasized by Secretary of Defense Harold Brown and National Security Adviser Breszinski; now, under Reagan and Weinberger, the central idea had become one of prevailing in a protracted conflict. But the basic underlying idea had not changed; America continued to found its hopes for peace on the concept of armed deterrence, protecting Europe and itself with what was called the nuclear umbrella.

The proponents of deterrence could point to more than thirty years of nuclear peace, during which the United States and the Soviet Union had each been armed with nuclear devices and had not used them, in spite of the obvious provocation of such events as the Cuban missile crisis and the East-West confrontation in Vietnam. But the same thirty years had seen an arms race of unprecedented magnitude, and no one was entitled to feel safe. The nuclear arsenals had grown to unthinkable size; the United States and the Soviet Union were each armed with more than 20,000 nuclear warheads, more than 7,000 megatons of nuclear explosives.[29]

The trouble, said Jerrold, was that peace through mutual deterrence was unstable. Any minor perturbation, even an unintended one, might be sufficient to bring on the horrors. For a policy of mutual deterrence to be effective, two conditions are required. First, neither side can have reason to believe that the other side can defend itself successfully against a nuclear attack, for if it could, it would have no reason to fear retaliation. Second, neither side must have grounds to fear that its ability to retaliate against the other is at risk. But since each side controls only its own military posture, these requirements turn out to be contradictory, and the contradiction leads inevitably to an arms race. Acting unilaterally, each side can guarantee the effectiveness of deterrence only by developing striking power against which the other side cannot defend. Simultaneously, each side strives to defend itself, to preserve itself and its population and its retaliatory power. The nature of deterrence is such that it fails if either effort succeeds. And yet there was no policy to take its place.

Jerrold, appalled at the militant and threatening policies of the Reagan administration and hearing nothing but their echoes from Moscow, followed the arguments closely. The nuclear freeze was not the answer, he was sure; it left all the nuclear arms in place. Getting his head straight about nuclear weapons, in his familiar phrase, became of singular importance to him. He went at it with his customary single-mindedness: he collected opinions, circularized his friends for their views, made his recurrent lists on his blackboard, and discussed and debated with anyone who came to see him. He was as troubled as he had ever been in his long life.

Primer

Asked about his health in 1982 or thereafter, it was unlikely that Jerrold would have said he was "as good as new," as he had described himself four years earlier. He was bothered by various complaints: gastric ulcers gave him little peace and robbed him of the enjoyment of his table; an increasingly severe loss of hearing took much of the pleasure out of viva voce argument and debate. He had a hearing aid but found it more annoying than helpful. The years appeared finally to be taking their toll. He was visibly slowing down, and his physical resources, he said, were depleted. For the first time, he was beginning to look his age.

Still, his mind remained sharp and clear, and he continued to think hard about the dilemmas of nuclear weapons, deterrence, arms control, and coexistence. Only a few people remained engaged with him on a day-to-day basis. Saville Davis, who had been at various times the national news editor, foreign correspondent, managing editor, and deputy editor-in-chief of the respected *Christian Science Monitor*, and whom he had known for more than twenty years, came frequently to join in the effort to think these matters through.[30] So did Myles Gordon, on the staff of EDC, who had worked closely with Jerrold in recent years on energy problems. Once a week, typically, Jerrold would call Rabi in New York to talk about these difficult issues; he always felt stimulated by his conversations with his old mentor. George Rathjens, a political scientist at MIT, was another with whom he found discussion valuable.

But for the most part, he went at it as he had gone at most other things—staying with it, worrying each piece of the problem like a dog with a bone, until he felt he understood it. Each day new lists would grow on the blackboard, recorded at appropriate moments by a devoted secretary; each day he would begin again.

A dozen years before Carter's Presidential Directive 59, fifteen years before Weinberger's exhortation to the graduates of the Naval War College, the *New York Times* had published in full an essay by the Soviet physicist Andrei Sakharov. Appearing on July 22, 1968, at the height of the political disarray and student unrest engendered by Vietnam, entitled "Progress, Coexistence and Intellectual Freedom,"[31] it had a stunning impact on Jerrold, who said at the time that it "should be the basis of a seminar in every college and university in the country."[32]

The Sakharov paper had let Jerrold see that coexistence was possible. Perhaps not until that moment had he fully believed it; he had no faith in the Soviet Union, no trust in its word, no tolerance for communism or totalitarianism in any of its forms. And yet Andrei Sakharov, a man he never met and whose physics he had never read, had managed in that short document to catch his attention and to crystallize his own ideas. Sakharov had enumerated the dangers: the threat of nuclear war, the failure to fulfill the Declaration of the Rights of Man, hunger and overpopulation, and the existence of police dictatorships. And he described a basis for hope: peaceful competition between the forces of socialism and those of capitalism could lead to eventual cooperation, a gradual convergence of goals, and a peaceful coexistence and collaboration between East and West. Jerrold acknowledged that one side did not have to love the other in order to share the planet.

Now, with the atmosphere and sound of nuclear confrontation everywhere, Jerrold tried to distill his ideas into a nuclear primer: a document that could be understood by anybody, that would lay out the issues simply, rationally and clearly. It was what the scientists had failed to do at Los Alamos in 1945, he said, but it was never too late to try again.

The first part of the primer, entitled "Common Sense and Nuclear Peace: An Essay by Jerrold R. Zacharias, Myles Gordon and Saville R. Davis," appeared as a special supplement to the April 1983 issue of the *Bulletin of the Atomic Scientists*.[33] It was written in hope, at a time when hopefulness was hard to find. The paper made seven principal points, summarized in an introduction:

1. . . . Detailed knowledge about the properties and the numbers [of bombs] is in fact secondary. It is enough to know that one bomb is terrible, that one hundred can be devastating and that tens of thousands exist. Arms control, arms limitation

and arms reduction, though moving in the right direction, are not sufficient. . . .
2. Nobody can be sure that a military action of any kind will indeed get out of hand; nor can anyone guarantee that a military action of any kind will not get out of hand. . . .
3. "There is no issue at stake in our political relation with the Soviet Union—no hope, no fear, nothing to which we aspire, nothing we would like to avoid—which could conceivably justify the resort to nuclear weaponry."[34]
4. We must find new ways to prevent and resolve conflict among nations. Our military thinks in terms of military solutions. Our State department thinks in terms of military solutions. Even the anti-missile groups are counting missiles and warheads. . . .
5. . . . We have to learn to live with and tolerate people who are different and whom we may not like. . . . We must learn to bargain and barter about the things that really matter. And they must do the same.
6. Nations, like people, are competitive by nature. The question is, on which turf do we want to compete? Military turf, which is likely to end in death for both the winners and the losers? Or economic turf, which might just lead to greater happiness and prosperity for all concerned? . . .
7. The interests of the United States and the Soviet Union are not always at odds. What is good for them is not always bad for us—and vice versa. . . . Good treaties based on commonality of interest make good neighbors.[35]

In about a dozen pages, these points were expanded and explained, questions posed, answers suggested. It was an eloquent call for sanity. "We have lived with nuclear explosives for forty years," they wrote. "We must learn to do so for forty thousand."

It is tempting to look on the primer as the distillation of a long life's experience, and in one sense that was true. It reflected Jerrold's unwavering belief that an educated public need not rely on the experts but could recognize and accept common sense. The piece was written, they wrote, for "the many people who, though apprehensive about the present nuclear dangers, are intimidated by the millions of words written on the issues, the complex arguments and the countervailing experts. It is important to recognize that every person has a right, even a duty, to think through these issues and to form opinions. No general, no defense scientist, no Secretary of Defense has ever fought a strategic or a tactical nuclear war. Regardless of credentials and experience, we are all operating in this arena on the basis of conjecture. . . . We hope that people will stick with us and reach their own conclusions."[36]

It was the optimistic credo of a man who believed with all his heart in democracy and who based all his hope on it—that every person has a right, even a duty, to think through the issues and to form opinions. It was indeed the distillation of his life's experience and a fine lesson to hand on.

Memorial

Memorial

Jerrold Zacharias died on July 16, 1986, at the age of eighty-one. The official medical cause was coronary artery disease, but anyone who knew him in those last months would have said that he simply wore out. His principal joy had been in work, and he had worked until he seemed too tired to work further. His death seemed to take people by surprise. It was hard to associate this man who had been so energetic with the sadness of dying, hard to imagine his quizzical voice stilled, hard to associate him, who had been the source of so many things, with ending.

There was an extraordinary flood of letters to Leona and the family, among which a common theme echoed over and over: "he changed my life," they said, and he had, by offering them new and unexpected opportunities and by making them believe that what they had to offer could make a difference. But the same words could have been said as well by thousands of others who had never met him but because of him had overcome their fear of learning. He had changed their lives too.

That October there was a memorial service at MIT. Al Hill presided, somewhat grizzled now and bearded, uncomfortable with the unaccustomed formality of the occasion, offering sympathy and a measure of comfort to Leona, Susan, and Johanna and their families.

Jerry Wiesner spoke of his affection and admiration for Jerrold, recalling past ventures: "One distinguished friend . . . not liking what the two of us were doing, once said to me, not as a compliment, 'You guys are just alike and you're going to be in trouble if you go on behaving like Zacharias.' I always wished I was just like Zach because he did things better than I did. . . . This warning was one of the great compliments paid to me in the course of my lifetime. . . . We were to each other the brother neither of us had been given by our parents. He was in every sense my older brother." He recalled Jerrold's many accomplishments, from the Radiation Lab on: "I believe that no single individual has contributed more to the development of MIT except William Barton Rogers, who started the place. We should all remember, as we recall Zach, that this was the place he was proudest of."

Ed Purcell spoke next, quietly reflecting on the forty-six years of their friendship that had begun, barely twenty yards from where he spoke, in the building that had housed the early Radiation Laboratory. He too described Jerrold's life as a scientist, as a technologist, and above all as an educator: "As you look back over the whole thing as I do, the theme of it all is education. In the very broadest sense, he was our teacher. In his unique way, he was one of the greatest teachers any of us ever had the privilege to run into."

Philip Morrison spoke last, finding a metaphor in the Gospel of St. Matthew: Jerrold as fisher of men, expanding it to include women and children as well: "I tried to think of people, individual persons, not institutions or grand things . . . and I could think of individual people in Rio, in Lagos, in Bombay, in Somalia, in Houston, Atlanta, San Francisco, Manhattan, in London and Rome and Israel and in a few other places where there were persons who had been caught . . . in the same net with me. Who were drawn out in that way from the wild waters of many purposes and who now had a large part of their lives informed by a single purpose, by a kind of taste and energy and broad direction which we had not experienced and, indeed, did not realize until the mesh included us."

The net that had caught people up in Jerrold's enterprises never really existed, of course, except in the mind, and those who were caught in it were always free to return to their many purposes. They did so, however, enriched with a new sense of what they could accomplish and how they might set about it.

But first, they all said, you have to get your head straight.

Appendix A

Project Hartwell

Scientific Personnel

Paul Adams
Head of the Navigation Division of Federal Telecommunications Laboratories. During and since the war has done extensive work on direction-finding systems and air-navigation systems.

Luis W. Alvarez
Professor of Physics, University of California, Berkeley. During the war, was head of the Special Systems Division of Radiation Laboratory, MIT, and later worked on atomic bomb development at Los Alamos.

Lloyd V. Berkner
Department of Terrestrial Magnetism, Carnegie Institution of Washington. During the war, was Director, Electronic Material Branch, Bureau of Aeronautics.

Harvey Brooks
Gordon McKay Professor of Physics, Harvard University. During the war, worked for OSRD at Harvard Underwater Sound Laboratory.

Bernard F. Burke
Research assistant, MIT

Edward E. David
Research associate, MIT

Charles R. Denison
Engineering consultant on port development and construction; recent project engineer, Port of Boston Authority; and port development research engineer, U.S. Maritime Commission; wartime colonel, Corps of Engineers, engaged on wartime port work in U.S. and Europe.

Robert H. Dicke
Associate Professor of Physics, Princeton University. During the war, was a staff member of Radiation Laboratory, MIT

Harry Dreicer
Student, MIT

Note: The biographical information is reproduced from the Hartwell Report.

Carl Eckart
Professor of Geophysics, Scripps Institute of Oceanography; Director, Marine Physical Laboratory, University of California. During the war, was associate director of the Division of War Research, University of California, at San Diego.

Francis L. Friedman
Assistant Professor of Physics at MIT. During the war, served as head of the Division on Radar Fire Control, Radiation Laboratory, MIT, and as consultant to Division on Fire Control, NDRC.

William H. Groverman
Commander USN. Head of Undersea Warfare Branch, Office of Naval Research. During the war, commanded the destroyers USS *Philip* and USS *DeHaven* and served on staff of Commander Destroyers Atlantic Fleet as Anti-Submarine Warfare and Combat Information Center Office.

Albert G. Hill
Professor of Physics and Director, Research Laboratory of Electronics, MIT During the war, was Chairman of the Radio Frequency Components Group and later the Transmitter Components Division of Radiation Laboratory, MIT.

Malcolm M. Hubbard
Assistant Director, Laboratory for Nuclear Science and Engineering, MIT. During the war, was in charge of Component Engineering at Radiation Laboratory, MIT.

Frederick V. Hunt
Professor of Physics, Chairman of Department of Applied Physics and Engineering Science, Harvard University. During the war, was director of the Harvard Underwater Sound Laboratory.

J. Wallace Joyce
Head, ASW Section, Radar Branch Electronics Division, Bureau of Aeronautics.

Winston E. Kock
Bell Telephone Laboratories. During the war, worked on microwave antenna development at BTL; Research Engineer on acoustical problems since 1948.

Charles C. Lauritsen
Professor of Physics, California Institute of Technology. During the war, led a group in torpedo and rocket development, and took an active part in the atomic bomb project at Los Alamos.

J. C. R. Licklider
Associate Professor of Electrical Engineering, MIT. During the war, worked at Psycho-Acoustic Laboratory of Harvard.

Harold S. Mickley
Associate Professor of Chemical Engineering, MIT During the war, was project leader of a torpedo power plant development program under NDRC.

Philip M. Morse
Professor of Physics, MIT, and consultant to the Weapons Systems Evaluation Group. During the war, established and directed the Operational Research Group of OSRD.

Arnold Nordsieck
Professor of Physics, University of Illinois. During the war, was a member of the scientific staff of Columbia Radiation Laboratory and of Bell Telephone Laboratories, specializing in microwave electronics.

John A. Pierce
Research Fellow of Harvard University. During the war, was head of LORAN Division of the Radiation Laboratory, MIT.

Ralph K. Potter
Director of Transmission Research, Bell Telephone Laboratories.

Edward M. Purcell
Professor of Physics, Harvard University. During the war, was Chairman of the Advance Development Group, Radiation Laboratory, MIT.

Richard B. Roberts
Staff Member, Department of Terrestrial Magnetism, Carnegie Institution of Washington. During the war, worked on proximity fuses; head of fire-control group and guided-missile group (Bumblebee) at Applied Physics Laboratory, Silver Spring, Md.

Merle A. Tuve
Director, Department of Terrestrial Magnetism, Carnegie Institution of Washington. During the war, was Director, Applied Physics Laboratory, Johns Hopkins University, and Fire Control Division, NDRC.

Foster L. Weldon
Staff of Naval Ordnance Laboratory, working on mine warfare. Now with Weapons Systems Evaluation Group.

Jerome B. Wiesner
Professor of Electrical Engineering and Association Director, Research Laboratory of Electronics, MIT. During the war, was leader of the Cadillac Project, Radiation Laboratory, MIT, and later head of the Electronics Division at Los Alamos.

Jerrold R. Zacharias, Chairman
Professor of Physics and Director, Laboratory for Nuclear Science and Engineering, MIT. During the war, served as head of the Transmitter Components Division, Radiation Laboratory, MIT, and later was leader of the Engineering Division at Los Alamos.

Organization of Project Hartwell

A. Atomic Weapons

L. W. Alvarez
F. L. Friedman
C. C. Lauritsen
J. R. Zacharias

B. Weapons and Weapons Systems

L. V. Berkner
R. Brooks
E. E. David
I. A. Getting
W. H. Groverman
M. M. Hubbard
J. W. Joyce
C. C. Lauritsen
J. C. R. Licklider
E. M. Purcell
M. E. Tuve

C. Shipping and Ports

C. R. Denison
W. H. Groverman

D. Detection and Identification

P. Adams
L. W. Alvarez
C. Eckart
H. T. Friis
A. G. Hill
F. V. Hunt
W. E. Kock
J. C. R. Licklider
J. A. Pierce
R. K. Potter
E. M. Purcell
M. A. Tuve
J. B. Wiesner

E. Harbor Defense

L. V. Berkner
B. F. Burke
A. G. Hill
A. Nordsieck

F. Mines and Mine Countermeasures

B. F. Burke
A. G. Hill
A. Nordsieck
R. B. Roberts
F. L. Weldon

G. Communication

E. E. David
H. T. Friis
R. K. Potter
J. B. Wiesner

H. Oceanographic Problems

C. Eckart

I. Power Plants

H. S. Mickley

J. Paramagnetic Resonance Absorption in a Magnetometer for MAD (Magnetic Anomaly Detection)

R. H. Dicke
E. M. Purcell

K. History and Sources

W. H. Groverman
M. M. Hubbard

Appendix B

Memorandum on Movie Aids

TO: Dr. James R. Killian, Jr.
FROM: Dr. Jerrold R. Zacharias
DATE: March 15, 1956
SUBJECT: Movie Aids for Teaching Physics in High Schools

In an effort to improve the teaching of high school physics I want to propose an experiment involving the preparation of a large number of moving picture shorts.

In order to present one subject, say physics, it is proposed that we make 90 films of 20-minutes duration, complete with text books, problem books, question cards and answer cards. Each of these points requires some discussion but before taking up the detailed mechanism it is necessary first to look at the subject matter. Success or failure depends to a large extent on having the entire apparatus of the experiment really right. Like a high fidelity phonograph, one must have besides the machine a good piece by a good composer played by an artist. The room must be good, not too noisy, and the people have to want to listen, but that all depends upon the piece.

Physics is a new science, so new that it is hard to remember that the concept of atomic number was completely accepted only just before Moseley's death in World War I. The tidy notions of classical mechanics still predominate in the way we try to introduce the subject of physics and it usually happens that these notions are dull as presented because it is so difficult to present arresting experiments to liven them up. Because one needs to understand many of the basic ideas and mathematical methods before gaining a profound comprehension, it always seems logical to begin with the basic ideas of vectors, velocities, statistics, hydrostatics, force, mass, etc. Now I think that physics can be divided up into the following parts:

1. The particles and the bodies.
2. Between these bodies we have laws of force.
3. Under the action of these forces the things move with laws of motion (which we think we really know and for which Newton's laws of motion are respectable approximations for bodies which are not too big, not too small, not too slow and not too fast.)
4. Mathematical methods which are used for comparing the theory with experiment.

So I propose, to be provocative, that we start with the particles. Some of these are heavenly, some are baseballs and some are molecular. Let's start with all three. But at this point I want to discuss only the molecules—for obvious reasons.

During the past few years it has become easy to detect neutral atoms or molecules one at a time. It is therefore possible to make lecture demonstration apparatus which will show the properties of molecules in such a simple way that an audience can be shown all of their properties which make us believe that matter exists in a particular form. The basic apparatus consists of an evacuated glass tube about 1/2-inch in diameter and ten or so feet long. At one end is a box with a small hole which contains some material the molecules of which are to be examined. These molecules evaporate from the surface of the material and effuse from the hole in various directions. At the other end of the tube is a detector for the molecules which can count them as they arrive at the other end, and can count them fast or slow depending on the rate at which they arrive. In the mid section of the tube there are a variety of slits or wires as obstacles, plates, magnets, or other things depending on the property to be demonstrated, such as:

1. Molecules arrive at the detector at a random rate. This is the property of particles and not of continuum. The arrival of one is not affected by the arrival of another.
2. They travel in straight lines, cast shadows, do not go around obstacles except that
3. Like baseballs they fall under the action of gravity
4. But how far they fall depends on how long a time they spend in falling and that depends on
5. How fast they were going in the first place and
6. Some are going fast and some are going slowly but on the average
7. They are going with a velocity which is about a hundred times as fast as a man can run which is equal to the
8. velocity of sound in a gas that consists of these molecules.
9. The molecules are not easily deflected by magnets or electric fields even though a water stream out of the faucet can be easily deflected by a comb that has been rubbed against your suit. We will do more with this later.
10. They carry energy and linear momentum but these effects are easier to demonstrate with a very short beam of a similar type.
11. It is difficult to count the number in a pound of molecules but with some idea of their size we can get a good estimate of the number using a surface film. But to get the idea of size
12. We scatter the beam by introducing some more molecules in their paths and from the free path we get the diameter
13. Which shows that almost all molecules are about equal in size within a factor of ten but that this size is about a billionth of a foot (1Å is 10^{-10} meters) or in a cubic inch there are so and so many.
14. Now in the detection process we made them into ions with positive or negative charges depending on whether they were, say sodium or chlorine (this can be shown).
15. This can now blossom into ions and electrons but there are other phenomena to look at.
16. They emit and absorb light and when they do they can emit the same color that they can absorb (sodium lamps)

17. But when they absorb the light energy they also absorb the linear momentum of the light (light pressure).

etc.   etc.

The reason for giving the long list of experimental demonstrations about molecules is that one can then build up a structure of physics with at least one new interesting experiment for every lecture. These demonstrations can not be made with wax and string.

In any case the material to be presented is going to require imaginative but painstaking work. A summer study is probably the best way to get the outline and the beginnings of the test. Plans for the demonstrations could be laid out and perhaps arrangements could be made to have some things built and tested. Some sample films should be made to see whether we should alter the methods of the professional short subject shooters.

A word or two about texts and answer cards. One should try to foresee a great many simple-minded questions. They are after all the most important but they are sometimes hard to answer. In the first years of use of the films the edge in knowing the right answers should be given to the teachers. They have to appear wise and if they are asked the questions by the students— and questions can be planted—they can give good answers if they have read over the lesson. It may be that there should be a special text to go with the student text which is prepared to lend an apparent dignity to the teachers.

The financial side of such an enterprise becomes obvious on considering the amount of film that will have to be thrown away before a first-rate reel is accomplished. Copies of the films are easy. But the efficiency of Flaubert was not high.

Appendix C

Steering Committee of the Physical Science Study Committee, February 21, 1957

Paul Brandwein   Science Editor, Harcourt, Brace & Co.

Vannevar Bush   Chairman of the Corporation, MIT

Frank Capra   Film Producer

Robert Carleton   Executive Secretary, National Science Teachers' Association

Henry Chauncey   President, Educational Testing Services, Inc.

Bradley Dewey   Member, MIT Corporation

Nathaniel Frank   Chairman, MIT Physics Department

Francis Friedman   Associate Professor of Physics, MIT

Mervin Kelly   President, Bell Telephone Laboratories

James Killian   President, MIT

Edwin Land   President, Polaroid Corporation

Elbert Little   Executive Director, PSSC

Morris Meister   Principal, Bronx High School of Science

Walter Michels   Chairman, Physics Department, Bryn Mawr

Edward Purcell   Professor of Physics, Harvard University

I. I. Rabi   Institute Professor (Visiting), MIT

Jerrold Zacharias   Professor of Physics and Chairman of the Committee

Steering Committee of the Physical Science Study Committee, February 21, 1957

Appendix D

Steering Committee of the African Education Program, 1961

William O. Brown	Director, African Studies, Boston University
Jerome S. Bruner	Professor of Psychology, Harvard University
Louis Cowan	Professor of Communications, Brandeis University and President, Cowan Foundation
Saville Davis	Managing Editor, *Christian Science Monitor*
Amos de Shalit	Professor of Physics, Weizmann Institute, Israel
Francis L. Friedman	Professor of Physics, MIT
James R. Killian, Jr.	Chairman of the Corporation, MIT
William Ted Martin	Professor of Mathematics. MIT
Gilbert Oakley, Jr.	Vice-President, ESI
William E. Spaulding	President, Houghton Mifflin Company
Stephen White	Special Projects, ESI
Jerrold R. Zacharias	Chairman

Notes on Sources

The research for this book may fairly be said to have begun before the project itself was initiated or even thought of. In 1984 I was invited by Arthur Singer, vice-president of the Alfred P. Sloan Foundation (no doubt it was Jerrold Zacharias who suggested my name to Singer), to develop and produce a series of videotapes dealing with the life of I. I. Rabi. These were duly made at Brandeis University later that year. Intended for archival purposes, they took the form of Rabi's unrehearsed conversations with selected groups of colleagues on particular themes. Three such videotapes were made, each running about 3 hours, and transcripts were subsequently prepared.

The first of these conversations, on the question of how scientists may offer useful advice to a government that wants it, focused principally on the PSAC experience. In addition to Rabi, the participants were Edwin H. Land, Edward M. Purcell, Richard Garwin, and Jerrold Zacharias, all former members of PSAC. The second conversation provided reminiscences of and reflections about doing physics in the 1930s. The participants were Rabi, Julian Schwinger, Norman Ramsey, Sidney Millman, and Zacharias, all of whom had worked in Rabi's lab at Columbia during that interesting period. The third was a more introspective conversation about the nature of the questions with which physics is or should be concerned; the discussants were Edward Purcell, Robert Palmer, Victor Weisskopf, and, once again, Jerrold Zacharias.

These videotaped conversations are accessible to scholars at the Smithsonian Institution and at the MIT Archives. They are referred to in the text as Brandeis/Sloan I, II, and III.

Two additional videotapes have proved valuable. In 1982, under the auspices of the Sloan Foundation, Arthur Singer arranged for a number of former participants to meet and discuss before the camera their recollections of Project Charles. Included were James Killian, Jerome Weisner, Paul

Samuelson, Malcolm Hubbard, Jerrold Zacharias, and a number of others. This document is referred to in the text as Sloan/Charles.

Also in 1982, a few of those who had worked with Zacharias and Philip Morrison on educational problems persuaded the two of them to discuss on camera their thoughts about their experiences; the conversation was video-taped by Henry Felt and John M. B. Churchill, and a copy was kindly provided to me by Mr. Felt. It is referred to in the text as Felt/Churchill.

The MIT Archives contain other very useful source material. In particular, the annals of the MIT Oral History Project are located there; in the early 1970's, a group of students under the direction of Prof. Charles Weiner succeeded in capturing much of the history of PSSC by interviewing a number of its developers, including Zacharias. These interviews, together with many other PSSC documents also deposited in the Archives, have proven immensely valuable. They are referred to in the text either as MIT/Oral or as MIT/PSSC.

Other valuable interviews are to be found in the Center for the History of Physics, located at the Neils Bohr Library of the American Physical Society. In particular, an interview of Jerrold Zacharias recorded by Paul Henricksen has been especially valuable; other interview material located in the center has also been very useful. This resource is abbreviated as APS.

Jerrold Zacharias's own papers constitute an essential resource. These have now been placed in the MIT Archives, where they await sorting and classification. They are referred to in the text by the abbreviation JRZ. In particular, these papers contain preliminary drafts of parts of an autobiographical memoir on which Jerrold was working at the time of his death. He was assisted in this by Myles Gordon, now vice-president of Education Development Center, who has graciously shared with me his recollections of working with Jerrold. The manuscript is referred to in the text with the abbreviation JRZ/Gordon.

In general, Jerrold's papers contain little of a personal nature beyond his reminiscences in JRZ/Gordon, but they do contain a number of handwritten screeds as well as some tape-recorded speeches, a few of which offer no indication of time or place. Many of the screeds were never put into final form, never published and often were abandoned halfway through. I have not hesitated to use them, nevertheless, as they show how Jerrold thought about their subject matter at the time; but I have indicated their nature in referring to them and their provisional quality should be noted. Other documents have a somewhat ambiguous provenance; for example, among Jerrold's papers there is to be found the transcript of an interview with Jerrold concerning the foundation of the Brookhaven National Laboratories. It is dated January 27, 1983, but I have not been able to identify either the interviewers or the occasion for the interview. Nevertheless, since it is corrected in Jerrold's hand, I believe it to be a reliable source. It is abbreviated as AUI-BNL in the text.

A word of thanks is in order for those who have custody of the various archives I have consulted and who have invariably proved helpful. It seems to me characteristic of that profession that its practitioners cheerfully go well beyond what one might reasonably expect from them, and I am grateful to the library staff of Brandeis University, MIT, and the Neils Bohr Library of the American Physical Society for their generous assistance.

Notes

Abbreviations

Sloan/Brandeis I, II, III: Interviews centered on the life and times of I. I. Rabi, videotaped at Brandeis University on December 16, 1983, March 29, 1984, and November 15, 1985. Deposited in the Smithsonian Institution and the MIT Archives.

Sloan/Charles: Videotape of various participants in Project Charles, 1982. Deposited in the Smithsonian Institution.

Felt/Churchill: Videotaped conversation between Jerrold Zacharias and Philip Morrison, 1982, deposited in the MIT Archives.

MIT/Oral: Interview materials collected under the direction of Charles Weiner, relating to the origins of the Physical Science Study Committee, deposited in the MIT Archives.

MIT/PSSC: Documents and manuscripts pertaining to the early period of PSSC, deposited in the MIT Archives.

APS: The Neils Bohr Library of the American Physical Society, New York.

JRZ: The personal papers of Jerrold Zacharias, deposited in the MIT Archives.

JRZ/Gordon: Draft of parts of an autobiographical memoir, prepared with the assistance of Myles Gordon. (JRZ).

AUI-BNL: Transcript of an interview dated January 27, 1983 (JRZ)

Chapter 1

1. Folk song (JRZ).

2. Isadore Zacharias chose that number for his house arbitrarily and whimsically, because he had a servant named Columbus. The house, which is still standing, now carries the number 1252.

3. He reported to the census taker in 1880 that he had been born in Prussia; but in the censuses of 1900 and 1910, he reported having been born in New York City. He and his younger brother were both listed on the Jacksonville voters' roll in 1885.

4. The other Zacharias children were Pauline, born in 1875; Lawrence, 1879; Rhoda, 1881; Herbert, 1883; Percy, 1887; and Ellis, 1890.

5. Encyclopedia Britannica, 11th edition (1911).

6. Rebecca Brown, born in 1894, remembered it this way in a personal interview in 1986.

7. Beryl Rubinstein lived with the Zacharias family for a dozen years and subsequently had a distinguished career both as a performer and a composer. In 1921 he settled in Cleveland, Ohio, where he eventually became director of the Cleveland Institute of Music. He died in 1952.

8. J. R. Zacharias, review of three books, *National Elementary Principal*, 51 (April, 1972):73–75.

9. Interview by Paul Henriksen, June 21, 1982, APS.

10. Audiotape dated January 26, 1971, Rainer Weiss (JRZ).

11. JRZ.

12. J. R. Zacharias, Panel Discussion: Science, Technology and Education, *Celebration - Fiftieth Anniversary of Pupin Physics Laboratories* (New York: Columbia University, November 4, 1977).

13. JRZ/Gordon.

14. L. Hoddeson, and G. Baym, "The Development of the Quantum-Mechanical Electron Theory of Metals," *Reviews of Modern Physics* 59 (1987):287–321.

15. Brandeis/Sloan II.

Chapter 2

1. Quoted in Charles Weiner, "Physics in the Great Depression", *Physics Today* (October, 1970):31–38.

2. See John S. Rigden, *Rabi: Scientist and Citizen*, for the story of how this position came to be offered to Rabi.

3. Spencer R. Weart, "The Physics Business in America, 1919–1940: A Statistical Reconnaissance" in Nathan Reingold, ed., *The Sciences in the American Context; New Perspectives* (Washington, D. C.: Smithsonian Institution Press, 1979):295–358.

4. Paul Henriksen interview, June 21, 1982 (APS).

5. Spencer R. Weart, "The Physics Business."

6. JRZ/Gordon.

7. The dissertation appeared in *Physical Review* 44 (1933):116–122. Jerrold's first publication resulting from this work was a brief note of which he was a coauthor: A. Dingwell, J. R. Zacharias, and S. Siegel, "The Contamination of Nickel Crystals Grown in a Molybdenum Resistance Furnace," *Transactions of the Electrochemical Society* 63 (1933):395–400.

8. JRZ/Gordon.

9. Brandeis/Sloan II. Arnold Sommerfeld, Wolfgang Pauli, and Erwin Schrodinger were theoretical physicists acknowledged to be among the principal architects of the quantum theory. "Copenhagen," in this context, implies Niels Bohr and the Bohr Institute. Hans Bethe, who came to Cornell University in 1937, is a member of the somewhat younger generation that included Rabi.

10. G. Breit and I. I. Rabi, "Measurement of Nuclear Spin," Physical Review 38 (1931):2082–2083.

11. V. W. Cohen, "The Nuclear Spin of Caesium," Physical Review 46 (1934):713.

12. JRZ/Gordon.

13. F. G. Brickwedde, "Harold Urey and the Discovery of Deuterium," Physics Today 35 (September, 1982):34–42.

14. R. Frisch and O. Stern, "Uber die magnetische Ablenkung von Wasserstoff-molekulen und das magnetische Moment des Protons I," Zeitschrift fur Physik, 85 (1933):4–16; I. Estermann and O. Stern, part II, ibid. 17–24.

15. Except for Schwinger, the list consists of physicists whose achievements were principally in experiment, although all of them were capable of good theoretical work. Schwinger, who won the Nobel prize in physics in 1965, was a pure theorist and may therefore seem like the odd man out in this collection. But he was very much a Rabi protégé, and Rabi would have insisted on his inclusion here.

16. Brandeis/Sloan II.

17. Jerrold liked to refer to the summer when they had measured the magnetic moment of the element indium as "indium summer." Millman remembered that small joke with pleasure. The work was subsequently published; see S. Millman, I. I. Rabi, and J. R. Zacharias, "On the Nuclear Moments of Indium," Physical Review 53 (1938):384–391.

18. Sidney Millman, letter written on the occasion of a sixtieth birthday celebration for JRZ, June 3, 1965 (JRZ).

19. The magnetic field had to be inhomogeneous, that is, varying from place to place in strength and direction. But the direction and strength of the field at any point had to be known as precisely as possible.

20. JRZ/Gordon.

21. Francis Bitter, Magnets: The Education of a Physicist (New York: Doubleday Anchor, 1959).

22. JRZ/Gordon.

23. The slit problem was solved in a simple and ingenious way. First, a larger slit—human hair thickness—was made by sawing a cut into the end of a closed glass tube. The sides of this slit were then collapsed onto a thin copper foil by heating. Afterwards, the copper was dissolved out in acid.

The source of current was a problem that would have taxed their limited financial resources; however, Rabi was able to borrow twelve 160 amp-hr storage batteries from the navy, and these worked very well.

The electrical conductors were copper tubes only a tenth of an inch in diameter and about six inches long. They were mounted in an aluminum block, insulated by thin mica sheets, and they were cooled by carrying a flow of water.

Detection was accomplished by allowing the beam of hydrogen atoms to impinge on a layer of molybdenum oxide "soot" deposited on a glass plate. The soot is normally yellow; atomic hydrogen reduces it to the blue oxide. The place where the beam strikes the layer is thus made visible—another example of folk wisdom at work.

24. This theory, invented by P. A. M. Dirac, correctly predicted the value of the magnetic moment of the electron. That it could not do so for the proton or the neutron suggested strongly that those particles might not really be elementary but might instead have some internal structure.

25. I. I. Rabi, J. M. B. Kellogg, and J. R. Zacharias, "The Magnetic Moment of the Proton," *Physical Review* 46 (1934):157–163.

26. Notebooks (JRZ).

27. JRZ/Gordon.

28. Brandeis/Sloan II.

29. I. I. Rabi, "On the Process of Space Quantization," *Physical Review* 49 (1936):324–328.

30. JRZ/Gordon.

31. Brandeis/Sloan II.

32. Brandeis/Sloan II.

33. I. I. Rabi, J. R. Zacharias, S. Millman, and P. Kusch, "A New Method of Measuring Nuclear Magnetic Moments," *Physical Review* 53 (1938):318.

34. J. M. B. Kellogg, I. I. Rabi, N. F. Ramsey, and J. R. Zacharias, "An Electric Quadrupole Moment of the Deuteron," *Physical Review* 55 (1939):318–319(L); "An Electric Quadrupole Moment of the Deuteron," *Physical Review* 57 (1940):677–695.

35. Brandeis/Sloan II.

36. A. Nordsieck, "Electric Quadrupole Moment of the Deuteron," *Physical Review* 57 (1940):556(A).

37. JRZ/Gordon.

Chapter 3

1. *New York Times*, October 10, 1933, 11:5.

2. Reproduced in Charles Weiner, "A New Site for the Seminar: The Refugees and American Physics in the Thirties." in D. Fleming and B. Bailyn, eds., *The Intellectual Migration* (Cambridge: Harvard University Press, 1969).

3. Leo Szilard, "Reminiscences," in ibid.

4. Brandeis/Sloan II.

5. J. R. Zacharias, "The Nuclear Spin and Magnetic Moment of K40" *Physical Review* 61 (1942):270–276.

6. General sources for this section are James P. Baxter, *Scientists Against Time* (Cambridge: MIT Press, 1946); I. Bernard Cohen "American Physicists at War: from the First World War to 1942," *American Journal of Physics*, 13 (1945):333–346; Daniel J. Kevles, *The Physicists* (New York: Knopf, 1978)

7. From an aide-memoire from Lord Lothian to President Roosevelt, reproduced in E. G. Bowen, *Radar Days* (London:Adam Hilger, 1987).

8. Principal sources for this section are H. Guerlac, *Radar in World War II* (New York:Tomash/AIP, 1987); Robert M. Page, *The Origin of Radar* (New York: Anchor Books, 1962); and E. G. Bowen, *Radar Days*.

9. G. Marconi, "Radio Telegraphy," *Proceedings of the IRE* 10(1922):215–238. Robert M. Page, one of the early workers in radar at the U. S. Naval Research Laboratory, pointed out in his book on the history of radar that since no work was done as the result of Marconi's address, "no radar resulted from his suggestions." As Rabi had said in connection with the discovery of the resonance method, "The important thing was to do it."

10. G. Breit and M. A. Tuve, "A Test of the Existence of the Conducting Layer," *Physical Review* 28 (1926):554–575.

11. Bowen, *Radar Days*, p.141

12. This figure corresponds roughly to the case of an antenna about 1 square meter in area, the characteristic size of which is about ten times the wavelength, and to a target with a cross-sectional area of about 100 square meters at a distance of about 100 miles.

13. Guerlac, *The History of Radar in World War II*

14. It was actually a fairly crude shack. D. Kevles, in his book *The Physicists*, remarks, "By December, DuBridge's physicists were shivering in their unheated rooftop penthouse—one of them wore a coonskin coat against the cold." It is hard to imagine that this refers to anyone other than Jerrold.

15. JRZ/Gordon.

16. Jerrold recalled that although he had had some experience with high frequencies, he had relied entirely for the electronics on someone who knew much more about it. This was Dale Corson, a nuclear physicist who later became president of Cornell University.

17. The switching device that was later developed was referred to as a duplexer; the British referred to it as a TR (transmit-receive) box. Both terms are still in common use.

18. One of the members of the group, J. L. Lawson, discovered that a klystron tube (normally used as a power amplifier), if used as a preamplifier, served to isolate the

receiver crystal satisfactorily. Unfortunately, the klystron introduced a great deal of noise into the system; therefore, it could be only a temporary expedient at best.

19. JRZ/Gordon.

20. Paul Henriksen, interview, June 21, 1982 (APS).

21. Ibid.

22. The following nomenclature is used: wavelengths of about 10 cm. are referred to as S-band; 3 cm. radar lies in the X-band; 1 cm. radar lies in the K-band. All of these are part of the microwave spectrum.

23. Henriksen, interview.

24. JRZ/Gordon.

25. Vannevar Bush, *Pieces of the Action* (New York: Morrow, 1970).

26. Henriksen, interview.

27. I. I. Rabi, speech, December 5, 1977 (JRZ). The anecdote is presented almost verbatim in John Rigden's biography, *Rabi: Scientist and Citizen* (New York: Basic Books, 1985).

28. Killian's telling of the anecdote is recorded on Sloan/Charles. The carousel represented a legitimate expenditure; it was an inexpensive way to provide a rotating mount for an experimental radar antenna.

29. JRZ/Gordon

30. Jerrold nominated him for that position. Leona remembered Jerrold's working hard in their living room to persuade Berkner to accept the nomination: "It was pure snake oil," she said.

31. Daniel S. Greenberg, *The Politics of Pure Science* (New York: New American Library, 1967).

32. JRZ to D. R. Inglis, January 28, 1942 (JRZ).

33. David Hawkins, Project Y: The Los Alamos Story. Part I: Toward Trinity (Los Angeles: Tomash, 1983).

34. That August, Parsons would achieve a kind of fame by riding in the bomb bay of the *Enola Gay* to arm the first atomic bomb for its drop on Hiroshima.

35. AUI-BNL

Chapter 4

1. Paul Henriksen, interview, June 21, 1982 (APS)

2. AUI-BNL

3. Some of these "temporary" buildings still stand at MIT, having sustained continual heavy use for more than forty years. They are hardly architectural ornaments on the MIT campus, but they have been invaluable assets nevertheless. Perhaps they could be considered intellectual ornaments instead.

4. The task was carried out under the supervision of L. N. Ridenour, a physicist who had been at the University of Pennsylvania before the war.

5. Quoted in J. A. Stratton, "R.L.E.—The Beginning of an Idea," in *R.L.E.: 1946+20,* a publication of the Research Laboratory of Electronics, MIT (May 1966)(JRZ).

6. The National Science Foundation, charged with the support of basic research in the civilian sector, was not established by the Congress until 1950, although enabling legislation had been proposed as early as 1945.

7. The bill referred to was the May-Johnson bill, which would have put all post-war nuclear research under the Manhattan District.

8. Bruce S. Old et al., "The Evolution of the Office of Naval Research," *Physics Today* (August, 1961)14:30-35.

9. R. D. Conrad, From a Navy Day address at the University of Illinois, Urbana, October 27, 1946, quoted in the *ONR Originator,* October 28, 1946. (JRZ).

10. A. G. Hill, interview by Paul Henriksen, June 25, 1982 (APS).

11. Daniel J. Kevles, *The Physicists,* (New York: Knopf,1977).

12. Henriksen, interview (APS); Henry D. Smyth, *Atomic Energy for Military Purposes: The Official Report on the Development of the Atomic Bomb under the Auspices of the United States Government, 1940-1945* (Princeton: Princeton University Press, 1945).

13. This machine was known as the Markle Cyclotron. It produced deuterons at an energy of 14 million volts, a respectable figure when it was built.

14. Robley E. Evans, taped interview (APS)

15. AUI-BNL

16. They were Columbia, Cornell, Harvard, Johns Hopkins, MIT, Princeton, Pennsylvania, Rochester and Yale. Harvard was not represented at the first meeting but joined in almost immediately thereafter.

17. Colonel, later General, K. D. Nichols had been second in command to General Leslie R. Groves in the Manhattan Project. He subsequently became general manager of the Atomic Energy Commission.

18. AUI-BNL.

19. Ibid.

20. Ibid.

21. Because the abbreviation then in use for a billion electron-volts was BeV, the machines were known as bevatrons. They represented large-scale and expensive undertakings. At that time, there were two under construction, one at Berkeley and one at Brookhaven.

22. Excerpt from minutes of tenth meeting of Naval Research Advisory Committee, September 19, 1949 (JRZ).

23. Ibid.

24. Ibid.

25. Lee A. DuBridge to Alan Waterman, October 5, 1949 (JRZ)

26. Victor F. Weisskopf, "Growing Up With Field Theory: The Development of Quantum Electrodynamics," in L. M. Brown and L. Hoddeson, eds., *The Birth of Particle Physics* (New York:Cambridge University Press, 1983), pp. 56–81.

27. W. V. Houston, "A New Method of Analysis of the Structure of Ha and Da ," *Physical Review* 51 (1937):446; R. C. Williams, "The Fine Structure of H and D Under Varying Discharge Conditions," *Physical Review* 54 (1938):558–567; S. Pasternack, "Note on the Fine Structure of Ha and Da ," *Physical Review* 54 (1938):1113(L).

28. The Dirac theory of the electron predicted a precise value for the electron magnetic moment. The discrepancy in question was a tiny departure from that value.

29. J. E. Nafe, E. B. Nelson, and I. I. Rabi, "The Hyperfine Structure of Atomic Hydrogen and Deuterium," *Physical Review* 71 (1947):914(L). The discrepancy sounds small, as indeed it is, but Nafe, Nelson, and Rabi pointed out that the "difference is five times greater than the claimed probable error in the natural constants." When Nagle, Julian, and Zacharias published their confirming results, they described them as "in violent disagreement with those calculated."

30. D. E. Nagle, R. S. Julian, and J. R. Zacharias, "The Hyperfine Structure of Atomic Hydrogen and Deuterium," *Physical Review* 72 (1947):971.

31. JRZ/Gordon.

32. Hill was the associate director of the Research Laboratory of Electronics, where Jerrold had set up his lab. Wiesner, after a period of handling the instrumentation for the first postwar tests of the atomic bomb, had also come to MIT, where, at Jerrold's urging, the department of electrical engineering had offered him an assistant professorship. Wiesner became a member of RLE and some years later its director. His subsequent career followed a brilliant and remarkable trajectory.

33. W. E. Lamb and R. C. Retherford, "Fine Structure of the Hydrogen Atom by a Microwave Method," *Physical Review* 72 (1947):241-243; see also W. E. Lamb, "The Fine Structure of Hydrogen," in Brown and Hoddison, eds., *The Birth of Particle Physics*, pp. 311–328, where the history is discussed.

34. See, for example, S. S. Schweber, "A Short History of Shelter Island I," in R. Jackiw et al., eds., *Shelter Island II: Proceedings of the 1983 Shelter Island Conference on the Fundamental Problems of Physics* (Cambridge: MIT Press, 1985).

35. Hans Bethe, of Cornell University, provided a first theoretical explanation of the Lamb shift only days after the close of the conference. Richard Feynman and Julian Schwinger (of Cornell and Harvard, respectively) separately provided new and comprehensive formulations of quantum electrodynamics, for which they were awarded the Nobel Prize in 1965 together with S. Tomonaga of Japan who had done equivalent work. (Bethe won the Nobel Prize in 1967 but for work of an entirely different nature.)

36. Feld moved on into the new field of meson physics after a few years.

Chapter 5

1. Daniel Yergin, *Shattered Peace: The Origins of the Cold War* (New York:Penguin Books, 1990), p.398. By contrast, the military budget was to rise to $22.3 billion in 1951, to $44 billion in 1952, and to $50.4 billion in 1953. The difference, of course, was the Korean war.

2. Bernard Brodie, "War Department Thinking on the Atomic Bomb," *Bulletin of the Atomic Scientists* 3 (June 1947):150–155,168.

3. Ibid.

4. Bernard Brodie and Eilene Galloway, "The Atomic Bomb and the Armed Services," Library of Congress Public Affairs Bulletin 55 (Washington, D. C., 1947); Bernard Brodie, "A Critique of Army and Navy Thinking on the Atomic Bomb," *Bulletin of the Atomic Scientists* 3 (August 1947) pp. 207–208; Bernard Brodie, "The Development of Nuclear Strategy," *International Security* (Spring 1978); Barry H. Steiner, "Using the Absolute Weapon: Early Ideas of Bernard Brodie on Atomic Strategy," *Journal of Strategic Studies* 7 no. 4 (December, 1984):365–393; David A.Rosenberg, "The Origins of Overkill: Nuclear Weapons and American Strategy, 1945–1960," *International Security* 7 (Spring 1983): 3–71.

5. Although the air force did not become an independent entity until 1947, it had achieved substantially complete autonomy by the end of World War II while remaining officially part of the army. The navy (and the marines), on the other hand, maintained their own aviation units. In most of the interservice disputes about roles and responsibilities, the army aligned itself with the air force in opposition to the navy.

6. Quoted in R. G. Albion and R. Reed, Makers of Naval Policy 1798–1947 (Annapolis:1980), chap. 27. Major General Curtis E. LeMay was at that time the commanding general of the XXth and XXIst Bomber Commands, which had just carried out the bombing of some sixty Japanese cities, including the destructive firebombing of Tokyo. By 1949, LeMay was commanding general of the Strategic Air Command; in this role he had a profound effect on air force policies and attitudes. It is worth emphasizing that LeMay's remark, and McCain's letter, antedated the use of atomic weapons; after August, 1945, the stakes became considerably higher.

7. See, for example, Daniel Kevles, *The Physicists* (New York: Knopf, 1979), or Daniel S. Greenberg, *The Politics of Pure Science* (New York: New American Library, 1967).

8. "Atom Unit Reports on Studies of Uses: First Accounting to Congress by U. S. Commission Tells of Research on Planes," *New York Times,* February 1, 1947; "6 Aides Take Oaths in Air Force Set-Up: Atomic Division Is Organized," *New York Times*, September 27, 1947.

9. This effort included first, an around-the-world nonstop flight by a B-50 bomber, refueled in flight (March 1, 1948), and next, a 9,600-mile flight by a B-36 carrying a payload equivalent to an atomic bomb for 5,000 of those miles (March 14, 1948). See Paul Y. Hammond, "Super Carriers and B-36 Bombers: Appropriations, Strategy and Politics," in Harold Stein, ed., *American Civil-Military Decisions: A Book of Case Studies* (Tuscaloosa: University of Alabama Press, 1963).

10. Hammond, "Supercarriers."

11. The report actually stated "It is to be expected that crashes may occur, and the site of the crash will be uninhabitable." See J. Tierney, "Take the A- Plane: The $1,000,000,000 Nuclear Bird That Never Flew," *Science82,* (January-February 1982):46ff.

12. Joint Hearings before the Preparedness Investigating Subcommittee of the Committee on Armed Services and the Committee on Aeronautical and Space Sciences, Eighty-sixth Congress, second session, January 29, 1959, p.22

13. Paul Y. Hammond, "NSC-68: Prologue to Rearmament," in W. Schilling, P. Hammond and G. Snyder, *Strategy, Politics and Defense Budgets* (New York: Columbia University Press, 1962); see also Samuel P. Huntington, *The Common Defense* (New York: Columbia University Press, 1961).

14. *New York Times,* October 29, 1949.

15. Forrestal diaries, quoted in Hammond, "Super Carriers."

16. Albion and Reed, *Makers of Naval Policy.*

17. Harry S. Truman, *Years of Trial and Hope* (Garden City, N. Y.: Doubleday, 1956), p. 53.

18. A. Stevens, *New York Times,* October 28, 1949.

19. H. A. Baldwin, *New York Times,* November 3, 1949.

20. R. G. Hewlett and F. Duncan, *Nuclear Navy* (Chicago: University of Chicago Press, 1974).

21. Hewlett and Duncan, *Nuclear Navy,* p.77.

22. Quoted in J. R. Marvin and F. J. Weyl, "The Summer Study," *Naval Research Reviews,* (August 1966).

23. J. R. Zacharias, "Scientist as Advisor," speech at the Littauer Center of Public Administration, Harvard University, March 29, 1961 (JRZ).

24. The project was named after the Lexington restaurant Hartwell Farms, where some of the negotiation had taken place.

25. Massachusetts Institute of Technology, "Project Hartwell: A Report on the Security of Overseas Transport," contract no. N5 ori 07846, December 21, 1950.

26. JRZ/Gordon.

27. Marvin and Weyl, "Summer Study."

28. Except where otherwise indicated, quotations are drawn from the final report of Project Hartwell.

29. David A. Rosenberg, "U. S. Nuclear Stockpile, 1945 to 1950," *Bulletin of the Atomic Scientists* (May, 1982):25–30.

30. David A. Rosenberg, "American Atomic Strategy and the Hydrogen Bomb Decision," *Journal of American History*, 66 (1979):62–87. The naval officers included Admirals Daniel V. Gallery, Arleigh Burke, and Ralph Ofstie.

31. JRZ/Gordon.

32. Quoted in Daniel Greenberg, *Politics of Pure Science.*

33. JRZ/Gordon.

34. Alice K. Smith, *A Peril and a Hope: The Scientists' Movement in America 1945–47* (Cambridge: MIT Press, 1965).

35. Many scientists had lobbied against the May-Johnson bill, which they felt gave the military too much control of atomic matters. The bill was defeated and in its place the MacMahon Act was passed establishing a fully civilian Atomic Energy Commission responsible directly to the president for all matters relating to atomic energy.

36. Brandeis/Sloan I

37. Ibid.

38. JRZ/Gordon. *Victory Through Airpower* was the name of a book by A. de Seversky (New York: Simon and Schuster, 1942), which appeared early in the war and claimed that the United States could dominate the world through airpower. The book achieved some notoriety, especially for a map showing the center of future American dominance to be somewhere in Kansas or Oklahoma and reaching out to all parts of the world.

39. Sloan/Charles

40. Quoted in Kevles, *The Physicists.* The original quotation appears in G. A. W. Boehm, "The Pentagon and the Research Crisis," *Fortune* 57 (February, 1958):160.

41. J. R. Zacharias, speech to MIT Quarter Century Club, January 26, 1971 (JRZ).

42. K. C. Redmond and T. M. Smith, *Project Whirlwind* (Bedford, Mass.:Digital Press, 1980), esp. chaps. 10, 11.

43. J. R. Killian, Jr. and A. G. Hill, "For a Continental Defense," *Atlantic Monthly* (November, 1953).

44. George E. Valley, "How the SAGE Development Began," *Annals of the History of Computing*, 7 (1985):196.

45. Valley was also an alumnus of the Radiation Lab, where he had worked in the Airborne Systems division. Later he became a consultant to the air force and in 1957 became the chief scientist of the Air Force.

46. Quoted in Valley, "How the SAGE."

47. Sloan/Charles

48. Hubbard was at this time the assistant director of LNS&E, under Zacharias.

49. Sloan/Charles

50. Like Ridenour, Getting and Griggs were alumni of the Radiation Lab. Getting was a group leader in LNS&E, under Zacharias, in charge of the MIT synchrotron. He had differed sharply with Jerrold on whether the synchrotron should be located on the campus or at a field station such as Fort Devens. At the time of Project Charles, he was on leave from MIT to the air force. Griggs was a geophysicist from California who later succeeded Ridenour as air force chief scientist. Griggs later gave highly controversial testimony in the Oppenheimer proceedings.

51. Massachusetts Institute of Technology, "Problems of Air Defense," Final Report of Project Charles, contract no. DA36-039sc-5450, AUGUST 1, 1951.

52. Sloan/Charles.

53. JRZ to J. Stratton, October 19, 1951 (JRZ).

54. In addition to Buckley, the initial membership included, among others, Detlev W. Bronk, representing the National Academy of Sciences; Alan T. Waterman, as director of the fledgling National Science Foundation; William Webster, as chairman of the Research and Development Board of the Department of Defense; James B. Conant; Lee A. DuBridge; James R. Killian, Jr.; and J. Robert Oppenheimer.

55. J. R. Killian, Jr., *Sputnik, Scientists and Eisenhower* (Cambridge: MIT Press, 1977), pp. 66–67

56.** Final Report, Project VISTA, Historical Files. Box 65, Folder 1 (California Institute of Technology, 1952)

57. David C. Elliott, "Project Vista and Nuclear Weapons in Europe," *International Security* 11 (1986):163–183.

58. In fact, the study's proposals turned out to have serious foreign policy implications; implementation of them would ultimately require a debate in the White House National Security Council.

59. The study seems to have been unique in that it had no official code name. It was often referred to, however, as Project Lincoln or as the Lincoln Summer Study.

60. Zacharias, "Scientist as Advisor."

61. Ionospheric scatter communications was an important new technique. It had its origin in a State Department-sponsored study (Project Troy) in 1950, which was aimed at solving the radio transmission problems of the Voice of America, the department's overseas broadcasting agency. Jerome Wiesner, Edward Purcell, and Lloyd Berkner contributed to this study. A little later, tropospheric scatter techniques were discovered at Bell Labs. It was Wiesner who recognized the importance of these developments for air defense.

62. Zacharias, "Scientist as Advisor."

63. Ibid.

64. Frank J. O'Brien, letter written on the occasion of the sixtieth birthday of Jerrold Zacharias (JRZ).

65. Louis D. Smullin, letter written on the occasion of a sixtieth birthday celebration for JRZ, March 19, 1965 (JRZ).

66. S. E. Ambrose, *Eisenhower: The President* (New York: 1984), 2:141.

67. Jerrold, whose testimony was part of the document, said that he had been given less than twenty-four hours to object to the release of the proceedings had he wished to do so.

68. All testimony cited in this section has been taken from the transcript of the Oppenheimer hearing published by the Government Printing Office, 1954; useful secondary sources are C. P. Curtis, *The Oppenheimer Case: The Trial of a Security System* (New York: Simon and Schuster, 1955); J. W. Kunetka, *Oppenheimer: The Years of Risk* (Englewood Cliffs, N. J.: Prentice-Hall, 1982); P. M. Stern, *The Oppenheimer Case: Security on Trial* (New York: Harper and Row, 1969); H. F. York, *The Advisers: Oppenheimer, Teller and the Superbomb* (San Francisco: Freeman, 1976).

69. David T. Griggs was a professor of geophysics at UCLA and a consultant on nuclear weapons to the Rand Corporation. He had been at the Radiation Lab in 1941–1942, where he had become acquainted with George Valley; at the time of Project Charles and of the summer study of 1952 he was chief scientist of the air force.

70. Hill remembered that he was interrupted by one of Oppenheimer's attorneys who "bounced up and protested, 'You are attacking the credibility of our witness!'"

71. C. Murphy, "The Hidden Struggle for the H-Bomb," *Fortune* (May, 1953).

72. See, for example, a staff-authored article entitled "The Truth about Air Defense," *Air Force Magazine* (May, 1953), where a figure of $20 billion is used.

73. The full report is reproduced in R. C. Williams and P. L. Cantelon, *The American Atom* (Philadelphia: University of Pennsylvania Press, 1984).

74. Quoted in ibid.

75. Atomic Energy Commission, "In the Matter of J. Robert Oppenheimer: Texts of Principal Documents and Letters of Personnel Security Board, General Manager, Commissioners" (Washington, D. C., Government Printing Office, May 27, 1954–June 29, 1954).

Chapter 6

1. J. R. Zacharias, from an unlabeled, undated tape of a talk (JRZ).

2. Positronium is a short-lived bound system consisting of an electron and its positively charged antiparticle, the positron. Jerrold often remarked that Deutsch should have had a Nobel Prize for that discovery.

3. For a full description of the history of the atomic clock, see Paul Forman, "Atomichron: The Atomic Clock from Concept to Commercial Product," *Proceed-*

ings of the IEEE, 73 (July, 1985):1181-1204; Norman F. Ramsey, "History of Atomic and Molecular Standards of Frequency and Time," *IEEE Transactions on Instrumentation and Measurement*, IM-21 (May, 1972):90-99; idem., History of Atomic Clocks," *Journal of Research of the National Bureau of Standards* 88 (1983):301-320.

4. Brandeis/Sloan II

5. J. R. Zacharias, "Undergraduate Education and Atomic Clocks," in *R.L.E.: 1946+20*, MIT (1966) (JRZ).

6. JRZ/Gordon.

7. *R.L.E.: 1946+20.*

8. J. R. Zacharias, taped lecture, January 18, 1971 (JRZ).

9. R.D. Haun obtained his Ph.D. under Jerrold's direction, using a modified "small clock" to make measurements on cesium. He went on eventually to become a research director at Westinghouse. Yates returned to his position in England at the end of his year's leave.

10. See Forman, "Atomichron," for a fuller explanation of the split-field method. When the 1988 Nobel Prize was awarded to Norman Ramsey, most physicists understood that it was for a lifetime of significant contributions to physics. The only one of these contributions mentioned in the Nobel citation, however, was the invention of the split-field idea.

11. I. Estermann, O. C. Simpson, and O. Stern, "The Free Fall of Atoms and the Measurement of the Velocity Distribution in a Molecular Beam of Cesium Atoms," *Physical Review* 71 (1947):238-249. See also I. Estermann, S. N. Foner, and O. Stern, "The Mean Free Paths of Cesium Atoms in Helium, Nitrogen, and Cesium Vapor," *Physical Review* 71 (1947):250-257.

12. Zacharias, lecture, January 18, 1971.

13. A highly accurate description of the shape of a nucleus is that it is either a very slightly flattened sphere or a very slightly elongated one. A quadrupole moment, such as Kellogg, Rabi, Ramsey, and Zacharias had discovered in deuterium in 1939, indicates a very small departure from that relatively simple shape. If a nucleus also happens to have an octupole moment, then its shape differs even further but by a still smaller amount. Such tiny variations in shape are extremely difficult to measure.

14. Zacharias lecture, January 18, 1971.

15. Roy M. Cohn, a young lawyer, and his friend G. David Schine, heir to a hotel fortune but otherwise unqualified, did highly publicized legwork for Senator Joe McCarthy's committee (Permanent Subcommittee on Investigations of the Senate Committee on Government Operations). They were widely feared for their recklessness and irresponsibility and the great harm of which they were capable.

16. At that time, Wiesner was a director of the National Company, the small firm in Malden that was supporting the commercial development of the atomic clock. Jerrold was a consultant to the company with royalty rights, but it was premature

to think about royalties. The clock had not yet been put on the market, and it was anybody's guess as to what kind of commercial success it would have or whether it would mean anything to them financially.

17. J. R. Zacharias, interview by S. S. Schweber, recorded October 23, 1985.

18. Jerrold always claimed that Purcell deserved a second Nobel Prize for the discovery of microwave radiation from hydrogen, at 21 centimeter wavelength, which has provided an important window for radio astronomy.

19. Dicke and Pound were experimental physicists, Guillemin an electrical engineer; all three were alumni of the Radiation Lab. Zimmerman, also an electrical engineer, was Wiesner's deputy at RLE and would eventually succeed him as director.

20. Trevor Gardner, Testimony before the Subcommittee on Military Operations of the House Committee on Government Operations, 83rd Congress, 1st session, June 10, 1954.

21. J. R. Killian, *Sputnik, Scientists and Eisenhower* (Cambridge: MIT Press, 1977), p.68.

22. *Time*, April 29, 1957, pp. 84–90.

Chapter 7

1. MIT/Oral.

2. JRZ to J. A. Stratton and G. Harrison, November 7, 1955 (JRZ).

3. MIT/Oral.

4. Ibid.

5. J. R. Zacharias, taped lecture, 1971. By 1956, more than twenty-five Ph.D.s had been achieved under Jerrold's direction, and another five or six were in progress and would be awarded within the next three years, an impressive record by any measure.

6. "African Summer Study," (Newton, Mass.: Educational Services, Inc., 1961).

7. See, for example, H. Krieghbaum and H. Rawson, *An Investment in Knowledge* (New York: New York University Press, 1969), for a full account of these teacher training institutes in the period 1954-1965. See also P. Marsh and R. Gortner, *Federal Aid to Science Education: Two Programs* (Syracuse, N. Y.: Syracuse University Press, 1963).

8. Krieghbaum and Rawson, *An Investment in Knowledge*, p.137.

9. MIT/Oral.

10. Ibid.

11. I. I. Rabi, speech at MIT, October 24, 1968 (JRZ).

12. MIT/Oral.

13. Ibid.

14. Land was also a member of the President's Science Advisory Committee (PSAC) and a member of the Greater Cambridge community, so that Jerrold already knew him well.

15. MIT/PSSC. The committee was sponsored by the American Institute of Physics, the American Association of Physics Teachers, and the National Science Teachers' Association.

16. Krieghbaum and Rawson, *An Investment in Knowledge*.

17. MIT/PSSC.

18. Morrison has been described by the historian Alice Kimball Smith as a man who "combined a rare sensitivity of spirit with a wide-ranging mind and a gift of language." It would be hard to find anyone to disagree. A telling and poignant excerpt from his description of his flight over Japan is reproduced in Smith's book, *A Peril and a Hope* (Cambridge: MIT Press, 1970).

19. MIT/Oral.

20. White recounted that, alone among the reporters in the Herald Tribune newsroom, he had taught himself to use a small slide rule. Impressed with this evidence of scientific talent, his editor had given him the assignment to cover the atomic bomb. The famous cover photograph of Oppenheimer's hat that appeared on the inaugural issue of the journal *Physics Today* was taken by White.

21. MIT/Oral.

22. General Report of the Physical Science Study Committee, March 25, 1957 (MIT/PSSC).

23. The ripple tank is an inexpensive device subsequently developed by PSSC and widely used in the classroom. It is used by the students themselves to demonstrate important general properties of waves by making visible the behavior of ripples on the surface of a layer of water.

24. MIT/Oral.

25. A fly's wing weighs about fifty micrograms—about two-millionths of an ounce. It makes a useful reference for very small weights and forces.

26. These are capitalized here because Jerrold always capitalized them. They were for him as fundamental as breathing.

27. J. R. Zacharias, "Research Scholars and Curriculum Development," address given at the General Meeting on English in New York, December 29, 1964.

28. E. M. Forster, *Two Cheers for Democracy* (London: Edward Arnold, 1951).

29. J. R. Zacharias, handwritten screed, n. d. (JRZ).

30. James Thurber, *Further Fables for Our Times* (New York: Simon and Schuster, 1956).

31. MIT/Oral.

32. Ibid.

33. *New York Times,* October 5, 1957, 1:6.

34. Dwight D. Eisenhower, *The White House Years: Waging Peace, 1956-1961,* quoted in J. M. Logsdon, *The Decision to Go to the Moon* (Cambridge: MIT Press, 1970).

35. Senate Space Committee, *Documents on International Aspects of Space,* quoted in Logsdon, *Decision.*

36. Eisenhower, *The White House Years,* p.226.

37. Quoted in Krieghbaum and Rawson, *An Investment in Knowledge,* p.228.

38. MIT/Oral.

39. Paul C. Reed, "Misguided Scientists," *Educational Screen and Audiovisual Guide* (January, 1958) (JRZ). Mickey Spillane was the author of the extremely successful Mike Hammer private-eye novels. Bill Board's Bumpkins and Uncle Jim's Animal Cousins, if they ever existed, are mercifully lost to memory.

40. Baez had suggested a way to film what is called Brownian motion; in the experiment, smoke particles, just barely visible under a microscope, are buffeted about by collisions with molecules that are themselves too small to see. It is a very important and persuasive demonstration of the existence of molecules. Baez also had made suggestions for using animation as a way of illustrating the behavior of molecules in a gas. Jerrold was bound to be favorably impressed with ideas that made molecules convincing.

41. J. R. Zacharias, "Today's Science - Tomorrow's Promise," *Technology Review* (July 1957):501.

42. Among these were a series on the Netsilik Eskimos and a series of films on fluid flow for the National Committee for Fluid Mechanics. Each of these series won many awards and prizes.

43. Webb served only a few months before becoming the first director of NASA, the new agency established to develop the U. S. role in space.

44. J. R. Zacharias, testimony before a subcommittee of the Committee on Appropriations, House of Representatives, "Highlights of Science in the United States," 87th Congress, 1st session, 1962.

45. Biological Sciences Curriculum Study; Chemical Bond Approach Project; Chemical Education Materials Study; Earth Sciences Curriculum Project; School Mathematics Study Group; University of Illinois Committee on School Mathematics.

46. MIT/Oral.

47. J. R. Zacharias, "Team Approach to Education," *American Journal of Physics* 29 (1961):347-349.

48. JRZ to Jon Schaffarzick, National Institute of Education, January 14, 1977 (JRZ).

49. Ibid.

50. See, for example, Anthony P. French, "Fifty Years of Physics Education," *Physics Today*, (November, 1981):51.

51. Dorothy M. Fraser, "Current Curriculum Studies in Academic Subjects," (Washington, D. C.: National Education Association, 1962).

52. This activity soon grew into Harvard Project Physics under the guidance of Gerald Holton and Fletcher P. Watson. It made use of historical case studies as the vehicle for teaching physics and was very successful; it was often seen as a less difficult alternative to PSSC.

53. JRZ to Schaffarzick.

54. French, "Fifty Years."

55. Zacharias, "Highlights."

Chapter 8

1. J. R. Zacharias, "Scientist as Advisor," speech at the Littauer Center of Public Administration, Harvard University, March 29, 1961.

2. "African Summer Study," (Newton, Mass.: Educational Services, Inc., September 1, 1961.) All unattributed quotations in this section are from this source.

3. Quoted in J. R. Zacharias, "Scientific and Engineering Education in Newly Developing Countries," in W. T. Martin and D. C. Pinck, eds., *Curriculum Improvement and Innovation: A Partnership of Students, School Teachers and Research Scholars* (Cambridge, 1966).

4. The report of the conference mentioned specifically the University of Illinois Arithmetic Project, the Stanford University Experimental Mathematics Project, the School Mathematics Study Group, and the University of Illinois Committee on School Mathematics. None of these was more than a few years old.

5. Jerrold used this excerpt several times in essays, always to good effect. Most notably, perhaps, he used it in the foreword to a posthumously published book by Francis Friedman: *The Classical Atom* (Reading, Mass.: Addison-Wesley, 1965).

6. Statement by the President's Science Advisory Committee, "Education for the Age of Science," White House, Washington, D.C., May 24, 1959.

7. Quoted in Emily Romney and Mary Jane Neuendorffer, *The Elementary Science Study—A History* (Newton, Mass.: Education Development Center, 1973).

8. "African Summer Study."

9. Thomas H. Huxley, "Scientific Education: Notes of an After-Dinner Speech (1869)", in C. Bibby, *T. H. Huxley on Education* (Cambridge, England: Cambridge University Press, 1961). This excerpt was also quoted in Dean Whitla and Daniel C. Pinck, *A Handbook of Suggestions for Introducing and Maintaining Innovative Science Activities* (Cambridge: Harvard University Press, 1974).

10. Much of the history of the Science Teaching Center (later named the Education Research Center) is recounted in Edwin F. Taylor and Betty Weneser, "The MIT Education Research Center." A draft dated December 1, 1977 remains in the JRZ papers, bearing Jerrold's handwritten comments and annotations.

11. MIT/Proposal to NSF, April 2, 1959 (JRZ).

12. H. C. Kelly (NSF) to R. P. Webber (MIT), June 8, 1959 (JRZ).

13. Singer had arrived at MIT with a fresh master's degree in business administration in 1955. When he and Jerrold met in 1961, Singer was working with Howard Johnson, then dean of the School of Industrial Management (and later president of MIT).

14. E. F. Taylor and B. Weneser, "The MIT Education Research Center," (1977). The word friendly was inserted by Jerrold himself, to whom Taylor had given the manuscript for comment.

15. The remarks of Killian, Stratton, and Zacharias are reproduced in the Quarterly Report 1962-63 (Newton, Mass.: Educational Services, Inc., 1963).

16. Philip Morrison, "Insight and Taste," *American Journal of Physics* 31 (1963):477–479.

17. David Webster was an early participant in ESS. The Mealworms unit was a study of small animal behavior, suitable for small children to carry out. Mealworms, left alone in a box with some bran, always manage to find the bran. How do they do it?

18. Two essays in particular express the ESS philosophy extremely well: D. Hawkins, "Messing about in Science" and "On Living in Trees," both in *The Informed Vision* (New York: Agathon, 1971).

19. Romney and Neuendorffer, *The Elementary Science Study—A History.*

20. National Science Resources Center (Smithsonian Institution–National Academy of Sciences) *Science for Children—Resources for Teachers* (Washington, D. C.: National Academy Press, 1988).

21. Including Philip Morrison and the author, who had been brought by Morrison onto the committee a year earlier.

22. J. R. Zacharias, "Testing in the Schools: a Help or a Hindrance," *Prospects* 5 (1975):33–41.

23. Savage had taught in Britain, had been associated for several years with ESS, and had recently worked with Fafunwa at Nsukka.

24. The term Westerners as used here includes Americans, a few English-based educators, and a few expatriates working in Africa.

25. Dyasi was a refugee from South Africa, a graduate of Fort Hare College and of the University of Illinois, and a veteran participant in APSP. He is currently director of the Teacher's Workshop at the College of the City of New York.

26. Zacharias, "Testing in the Schools."

Chapter 9

1. I. I. Rabi, taped transcript of a seminar at MIT, March 3, 1969 (JRZ).

2. Soon to become the U. S. commissioner of education.

3. The National Science Foundation, the Ford Foundation, the Alfred P. Sloan Foundation, the New World Foundation, the Charles F. Kettering Foundation, the American Council of Learned Societies, and the Newton, Massachusetts, public school system.

4. J. R. Zacharias, in H. Conant, ed., "Seminar on Elementary and Secondary School Education in the Visual Arts" (New York University, 1965).

5. Report of the President's Task Force on Education, White House, Washington, D.C., November 14, 1964.

6. J. R. Zacharias, "In Defense of Committees," reprinted in ESI Quarterly Report (Summer–Fall 1964).

7. Citation reprinted in Quarterly Report (Summer–Fall 1964).

8. J. R. Zacharias, Lowell Lecture, Massachusetts General Hospital, February 2, 1966.

9. In addition to its own regular meetings, the panel had helped to organize, during 1963–1964, nine special meetings and seminars on a wide variety of topics related to the theme "Innovation and Experiment in Education." Stephen White took part in two of them, Kevin Smith in another, Philip Morrison and David Hawkins in a third. Jerrold's concerns were well represented.

10. "Education of the Deprived and Segregated," Bank Street College of Education Report, New York, 1965.

11. Ralph Ellison, *Invisible Man* (New York: Random House, 1953).

12. Ralph Ellison, "What These Children Are Like," in "Education of the Deprived and Segregated."

13. MIT/Oral.

14. "Innovation and Experiment in Education," Progress Report of the Panel on Educational Research and Development to the U.S. Commissioner of Education, the Director of the National Science Foundation and the Special Assistant to the President for Science and Technology, Washington, D. C., March, 1964.

15. John Niemeyer, "Next Steps," in "Education of the Deprived and Segregated."

16. ESI Quarterly Report (Summer–Fall 1964).

17. MIT/Oral.

18. Anthony P. French, letter included in Zacharias sixtieth birthday souvenir book (JRZ). Unless otherwise indicated, all quotations in this section are from this source.

19. Patterson was director of Tufts University's Lincoln Filene Center for Citizenship and Public Affairs, and director of the junior high school component of the Social Studies Curriculum Program at ESI.

20. J. R. Zacharias, "Learning by Teaching," in A. M. Rzepecki, ed., *Science and the Modern Mind* (Detroit: Sacred Heart Seminary, 1966).

21. Ibid.

22. Ibid.

23. Ibid.

24. Cyrus Levinthal, letter written for sixtieth birthday souvenir book (JRZ).

25. O. Cope, and J. R. Zacharias, *Medical Education Reconsidered: Report of the Endicott House Summer Study on Medical Education* July 1965 (Philadelphia: Lippincott, 1966). All quotations from the report are from this source.

26. Bacteriology, biochemistry, internal medicine, obstetrics, pathology, pediatrics, pharmacology, psychiatry, surgery, and hospital administration.

27. At the time, William Ellery Channing Professor of Medicine, Harvard Medical School.

Chapter 10

1. Galbraith and Kaysen were Harvard economists; Galbraith had been ambassador to India in the Kennedy administration. Kistiakowsky and Long were physical chemists, the former from Harvard, the latter from Cornell. Neustadt was a Harvard political scientist, as was Kissinger at that time. Skolnikoff was a political scientist from MIT. Lindsay was president of Itek, Inc., to which Jerrold was a consultant; Lindsay was also a board member of EDC. Wyzanski was a prominent Boston jurist; Benjamin was a New York attorney who had been one of the founders of the United Nations Association.

2. G. B. Kistiakowsky and J. B. Wiesner, Prospectus for the Cambridge Discussion Group and invitation to first meeting, privately circulated February 4, 1966 (JRZ).

3. J. K. Galbraith to President Lyndon B. Johnson, April 19, 1966 (JRZ).

4. Carl Kaysen, private communication, March 16, 1989.

5. Robert S. McNamara to JRZ, April 16,1966 (JRZ).

6. Jason, according to the Pentagon Papers, was used to conduct "ad-hoc high-level studies using primarily non-IDA scholars."

7. *The Pentagon Papers*, Sen. Gravel Edition (Boston: Beacon Press, 1975), vol. 4.

8. *Pentagon Papers*, narrative.

9. There was irony in the fact that not long after Jerrold led his group of scientists to this pessimistic conclusion, the Navy Cross was awarded in Saigon to another Jerrold Zacharias, for a night bombing raid over Hanoi. This was navy Commander Jerrold Zacharias, his cousin, who was the son of his uncle Admiral Ellis Zacharias.

10. Quoted in G. Lewy, *America in Vietnam* (New York: Oxford, 1977).

11. Paul Dickson, *The Electronic Battlefield* (Bloomington: Indiana University Press, 1976).

12. I am indebted to Dr. Gregg Herken for bringing Igloo White to my attention and for supplying this valuable insight into its significance.

13. Deborah Shapley, "Jason Division: Defense Consultants Who Are Also Professors Attacked," *Science* 179 (1973):459–462, 505.

14. JRZ to editor of *Science* (JRZ).

15. Data from "The Great Society: What It Was, Where It Is," *New York Times,* December 9, 1968, 1:6

16. J. R. Zacharias, testimony on the 1981 National Science Foundation Authorization before the Subcommittee on Science, Research and Technology of the Committee on Science and Technology, U. S. House of Representatives, 96th Congress, 2nd session, February, 1980.

17. Report of the President's Commission on Campus Unrest (Washington, D. C.: U.S. Government Printing Office, 1970).

18. Arthur M. Schlesinger, Jr., *The Crisis of Confidence* (Boston: Houghton Mifflin, 1969).

19. Report of the President's Commission on Campus Unrest.

20. C. R. Whitney, "Columbia Split as Buildings Are Blocked," *New York Times,* May 8, 1970.

21. Renamed the Draper Laboratory in October 1969.

22. This and succeeding quotations are from the *New York Times,* May 1, 1969, 41:3.

23. Described by the *New York Times,* November 4, 1969, as "a loose organization of about 30 antiwar groups."

24. E. M. Forster, *Two Cheers for Democracy* (London: Edward Arnold, 1951).

25. Zacharias to Edward B. Fiske, education editor of the *New York Times,* March 11, 1976 (JRZ).

26. H. Kipphardt, *In the Matter of J. Robert Oppenheimer* (New York: Hill and Wang, 1969).

27. Published in full in *New York Times,* July 22, 1968.

28. A week or so later, Rabi addressed the entire MIT community on the subject of the Sakharov paper; he was pleased to suggest that the names Sakharov and Zacharias were probably cognates, perhaps the same names spelled differently. The idea tickled him.

29. I. I. Rabi, colloquium at the Education Research Center, MIT, 1968, transcript of seminar proceedings, March 3, 1969 (JRZ).

30. J. R. Zacharias, colloquium at the Education Research Center, 1968.

31. Richard L. Meehan, *Getting Sued and Other Tales of the Engineering Life* (Cambridge: MIT Press, 1981).

32. JRZ to MIT Committee on Educational Policy, memorandum, May 17, 1967 (JRZ).

33. Aldous Huxley, "Education on the Nonverbal Level," *Daedalus* (1962):279–293.

Chapter 11

1. J. R. Zacharias, "Professional Education of Scientists, Engineers, Physicians and Teachers," unpublished manuscript (1969) (JRZ).

2. The first president was Arthur Singer, who had made the trip to East Africa in 1960. He had left MIT a few years earlier and had gone to the Carnegie Corporation as a senior program officer. He served there for two years and then left to become a vice-president of the Alfred P. Sloan Foundation in New York.

3. From a document distributed to incoming freshmen at MIT in 1969 (JRZ).

4. Report of the President and Chancellor to the MIT Corporation (1972).

5. The first was the Committee on Educational Survey, known as the Lewis committee, formed in 1947 and reporting in 1949; the second was the Committee on Curriculum Content and Planning (CCCP), which Jerrold had chaired. Formed in 1962, the CCCP had reported in 1964.

6. J. R. Zacharias, taped lecture, January 18, 1971 (JRZ).

7. K. Hoffman, et al., "Creative Renewal in a Time of Crisis," Report of the Commission on MIT Education (November 1970).

8. "A Proposal for a Division for Education Research," 1971 (JRZ).

9. Zacharias, J.R., "Professional Education of Scientists, Engineers, Physicians and Teachers," unpublished manuscript, 1969 (JRZ).

10. MIT/Oral.

11. The then existing schools were Architecture and Planning, Engineering, Humanities and Social Science, Management, and Science.

12. The staff person assigned to the committee from Wiesner's office was Barbara S. Nelson, whose 1974 dissertation describes the committee proceedings in some detail and is the basis for much of this section. See B. S. Nelson, "The Creation of MIT's Division for Study and Research in Education," (Ph.D. dissertation, Harvard University, 1974).

13. J. R. Zacharias, "Professional Education."

14. S. B. Sutton, "Report on MIT's Division for Study and Research in Education," Part I, September 1978; Part II, 1979; Part III, March, 1983. Prepared for the Ford Foundation.

15. Report of the President and the Provost to the MIT Corporation, 1974.

16. Early in his engineering career, Wiesner had been an associate and an admirer of the great MIT mathematician Norbert Wiener, one of the pioneers in the field

of artificial intelligence. Walter Rosenblith, a biophysicist who succeeded Wiesner as MIT provost, had himself been a protege of Wiener and was also a strong advocate for research in that field.

17. Sutton, "Report."

Chapter 12

1. M. Parry, "The Reward" (Newton, Mass.: Education Development Center, Inc., 1974)

2. J. R. Zacharias, handwritten screed, n.d. (JRZ).

3. J. R. Zacharias, "The Case of the Missing Scientists," *National Elementary Principal* 59 (January, 1980): 14-17.

4. Felt/Churchill. The term "up the street" implies the Harvard School of Education.

5. Parry, "The Reward."

6. *Infinity Factory*, Program 104 (Newton, Mass.: Education Development Center, Inc., 1976).

7. These were the Carnegie Corporation, the Alfred P. Sloan Foundation, the John and Mary Markle Foundation, and the JDR 3rd Fund.

8. "Education of the Deprived and Segregated" (New York: Bank Street College of Education Report, 1965).

9. "By the Numbers," *Time,* February 23, 1976.

10. F. Harvey, et al., "Evaluation of Eight 'Infinity Factory' Programs" (June, 1976); "Infinity Factory II: Evaluation of a Pilot Program," (Newton, Mass.: Education Development Center, August, 1976).

11. J. R. Zacharias, speech to the National Association of Elementary School Principals, Dallas, 1976 (JRZ).

12. J. R. Zacharias, *National Elementary Principal* 55 (July-August 1976).

13. J. R. Zacharias, "Testing in the Schools: A Help or a Hindrance?" *Prospects,* 5 (1975): 33-43.

14. Zacharias, speech to the National Association of Elementary School Principals.

15. Banesh Hoffman, *The Tyranny of Testing* (New York: Crowell-Collier Press, 1962). Hoffman opened his book with a notable example of the difficulty, a question from a British school entrance examination: "Which is the odd one out among cricket, football, billiards and hockey?" A reasonable case can be made for choosing any one of the four. There are countless other examples; this one is hardly the worst. See also Morris Kline, *Why Johnny Can't Add* (New York:St. Martin's Press, 1973).

16. See the following in *National Elementary Principal* 54 (1975): J. R. Zacharias, "The Trouble with IQ Tests," pp.23-29; J. Schwartz, "A Is to B as C Is to Anything at All: The Illogic of IQ Tests," pp.38-41; and J. Butler, "Looking Backward:

Intelligence and Testing in the Year 2000," pp.67–75. The Zacharias and Schwartz articles have been reprinted in P. Houts, ed., *The Myth of Measurability* (New York: Hart Publishing Co., 1977).

17. Zacharias, "The Trouble with IQ Tests."

18. Correspondence between S. H. White and J. R. Zacharias, 1978 (JRZ).

19. Philip Morrison, "The Bell Shaped Pitfall," in Houts, ed., *The Myth of Measurability*.

20. Zacharias, speech to the National Association of Elementary School Principals.

21. From Eleanor Duckworth, "Evaluation of the African Primary Science Program" (Newton, Mass.: Education Development Center, Inc. 1970), quoted in Zacharias, "Testing in the Schools."

22. J. R. Zacharias, and Saville Davis, draft for a book, 1978 (JRZ).

23. Jerrold was emphatic that the parents' right does not extend to the question, "Is my kid doing better than your kid?" That question, he claimed, was not legitimate, but it is the principal reason for tests designed to place children in numerical order.

24. M. Lazarus and J. Schwartz, "Teacher and Classroom Involvement in an Alternative Approach to Large-Scale Achievement Testing (Project TORQUE)" (Newton, Mass.: Education Development Center, Inc., June 25, 1975).

25. Willard Wirtz, "On Further Examination," Report of the Advisory Panel on the Scholastic Aptitude Test Score Design, College Entrance Examination Board, New York (1977).

26. Zacharias and Davis, book draft. The testimony referred to was by Hyman G. Rickover, Testimony before the Subcommittee on Education, Arts and Humanities of the Committee on Human Resources, United States Senate, 95th Congress, 1st session, July 14, 1977.

27. Note that a child would probably be marked wrong for this kind of arithmetic. *About* is rarely considered an appropriate arithmetic word for schoolchildren.

28. The proceedings of the symposium, including Jerrold's remarks, were published in *Celebration of the Fiftieth Anniversary of the Pupin Laboratories* (New York: Columbia University, 1977).

29. Gerard Piel, "The Sorry State of School Science: A Study in Decline," *National Elementary Principal* 59 (January, 1980):33–36.

Chapter 13

1. J. R. Zacharias, "The Case of the Missing Scientists," *National Elementary Principal* 59 (January, 1980):14–17.

2. *Celebration of the Fiftieth Anniversary of the Pupin Laboratories* (New York: Columbia University, 1977).

3. Proposal to the National Institute of Education for Project "M" (Newton, Mass: Education Development Center, Inc., August 24, 1976).

4. Ibid.

5. In 1967, the NSF budget for science education was $141.7 million; by 1977, it had decreased to $56.3 million. After adjustment for inflation, this was equivalent only to $34.1 million in 1967 dollars.

6. J. R. Zacharias, testimony on the 1981 National Science Foundation Authorization, before the Subcommittee on Science, Research and Technology of the Committee on Science and Technology, U.S. House of Representatives, 96th Congress, 2nd session, February 1980.

7. Draft of a report by the Review Group on NSF Science Curriculum Activities of the House Committee on Science and Technology, June 22, 1975 (JRZ).

8. *Congressional Record*, April 8, 1975.

9. James J. Kilpatrick, *Washington Star-News*, April 1, 1975.

10. JRZ.

11. Zacharias, testimony.

12. J. R. Zacharias, "The Common Reader," *National Elementary Principal* 51 (April, 1972):73–75.

13. Letter to H. J. Greenberg, July 7,1969 (JRZ)

14. Zacharias, "The Case of the Missing Scientists"

15. J. R. Zacharias, quoted in F. Hechinger, "About Education," *New York Times,* January 29, 1980

16. Ibid.

17. Ibid.

18. John J. Sirica, Chief Judge of the United States District Court for the District of Columbia, had presided over the Watergate trials. He had insisted, successfully, in spite of Nixon's defiance, that the tape recordings of the President's conversations be delivered to the Court as evidence in the trials of the Watergate defendants. Even the President of the United States was shown to be subject to the rule of law.

19. Jerrold had made the remark, he recalled, to President Eisenhower during a PSAC meeting, around 1957; its context was a comparison of the technological rate of growth of the United States and that of the Soviet Union. Eisenhower had understood the mathematical term.

20. President Ford's initial approach to the problem of inflation was to offer a slogan: "Whip Inflation Now." He wore a button emblazoned with the initials: WIN.

21. Zacharias, testimony.

22. Radioactive strontium-90 (Sr^{90}), a fallout product of atmospheric nuclear explosions, had appeared in milk both in the Soviet Union and in the United States

(and presumably elsewhere as well.) Substituting for calcium in the bone chemistry of growing children, Sr^{90} is an effective and deadly carcinogen.

23. For a non-partisan account, see "Nuclear Arms Control: Background and Issues," Committee on International Security and Arms Control" (Washington, D. C.: National Academy of Sciences, 1985).

24. Two important reports on this subject are: "Long Term Effects of Multiple Nuclear Weapons Detonations," (Washington, D. C.: National Academy of Sciences, 1975); "The Effects of Nuclear War," (Washington, D. C.: Office of Technology Assessment, May, 1979). See also S. Drell and F. v. Hippel, "Limited Nuclear War," *Scientific American* 235 (November, 1976):27-37.

25. In a 1984 newspaper article entitled "Civil Defense is Crucial" Edward Teller had the following to say: "About 40 million Americans are likely to survive a worst-case large-scale nuclear attack, even without protective measures... civil defense planning could reduce the number of American dead from 150 million to 50 million." (*New York Times*, January 3, 1984).

26. Richard Halloran, "Pentagon Draws Up First Strategy For Fighting a Long Nuclear War," *New York Times*, May 30, 1982.

27. Richard Burt, *New York Times*, August 6, 1980.

28. C. W. Weinberger, speech at the Naval War College, June 25, 1982 (JRZ)

29. It is estimated that all of the bombs dropped in the five years of World War II added up only to about three megatons.

30. Saville Davis had been at the Endicott House Conference on African Education in 1960.

31. Subsequently annotated by Harrison E. Salisbury and published in book form under that title. (New York: W. W. Norton and Company, 1969).

32. Quoted in *Technology Review*, June, 1969.

33. J. R. Zacharias, M. Gordon, and S. R. Davis, "Common Sense and Nuclear Peace," *Bulletin of the Atomic Scientists*, 39(April, 1983), special supplement.

34. This statement was a remark by George Kennan, published in Kennan's *The Nuclear Delusion,* (New York:Pantheon, 1982).

35. Zacharias, et al., "Common Sense and Nuclear Peace"

36. Ibid.

Index

Printed in the United States
by Baker & Taylor Publisher Services